Romans at War

This volume addresses the fundamental importance of the army, warfare, and military service to the development of both the Roman Republic and wider Italic society in the second half of the first millennium BC.

It brings together emerging and established scholars in the area of Roman military studies to engage with subjects such as the relationship between warfare and economic and demographic regimes; the interplay of war, aristocratic politics, and state formation; and the complex role the military played in the integration of Italy. The book demonstrates the centrality of war to Rome's internal and external relationships during the Republic, as well as to the Romans' sense of identity and history. It also illustrates the changing scholarly view of warfare as a social and cultural construct in antiquity, and how much work remains to be done in what is often thought of as a "traditional" area of research.

Romans at War will be of interest to students and scholars of the Roman army and ancient warfare, and of Roman society more broadly.

Jeremy Armstrong is a Senior Lecturer in Ancient History at the University of Auckland, New Zealand. He received his BA from the University of New Mexico and his MLitt and PhD from the University of St Andrews. He works primarily on archaic central Italy, and most specifically early Roman warfare. He is the author of *War and Society in Early Rome: From Warlords to Generals* (2016) and the editor of a number of volumes, including *Rituals of Triumph* (2013) and *Circum Mare: Themes in Ancient Warfare* (2016).

Michael P. Fronda is Associate Professor in the Department of History and Classical Studies at McGill University. He received his PhD from The Ohio State University. He writes on Roman political and military history, interstate relations, and Roman and Pre-Roman Italy. He is the author of *Between Rome and Carthage: Southern Italy during the Second Punic War* (2010).

Routledge Monographs in Classical Studies

Titles include the following:

Power Couples in Antiquity
Transversal Perspectives
Edited by Anne Bielman Sánchez

The Extramercantile Economies of Greek and Roman Cities
New Perspectives on the Economic History of Classical Antiquity
Edited by David B. Hollander, Thomas R. Blanton IV, and
John T. Fitzgerald

The Bible, Homer, and the Search for Meaning in Ancient Myths
Why We Would Be Better Off With Homer's Gods
John Heath

Fantasy in Greek and Roman Literature
Graham Anderson

Piracy, Pillage, and Plunder in Antiquity
Appropriation and the Ancient World
Edited by Richard Evans and Martine de Marre

Romans at War
Soldiers, Citizens, and Society in the Roman Republic
Edited by Jeremy Armstrong and Michael P. Fronda

Discourse of Kingship in Classical Greece
Carol Atack

For more information on this series, visit: https://www.routledge.com/classicalstudies/series/RMCS

Romans at War

Soldiers, Citizens, and Society in the
Roman Republic

**Edited by
Jeremy Armstrong and
Michael P. Fronda**

Routledge
Taylor & Francis Group
LONDON AND NEW YORK

First published 2020
by Routledge
2 Park Square, Milton Park, Abingdon, Oxon OX14 4RN

and by Routledge
605 Third Avenue, New York, NY 10017

First issued in paperback 2021

Routledge is an imprint of the Taylor & Francis Group, an informa business

© 2020 selection and editorial matter, Jeremy Armstrong and Michael P. Fronda; individual chapters, the contributors

The right of Jeremy Armstrong and Michael P. Fronda to be identified as the authors of the editorial material, and of the authors for their individual chapters, has been asserted in accordance with sections 77 and 78 of the Copyright, Designs and Patents Act 1988.

The Open Access version of this book, available at www. taylorfrancis.com, has been made available under a Creative Commons Attribution-NonCommercial-NoDerivatives 4.0 International license. Funded by The University of Auckland Faculty of Arts.

Trademark notice: Product or corporate names may be trademarks or registered trademarks, and are used only for identification and explanation without intent to infringe.

British Library Cataloguing-in-Publication Data
A catalogue record for this book is available from the British Library

Library of Congress Cataloging-in-Publication Data
A catalog record has been requested for this book

ISBN 13: 978-1-03-208916-4 (pbk)
ISBN 13: 978-1-138-48019-3 (hbk)

DOI: 10.4324/9781351063500

Typeset in Times New Roman
by codeMantra

For Nate Rosenstein

Contents

Maps

Tables

Contributors

Jeremy Armstrong is a Senior Lecturer in Ancient History at the University of Auckland, New Zealand. He received his BA from the University of New Mexico and his MLitt and PhD from the University of St Andrews. He works primarily on archaic central Italy, and most specifically early Roman warfare. He is the author of *War and Society in Early Rome: From Warlords to Generals* (2016) and the editor of a number of volumes, including *Rituals of Triumph* (2013) and *Circum Mare: Themes in Ancient Warfare* (2016).

Cary Barber is Assistant Professor in the Department of History at California State University – San Bernardino. He recently received his PhD from The Ohio State University under the supervision of Nate Rosenstein. His work focuses on political history, Roman demography, and the middle Republic.

Lee L. Brice is Professor of History and Distinguished Lecturer at Western Illinois University. He has published books on military history and on institutional and geographic history, as well as articles and chapters on Greek numismatics, military history, ancient history and film, and teaching. He is series editor of *Warfare in the Ancient Mediterranean World* and senior editor of the journal *Research Perspectives in Ancient History*.

Jessica H. Clark (PhD Princeton, 2008) is Associate Professor of Classics at Florida State University. Her work focuses on the history and historiography of the Roman Republic, with a particular interest in the third and second centuries BC. She is the author of *Triumph in Defeat: Military Loss and the Roman Republic* (2014), and the co-editor of *Brill's Companion to Military Defeat in Ancient Mediterranean Society* (2018).

Fred K. Drogula is the Charles J. Ping Professor of Humanities and Professor of Classics at Ohio University. He received his BA from Kenyon College, MA degrees from Boston University and the University of Virginia, and PhD from the University of Virginia. He is the author of *Commanders and Command in the Roman Republic and Early Empire* (2015) and *Cato* *the Younger: Life and Death at the End of the Roman Republic* (2019). He works widely on Roman history in the fields of politics and warfare.

Michael P. Fronda is Associate Professor in the Department of History and Classical Studies at McGill University. He received his PhD from The Ohio State University. He writes on Roman political and military history, interstate relations, and Roman and pre-Roman Italy. He is the author of *Between Rome and Carthage: Southern Italy during the Second Punic War* (2010).

François Gauthier is visiting Assistant Professor in the Department of Classics at Mount Allison University. He received his PhD from McGill University. He is interested in the Roman army of the Republic from a socio-economic perspective, imperialism, and political culture. He has published several articles and book chapters on these topics.

Marian Helm is Assistant Professor in the Seminar für Alte Geschichte at Westfälische-Wilhelms Universität Münster. He received his PhD from Ruhr-Universität Bochum, Germany. His research focuses on the rise of early Rome, political and military history, and the historical geography of the Ancient Mediterranean.

Jeremiah McCall teaches history and game design at Cincinnati Country Day School. He received his PhD from The Ohio State University. He writes on political culture and military systems in the Republic, and the intersection of games and the learning of history. His books include *The Cavalry of the Roman Republic (2001), Gaming the Past (2011),* and, most recently, *Clan Fabius, Defenders of Rome (2018).*

Kathryn H. Milne is Associate Professor of History at Wofford College. She earned her MA (Hons) from the University of Glasgow, an MA from the University of Manchester, and her PhD from the University of Pennsylvania. She works widely on aspects of the Roman army, with a particular interest in the social and cultural dynamics which underpinned Rome's military systems.

Saskia T. Roselaar obtained her PhD from Leiden University in 2009. She is best known for her monograph *Public Land in the Roman Republic: A Social and Economic History of Ager Publicus in Italy, 396–89 BC* (Oxford University Press, 2010). Her research interests cover the socio-economic history of the Roman Republic, as well as citizenship and integration in the Roman world.

Nathan Rosenstein is Professor Emeritus in the Department of History at The Ohio State University. He is the author or co-editor of several books, most recently *Rome and the Mediterranean 290–146 BC: The Imperial Republic*. He is currently co-editor of *The Oxford History of the Roman World*.

John Serrati is an Adjunct Professor in the Department of Classics and Religious Studies at the University of Ottawa. He is also an Assistant Professor of Classics at John Abbott College in Montreal. He has published extensively on Hellenistic and Roman Sicily, as well as warfare and ancient economies. These themes are broadly explored in his latest volume: *Money and Power in the Roman Republic* (ed. with H. Beck and M. Jehne, 2016). His current research examines gender and Roman warfare, specifically looking at the role of women in the martial society of the Roman mid-Republic.

James Tan is a Lecturer in the Department of Classics and Ancient History at the University of Sydney. His research focuses on the intersection between the political and the economic history of the Roman Republic. He is the author of *Power and Public Finance at Rome, 264–49 BCE* (2017).

Peter VanDerPuy is currently a Lecturer in the Department of History at The Ohio State University. He obtained master's degrees from Cardiff University and Oxford University, before earning his PhD in Ancient History at The Ohio State University. He specializes in the history of the Roman Republic and his current scholarship deals with ancient agricultural communities, warfare, and the environment.

Jack Wells is Associate Professor in the History Department at Emory & Henry College. He is interested broadly in the religious and political history of the ancient Mediterranean. He has published on Roman temple plundering in Italy in the middle Republic. He received his PhD from The Ohio State University.

Preface and acknowledgments

The origins of this volume nominally go back to July 2017 and the first meeting of the Celtic Conference in Classics in North America, held in Montreal and co-hosted by McGill University and Université de Montreal. At that event, 17 papers were presented as part of a panel entitled "New Directions in Roman Military Studies." The papers covered a wide range of topics, from the origins of *imperium* to a comparative analysis of the "Eurasian Way of War" in Late Antiquity, from the command structure of the legions and their transformation in the late Republic, to the triumph, martial tropes in love elegy, sculpted depictions of battles, and the Roman battle cry. It was, on its own, an engaging and useful event, and the quality and (perhaps surprising) cohesiveness of many of the papers meant they were certainly worthy of being collected in an edited volume in their own right.

The real seeds of both the volume and the panel, however, go back much further. One of the impetuses for organizing the panel in the first place was to bring together several scholars – some junior, others more advanced – who were mentored by Prof. Nate Rosenstein at The Ohio State University, as graduate students or postdoctoral researchers, or whose careers otherwise benefitted from Nate's advice, encouragement, or support. Indeed, the timing and location of the 2017 CCC was particularly serendipitous, as the panel co-organizers (the co-editors of this volume) had previously learned of Nate's pending retirement in July 2018 and thought it would be a good platform for his former students and mentees to pay tribute to him. Nate himself generously accepted an invitation to the panel and gave an excellent overview of the state of the field of Roman Republican military studies. Several speakers took a few minutes during their papers to thank Nate for being such a wonderful (if demanding) professor and generous, kind, and supportive colleague, and to acknowledge him as truly one of the leading voices in the field of Roman military history in the last generation.

It was also during the conference that the co-organizers began to discuss seriously transforming the conference panel into an edited volume. About two-thirds of the panel's speakers gave papers dealing with the Roman Republic – not surprising given our connection to Nate. Additional papers were solicited to flesh out the volume from friends and colleagues of Nate who were unable to make the initial event. As co-editors, we envisioned a volume that offered a broad coverage of many facets of war and society in the Roman Republic, which would both highlight and reflect some of the main trends in the field, and also point to possible future directions. We wanted a volume that cohered around a central theme, or set of themes, rather than present itself as an arbitrary collection of essays – and we hope that this has been achieved. And while we did not envision this volume as a *Festschrift* per se, we planned from the beginning to dedicate the volume to Nate Rosenstein, and hope that the chapters contained herein, in some small way, acknowledge his influence. That we convinced Nate to contribute the concluding chapter is, we think, additionally satisfying.

A volume such as this requires the work of many hands. First, we want to thank Anton Powell (the founder of the CCC), as well as Bill Gladhill and Elsa Bouchard (the local organizers of the 2017 CCC), for accepting our panel. We also must thank all of the speakers at the original panel, in particular those whose papers are not included in this volume due to their focus on later periods: James Gersbach, Justin James, Alison Keith, Sara Phang, Brian Turner, and Conor Whately. They certainly contributed to the rousing success of the panel, and their participation throughout the event – their papers, their questions, and their comments – undoubtedly improved this volume. We also thank the peer reviewers for the volume (all six of them), who must unfortunately remain anonymous, but whose contributions were integral to its success. We would also like to thank Anthony Pavoni, Sally Mubarak, and Ashley Flavell, whose editorial assistance was vitally important. We also should acknowledge Amy Davis-Poynter, Elizabeth Risch, and Ella Halstead from Routledge, Lindsay Holman and the Ancient World Mapping Center for the production of typically excellent maps, and the United States Military Academy at West Point for the generous use of their map of Roman Italy.

The co-editors also extend a special appreciation to their families, who patiently endured discussions of Roman military history and oddly timed Skype conversations between New Zealand and Canada. In Canada, this includes Jennifer, Annamaria, Gianna, and Carmela. In New Zealand, Ashle, William, Theodore, and Etta.

And finally, we would like to end the acknowledgments with a final thank you to Nate Rosenstein, without whom none of this would have been possible – a point which holds on several levels. Nate's influence has not only shaped our careers in various ways but also the field we inhabit, and the discussions we engage in. We, and the field of Roman military history, are better for having had the privilege of working with him and are forever indebted to him.

Jeremy Armstrong and Michael P. Fronda, 2019

Note on texts, translations, and abbreviation

Unless otherwise noted, all ancient Greek and Latin texts and all translations are taken from the *Loeb Classical Library* series or, for fragmentary Roman historians, from Cornell, T. (ed.) (2013) *The Fragments of Roman Historians*. Abbreviations generally follow those in the *Oxford Classical Dictionary*, 4th edition.

RGDA   *Res Gestae Divi Augusti* (Loeb).

Maps

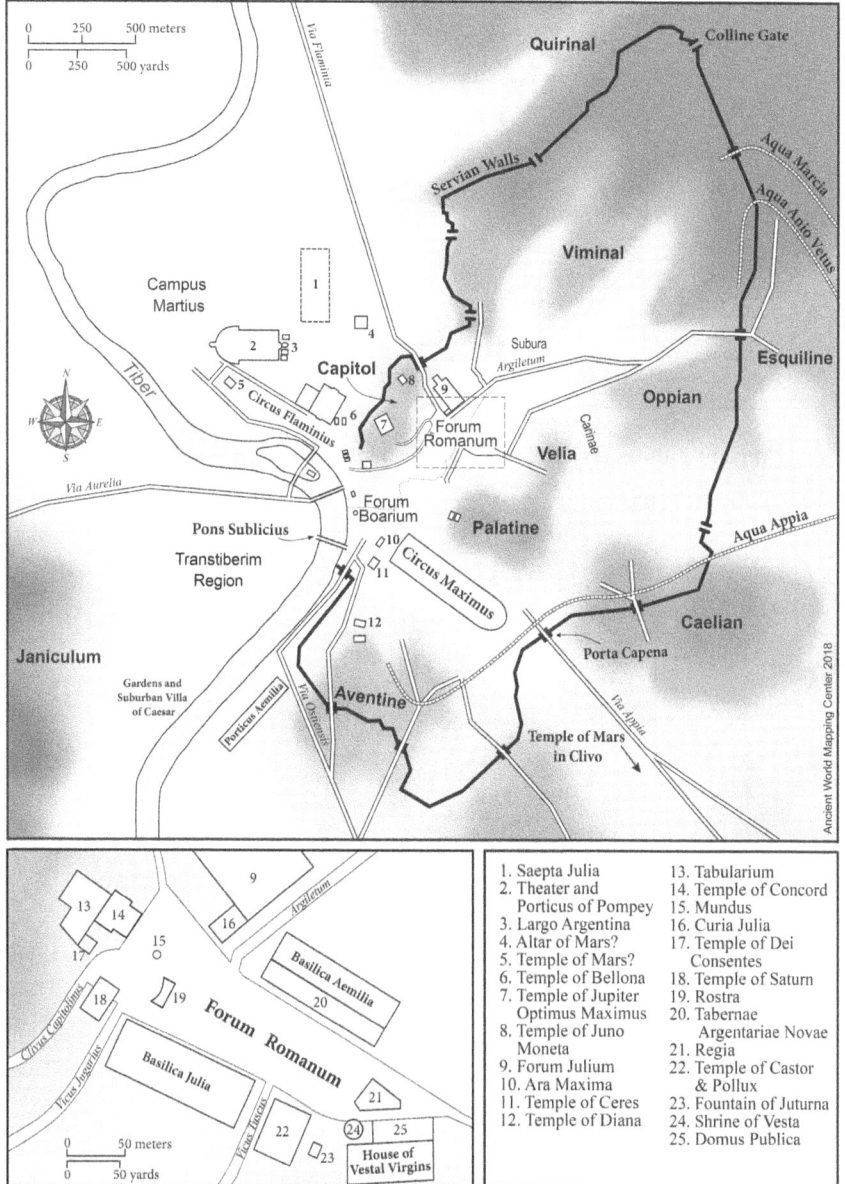

Map 1 The city of Rome. Produced by the Ancient World Mapping Center.

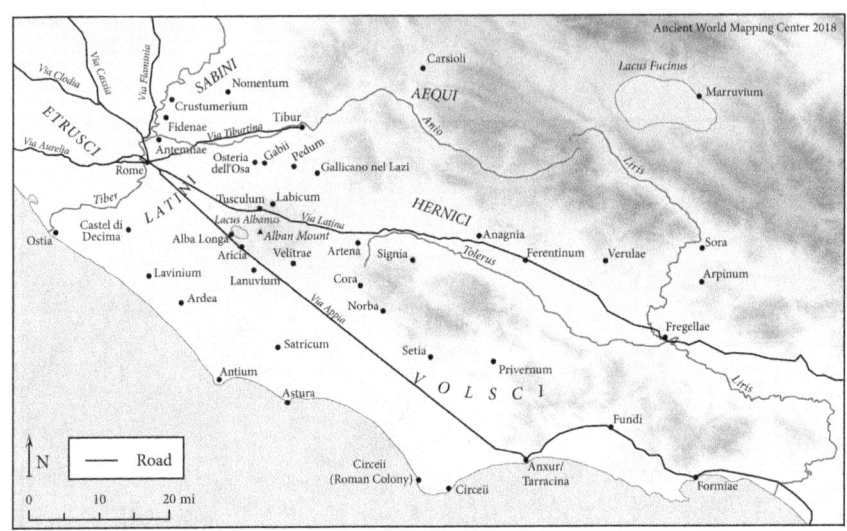

Map 2 Ancient Latium. Produced by the Ancient World Mapping Center.

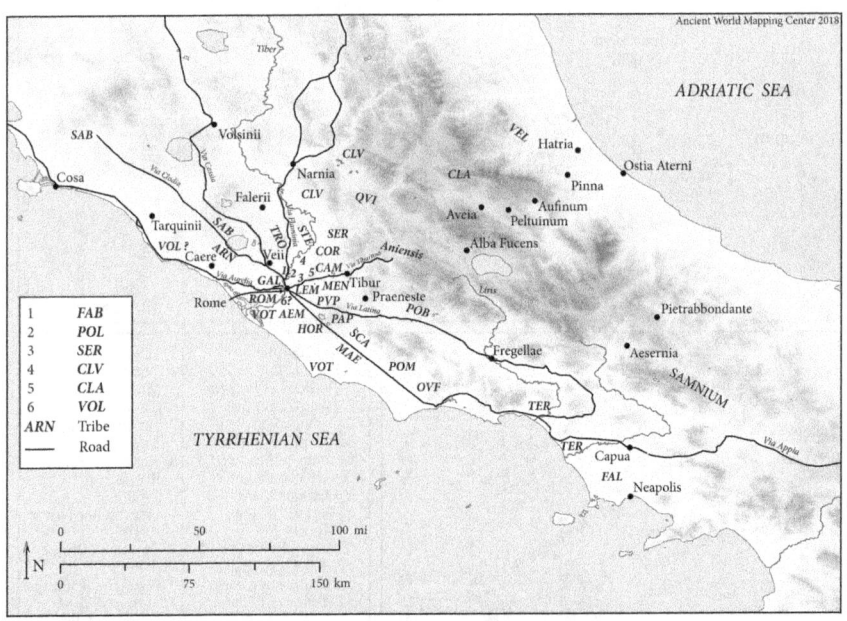

Map 3 Central Italy with Roman voting Tribes. Produced by the Ancient World Mapping Center.

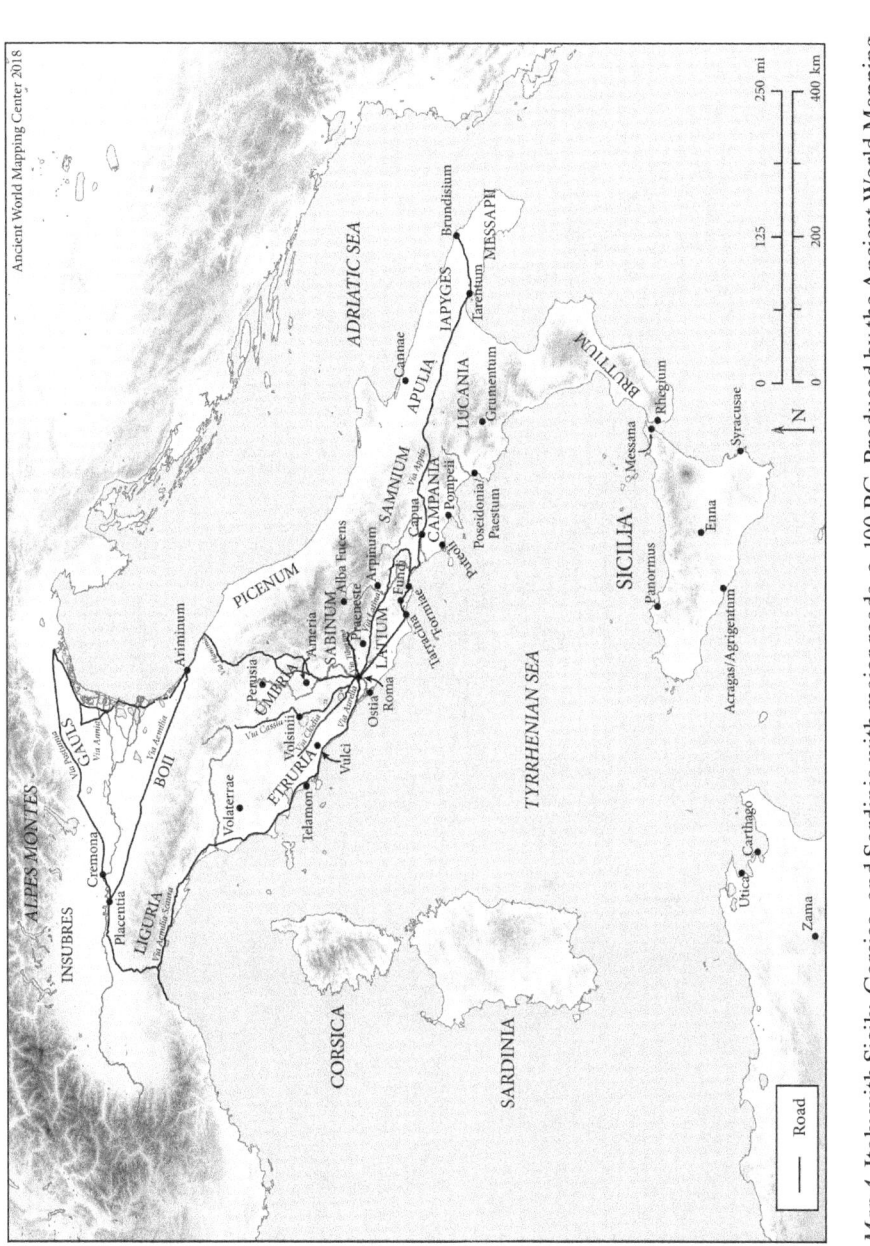

Map 4 Italy with Sicily, Corsica, and Sardinia with major roads, c. 100 BC. Produced by the Ancient World Mapping Center.

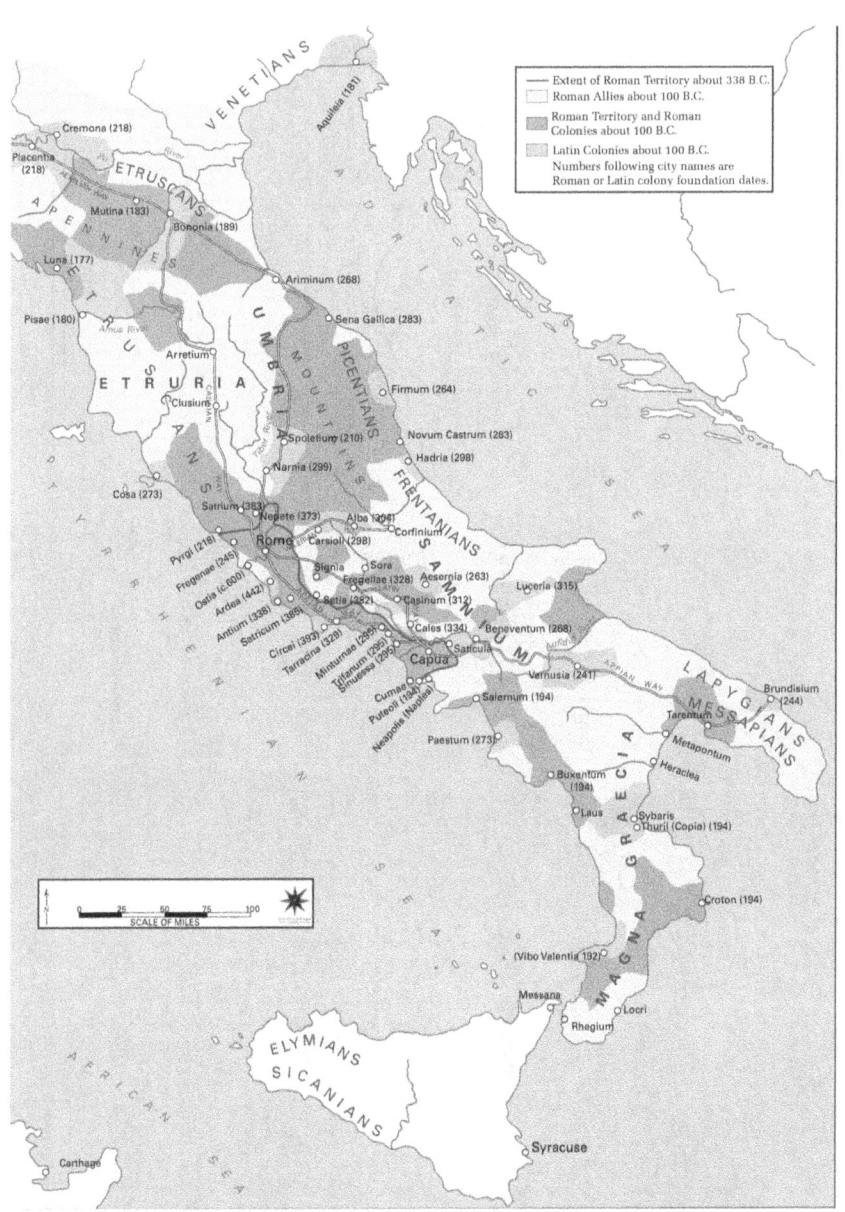

Map 5 Italy: Roman, Latin and Allied Territories, c. 100 BC. Map courtesy of the USMA, Department of History. Used with permission.

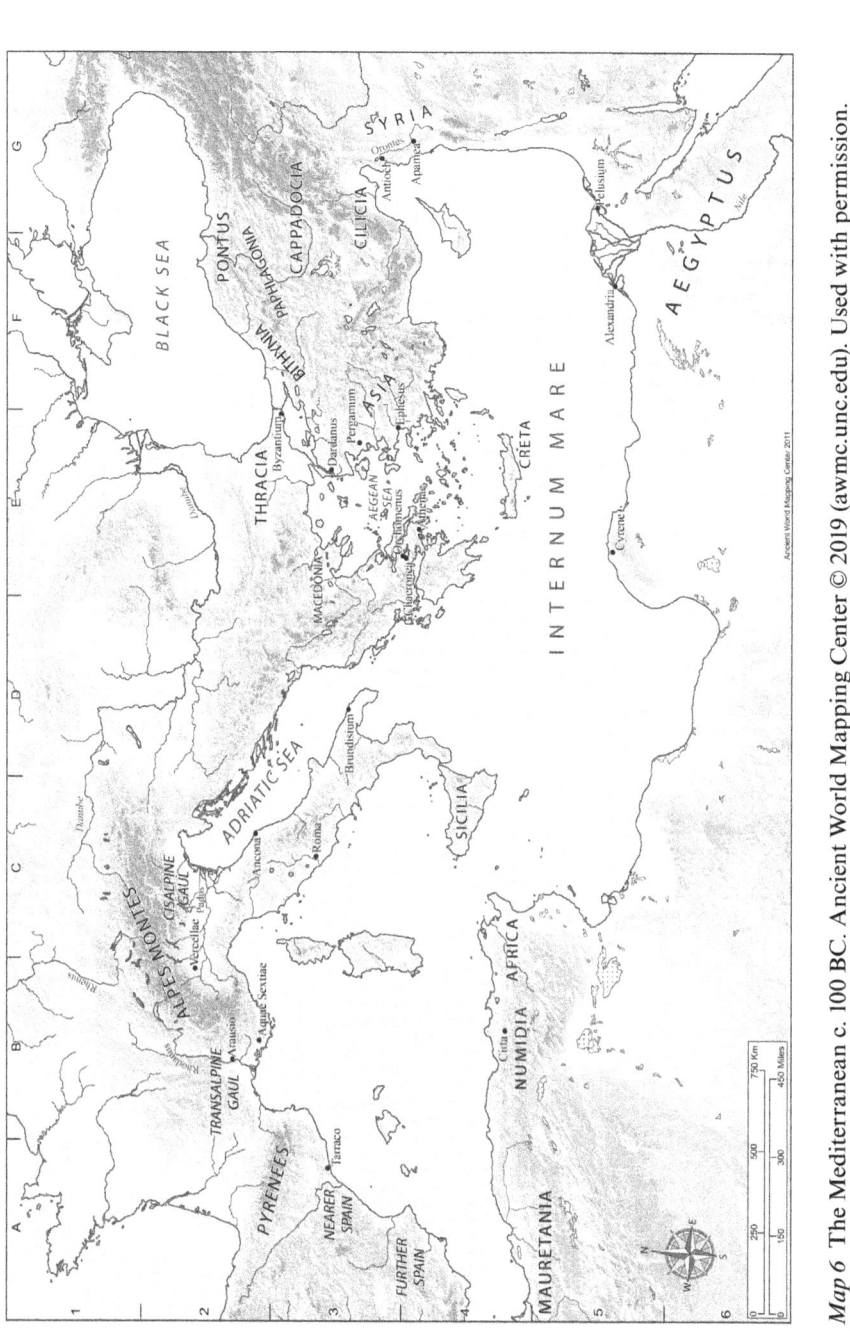

Map 6 The Mediterranean c. 100 BC. Ancient World Mapping Center © 2019 (awmc.unc.edu). Used with permission.

1 Writing about Romans at war*

Jeremy Armstrong and Michael P. Fronda

The study of the Roman army is almost as old as the genre of Roman history itself. Polybius, writing in the middle of the second century, offered the first – and still one of the most important – studies of the Roman army of any period. Taking up 24 chapters (19–42) within Book 6 of his *Histories*, his description and analysis of the army of the Republic both set the stage for, and helped to shape the trajectory of, the field of Roman military studies which came after him. He is, arguably, the father of the discipline as it exists today.[1]

The bulk of Polybius' discussion focuses on military praxis, organization, and equipment. He offered detailed descriptions of recruitment, marching orders, camp construction, and tactical arrangements. These topics were expanded upon in other sections, including his famous comparison of the Roman legion and Macedonian phalanx (18.28–32). His "nuts and bolts" approach to the army reveals both his familiarity with the practical details of military systems – Polybius having served as a *hipparchus* for the Achaean league in 170/169 – and his appreciation for organization and command structures. Polybius offered an educated officer's view of the Roman military system, which presented the army as a rational and practical tool of power – a set of systems and institutions designed to maximize the effectiveness of both the armed force and the power and influence of Rome's military elite. Accordingly, his analysis has long resonated with later aristocrats and military officers-turned-authors, both from antiquity and modernity. His detailed and precise descriptions of Rome's military order offered a paradigm for well-organized operations to which later generals could both compare and aspire. His description of the army did not include the sumptuous and dramatic individual moral *exempla* found in an author like Livy – although it did contain moral aspects – but represented a deeply practical and informed expression of military structures. Polybius' work did not focus on the ideal heroic warrior, but rather on the ideal *army*.

* The authors thank Brahm Kleinman, whose suggestions improved this introduction. All dates are BC unless otherwise noted.

1 For an overview of Polybius as military historian more generally, see Marsden (1974, 295), who concluded, "...at the very least, he began the breakthrough into more advanced, even modern, military history."

DOI: 10.4324/9781351063500-1

This chapter has been made available under a CC-BY-NC-ND license.

Polybius' focus on military systems is important for many reasons, but particularly because it is indicative of an entirely different approach to the study of ancient warfare and military forces from that found in earlier texts. While earlier writers, such as Herodotus, Thucydides, and Xenophon, certainly discussed and described large group actions and battles, their focus typically remained on the actions and influence of (often heroic) individuals – albeit sometimes considered collectively. Classical Greek historians generally only discussed military systems through their examination of the Spartan ἀγωγή, thought to produce the best soldiers. But even here, the emphasis was often on individual citizens and the purpose was usually to draw explicit contrasts with communities like democratic Athens. Military systems were rarely explored, in their own right and from a historical perspective, before the Roman period – at least within our extant sources.

The caveat "within our extant sources" is, however, a required one. Although Polybius is one of our first surviving sources to approach an army and warfare in this systemic way, he was most certainly working from, and building upon, a foundation of Hellenistic precedents which are now lost. The fourth century witnessed an expansion and "complication" of warfare which featured, among other things, ever larger and increasingly mercenary armies, composed of myriad different unit types, fighting for longer, and further from home. In this environment, military systems – including generalship, military organization, and logistics – became important topics of study. This can be seen in the fourth century with the work of Aeneas Tacticus, who wrote a number of military treatises, including his only extant one "How to Survive under Siege."[2] It also likely formed an important part of the now lost work of Hieronymus of Cardia, who was probably used by Polybius (although not mentioned by name) and certainly by Diodorus for organizational details, as well as many other Hellenistic writers.[3] The genre of military writing was clearly evolving. But Polybius still stands as a vitally important contributor in this area. First, because his work does survive, while those of his Hellenistic predecessors, for the most part, do not. Thus, he provided the model that became the core of the later discipline, which favored this systemic approach. Second, because he translated this approach to Republican Rome – framing, seemingly for the first time, Rome's army in these systemic terms.[4]

---

2 The exact identification and dates of Aeneas Tacticus are often debated but remain unknown. However, as an epitome of his work was made by Cineas, who worked in the court of Pyrrhus, he can be placed in the fourth century at least.

3 For the likely use of Hieronymus by Polybius, see Billows (2000, 286–306), with particular reference to Polyb. 8.10.11. See also Walbank (1967) *ad loc*. For use by Diodorus, see Diod. Sic. 2.1557f; see also Billows (2000, 300) for discussion; Hornblower (1981) for general discussion and later reception.

4 This is not to say that Polybius was *solely* concerned with systems. He knew that an army was only as good as the men who filled its ranks. Thus, he also implied that aspects of Roman military behavior (for example, discipline, encouragement of martial valor on and

As part of this innovative focus on the wider military system, Polybius explicitly acknowledged the link between the Roman military and Roman politics. In this, as with his discussion of military praxis, he was likely working from precedents dating back to the fourth century as well – although here, thankfully, his antecedents are somewhat more secure, with clear allusions to the work of Plato and Aristotle. The discussions of both the Roman "military system" and the Roman "political system" are included in Book 6, suggesting the inseparability of war-making and politics.[5] Indeed, Polybius "bookends" his discussion of the military system with politics – Chapters 11 through 18 describe the Roman constitution, and Chapters 43 through 56 compare the Roman system with others from around the ancient Mediterranean. This integrated and systemic approach is again likely a reflection of both his experience as an aristocratic, elected military leader, and the military context within which he operated: a Hellenistic world dominated by great kingdoms and empires, where armies acted as the military extension of a state's will, even if that "state" was a single king. And, again, this focus helps explain Polybius' enduring appeal. His experience and context resonated with many later authors, from both the late Republic and early Empire, as well as with those from more recent times. While modern readers may sometimes struggle to understand the individual motivations of Homer's Achilles or Ajax, we can appreciate the ordered relationships which existed between Rome's military, political, and social apparatus as described by Polybius. His model, based on systems, structures, and ideals common to large states, was able to transcend his specific context.

Thus Polybius, by focusing both on the practical aspects of the Roman military and war-making, as well as the cultural and imperial implications of these practices, paved the way for later writers to move beyond the individual, heroic ideals, and descriptions which seem to have dominated military literature of earlier periods, and move toward more systematic and relational approaches to ancient warfare. As a result, Polybius was seemingly the first historian (or at least the first whose work is extant) to explore

---

individual and corporate level, etc.) reflect a deeper Roman character, which, in turn, helps to explain the "success" of Rome and the superiority of their political system over others. No coincidence, then, that Book 6 ends with the brief account of "Horatius at the Bridge" and a description of the aristocratic funeral – both showing how martial valor was communicated to the people, inculcated among the young, and translated into political capital.

5 Polybius' discussion of the military is also invoked implicitly to explain Roman imperial success (itself a reflection of the superiority of the overall Roman system). After all, Polybius famously set out to explain how the Romans came to conquer the whole world in 53 years, an achievement hitherto unprecedented (Polyb. 6.2.3). According to Polybius, it was the Roman military-political system that ultimately accounted for Roman success, and for the Romans doing what no other great power before (Persians, Athenians, Spartans, Macedonians, etc.) were able to do.

what is now called "war and society" in antiquity – and almost certainly the first to do so in a Roman context. Returning to the point made above, although undoubtedly building upon an existing foundation of scholarship, Polybius was seminal in creating the framework for describing and defining the Roman army in its socio-political context which most subsequent writers have followed.

*   *   *

Like Polybius, the authors and editors of this work are explicitly building upon – and indeed are heavily indebted to – a much wider tradition of pre-existing scholarship on the subject of ancient warfare. However, unlike Polybius, we are thankfully not required to translate and transport this work to an entirely new culture. While he was seemingly forced to blaze a new scholarly trail in describing the Roman military system in the second century, in the twenty-first century, we are able to walk a very well-worn and established route. Indeed, the faint path of Roman Republican military history pioneered by Polybius has been transformed over the past two millennia into a massive thoroughfare, with many branches, each featuring rich and vibrant subfields of study.

Despite the tremendous developments that have occurred in Roman military history, many of the central aspects of the modern discipline still owe much to Polybius' work and aims, and indeed can be mapped onto the same major areas he focused on. As noted above, Polybius' "nuts and bolts" approach to the Roman army, with its focus on praxis, organization, and equipment, has long appealed to army officers-turned-academics, as many of the timeless and universal details involved in organizing masses of men likely resonated with their own experience in armed forces.[6] Recruitment procedures, ordered rows of tents, and strict marching orders would have been a common, shared experience for many who had served in armies during the nineteenth and twentieth centuries. Indeed, some of this resonance may have been either conscious or circular, as Polybius was often mined for useful principles and strategies by modern generals seeking to improve efficiency, or at least seen through the practical lens of contemporary military practice and doctrine.[7]

The most famous example of this for Roman Republican warfare is perhaps the work of Hans Delbrück, who not only wrote a hugely influential study of ancient warfare in the nineteenth century but also was a Prussian officer who saw service in the Franco-Prussian war. Delbrück consciously brought a deep and practical understanding of how modern armies functioned to

---

6 For discussion of classical warfare earlier in the Enlightenment, see: Earle (1971, 3–25, 260–86); Garlan (1975, 15–21); Dawson (1996, 169–91).

7 For example, officer-military theorists Ardant du Picq (d. 1890) and Alfred von Schlieffen (d. 1913) both attempted to apply lessons from classical antiquity to modern warfare.

his work on antiquity.[8] Indeed, his technique of *Sachkritik*, which judged sources and accounts based on their "practicality" (filtered through his own firsthand experience), has been critiqued as having "often degenerated into rejecting descriptions in Herodotus or Caesar through wooden comparisons with the experience and practice of the contemporary German army."[9] While this approach is obviously problematic in many respects, particularly for modern scholars trained to acknowledge and account for their modern biases, it is noteworthy that Delbrück felt enough of a connection to his material to be able to do this at all. While he may have struggled to bridge the massive cultural divide between antiquity and his own day, it is clear that men like Delbrück empathized with the issues Roman generals faced in organizing a group of men into a functioning fighting force. He also saw, in the writings of men like Polybius, reasonable and indeed imitable solutions to these issues. Ancient military history was seen as relevant and applicable to his contemporary military context and approached as such. Although these types of studies are no longer quite as common today, likely because the overlap between scholars and those with military backgrounds has shrunk, they can still be found in the works of scholars like Richard Gabriel (Royal Military College of Canada) and Donald Boose (U.S. Army War College).

Although Polybius was also concerned with military equipment, developments in military technology ensured that this area did not have exactly the same type of resonance as more organizational matters in modern, and early modern, historiography. While both ancient and more modern armies needed to organize large groups of men to fight, the equipment they used to fight was, obviously, vastly different. Despite this, however, descriptions of military equipment continue to hold an important place in Roman military studies. One part of this likely relates to the practical and very concrete nature of ancient military equipment – many pieces of which have been on display in both museums and private collections since the time of the Grand Tour. These items created a physical, and perhaps experiential, link to the past, which has long attracted both collectors and re-enactors as well as fueled publication areas like the *Osprey Military History Series*. In more recent years, new archaeological discoveries have allowed for further developments – for instance, the refinement of the picture of the legionnaire's kit described by Polybius and huge amounts of information on the panoplies of later periods.[10] The appearance of the *Journal of Roman Military Equipment Studies (JRMES)* in 1990– and its reappearance in 2016 after hiatus – along with

---

8 Hanson (2007, 7–8).
9 Hanson (2007, 7).
10 For example, Bishop and Coulston (1993), whose *Roman Military Equipment: From the Punic Wars to the fall of Rome* utilized then-recent archaeological evidence to displace earlier works such as Robinson (1975) and, even older, Couissin (1926). In turn, Bishop and Coulston published a second version of *Roman Military Equipment* (2006), revised greatly from the original in response to the growing corpus of relevant archaeological evidence.

the periodic iterations of the related Roman Military Equipment Conference (RoMEC) attest to the ongoing interest in Roman military equipment, with particular emphasis on material finds. Survey publications and more specialized studies of Roman arms, armament, and military materiel continue to appear regularly, and thus Polybius' focus on praxis remains evident in the modern field; although not, perhaps, to the same extent seen in his work, and almost always with a different aim and set of assumptions. Whereas Polybius typically focused on an (admittedly far from static) ideal form,[11] the modern field of military equipment studies fully embraces the variability and lack of standardization which the archaeological record displays.

The movement away from an ideal and the increased acceptance of the variability inherent in ancient warfare is visible elsewhere as well. For instance, there has been a gradual shift in modern scholarship away from discussions of the minutiae of specific military engagements, and particularly for the Roman Republic. Although interesting and instructive, single events and periods, no matter how well documented, are no longer understood to be indicative of the whole of Republican society and history.[12] With a few important exceptions – for example, John Lazenby's *Hannibal's War: A Military History of the Second Punic War* (1978, reprinted in 1998) – more "traditional" military histories focusing on specific wars have fallen out of favor, and relatively few scholarly publications are dedicated to the analysis of individual battles (an observation commented on by Nathan Rosenstein in the concluding chapter to this volume).

This is not to say that the nitty-gritty mechanics of battle in the Roman world have entirely disappeared from the scholarly landscape, and indeed Greg Daly's *Cannae: The Experience of Battle in the Second Punic War* (2002) is one example. However, recent work in this area is fundamentally different from the battle analyses which were common in earlier eras, as they typically reflect one of the major developments in military scholarship in the last 40 years: greater appreciation for the "fog of war," the chaos of the battlefield when viewed from the perspective of the individual soldier rather than the bird's eye view of the (armchair) general, and the practical realities of battle that give the lie to popular myths and theories of ancient warfare. This approach to military history has been inspired by John Keegan's *The Face of Battle* (1976) and is an area where, as his work so ably showed, one might profitably make use of comparanda from different periods. His methods

---

11 Indeed, Polybius recognized changes to Roman military equipment over time, namely their adoption of Greek-style armor and weapons for cavalry troopers (6.25). Polybius highlights this as evidence of what he saw as a particular Roman attribute: their willingness to adopt and adapt superior military practices from others.

12 The publication of Flower's *Roman Republics* (2010) marked a key turning point in the movement to break down the concept of "the Republic" as a single, monolithic entity and society. Instead, as Flower and others since have argued, Roman society was in a constant state of development throughout its history, but most notably during the Republic, and should at best be understood as a series of "Republics."

were first applied to the study of ancient warfare by scholars interested in the Greek world, especially hoplite warfare.[13] Subsequently, Keegan's approach influenced Roman military studies, though (as Philip Sabin has noted) the tradition remains relatively less developed for Roman warfare.[14]

Along these lines, the last generation of scholarship has seen a rising awareness of, and appreciation for, the premodern context of Roman warfare. This usually sees greater stress laid on the difficulties and messy realities of military campaigns in the ancient world, typified by such features as slow movement, limited communications across (and beyond) the battlefield, amateur quality leadership, and under-developed structures for recruitment and supply. Such an approach tends to be more open to comparative analysis while, correspondingly, more skeptical of ancient Greek and Roman authors who tend to glorify the Roman army and idealize its structures, organization, and functioning.[15] In particular, much recent scholarship has advanced our understanding of the profound logistical challenges for any ancient society, including the Roman state, to outfit and, especially, feed large armies – and the impact, therefore, of large armies on the local economy. Paul Erdkamp's *Hunger and the Sword* (1998) stands out as exemplary in this area, as well as Jonathon Roth's *The Logistics of the Roman Army at War (264 BC–AD 235)* (1999).[16]

Rome's remarkable capacity (by ancient standards) to levy large armies, to field multiple armies year after year for decades, and to absorb staggering defeats without crumbling has drawn particular scholarly notice. Indeed, the tradition can again be traced to Polybius, who situated his discussion of the Roman political and military systems chronologically in his narrative immediately after the battle of Cannae, a military disaster that would have crippled virtually any other ancient power.[17] Earlier in his narrative, Polybius famously claimed to present the total number of Roman citizens and Italian allied men of military age available for the levy, to impress upon his readers Rome's vast military resources. This lengthy passage (2.23–24) is a cornerstone of modern research on the Roman military and, more generally, Roman-Italian demography. The crucial modern study on this topic remains Peter Brunt's *Italian Manpower 225 BC–AD 14* (1971). This magisterial book has shaped and framed virtually all subsequent scholarship pertaining to the interlocking questions of the population of Rome and Italy in antiquity, the impact of that population (and the ability to marshal it) on Roman conquest, and the long-term demographic (and thus economic and political) ramifications of generations of warfare. To take but a single recent example, Nathan

---

13 For example, Hanson (1989, 1991).
14 Sabin (2000, 1–2); see also Sabin (1996); Koon (2011).
15 This trend again derives in part from important scholarship on the Greek world, for example, Hanson (1983).
16 See also Shean (1996); Erdkamp (2011a).
17 Polybius (6.52.5–7) comments explicitly on Rome's ability to weather defeats better than the Carthaginians because of the superiority of their system; cf. Polyb. 6.58.7–13.

Rosenstein's *Romans at War* (2014) argued that Roman and Italian agricultural rhythms and demographic patterns (relative age at marriage, number of children, etc.) were actually well synchronized with cycles of warfare even in the second century. His conclusions challenged the standard interpretation that near continuous overseas warfare resulted in rural depopulation and also led ultimately to the destabilization of politics in the late Republic.

However, it is Polybius' "war and society" approach, which brought together war, politics, and culture in a single, rational model, that is his most important historiographic legacy in the modern field of Roman military history. Although Roman warfare had presumably been associated with elite power and identity in Roman society well before Polybius set pen to papyrus, as indeed explored by several chapters in this volume (see especially those by Drogula and Serrati), his expression of Roman warfare and military expansion as being intimately connected to the nature and character of the Roman state fundamentally reshaped how it was viewed and understood – seemingly by the Romans themselves, as well as by modern scholars.[18] Indeed, bolstered by Enlightenment and early modern thinkers, most famously Clausewitz and his dictum that "war is the continuation of politics by other means," it has become virtually unthinkable in the modern discipline of Roman history to discuss warfare – especially in the Republican period – without linking it to its political context, if not its social context as well. Thus, most works focusing on the Roman army in the Republic tend to adopt a more holistic approach to the topic of Roman warfare. Although arguably initiated by necessity, this contextualized approach, which relies on the close relationship between Roman warfare and politics, has become perhaps the most important strand of in the development of Roman military studies in the last generation. The bibliography is vast, and the sheer number of related topics is impossible to summarize in a brief introduction. However, a few main lines of inquiry stand out.

The first centers on the direct influence and impact of the army – of soldiers and veterans – on the political system: the "politicization of the army." It is commonly argued that the Roman state was forced in the second century to lower property qualifications to serve in the legions in order meet a (real or perceived) shortage of military manpower, with the state increasingly funding the cost of service rather than recruiting only self-funded citizen-soldiers. The Roman army thus transformed from a militia to a professional force. Emilio Gabba (1976) argued that the evolution of the Roman army, in turn, had a transformative effect on Roman politics: as soldiers joined the legions increasingly in search of pay, booty, and (eventually) land upon discharge, their fortunes were tied more and more to their commanders, and eventually their loyalty to the state shifted to allegiance to their general, paving the path to civil war. In this interpretation, the so-called "Marian Reforms"

---

18 On the reception of Polybius in both antiquity and modern times, see the various chapters in Miltsios and Tamiolaki (2018).

mark a key moment in the professionalization of the army and, correspondingly, the breakdown of Republican institutions – even if (as Gabba already noted more than 40 years ago), the revolutionary nature of Marius' reforms is debatable.[19] This interpretation is commonly repeated,[20] though recently François Cadiou (2009, 2018) has argued forcefully against the notion of client armies in the late Republic, thus renewing the debate.

A second debate centers on the profound importance of military success in the Roman political system, where displays of martial valor and (for elites) successful military leadership were both a means to obtaining higher political honors and the most sought-after objectives upon obtaining those honors. The deep embeddedness of a martial ethos (shared by elite and common Roman alike), warfare, and politics was (as noted above) recognized by Polybius in antiquity. In modern scholarship, the relationship was promoted emphatically by William Harris in *War and Imperialism in Republican Rome, 327–70 BC* (1979), perhaps the most important work on war and society in the Roman Republic in the last 40 years.[21] Since Harris' publication, virtually every aspect of the connection between war and politics has been explored.[22]

Karl-Joachim Hölkeskamp has led the way in demonstrating how the Roman Republican elite self-fashioned and communicated their status, with particular emphasis on monuments (such as Duilius' column) and performative commemorations (such as the *pompa funebris*), which themselves focused heavily on martial achievement.[23] The work of Hölkeskamp and others reflects the impact of the "memory turn" on Roman military and political studies. Similarly, the "performative turn" is manifest in the rapidly expanding bibliography on the triumph, Rome's greatest martial celebration. Recent scholarship on the triumph, of which Mary Beard's *The Roman Triumph* (2007) is the most prominent, has demonstrated the deep interconnection not only of Roman warfare, politics, and political competition but also of religion and collective memory.[24] While scholars have tended to focus on Roman victories, in *Imperatores Victi: Military Defeat and Aristocratic Competition in the Middle and Late Republic* (1990), Nathan Rosenstein turned the subject on its head by looking at the defeated Roman generals. He concluded, perhaps surprisingly, that military defeat did not

---

19 See Gauthier (2016a, 2016b), arguing that Marius' "reforms" were simply ad hoc measures rather than lasting institutional changes.
20 See Keaveney (2007, 30–33, esp. 31 n. 250) for bibliography.
21 Although important, this work is also divisive for a number of reasons. Academically, its extreme view of Roman militarism has been forcefully challenged (as discussed below). Additionally, Harris' personal reputation has (re)shaped its reception in recent years.
22 See Rosenstein (2007) for an excellent summary of the intrinsic interrelationship of military courage (*virtus*), glory (*gloria*), praise (*laus*) and renown (*fama*), military command and political success in the middle and late Republic.
23 Hölkeskamp (2010); see also various articles collected in Hölkeskamp (2004).
24 Recent treatments of the triumph include Itgenshorst (2005); Bastien (2007); Pittenger (2008); Östenberg (2009); Lange and Vervaet (2014); Lange (2016); Popkin (2016).

cause significant political harm to aristocrats. The Republican political system had developed various structures, such as faith in the *pax deorum*, that allowed defeats to be explained without upsetting aristocratic consensus or destabilizing competition. Rosenstein's influence, in turn, is seen in more recent publications on Roman losses, such as Jessica Clark's *Triumph in Defeat: Military Loss and the Roman Republic* (2014).[25]

Perhaps the most prominent debate in the last 40 years, within the broader subject of war and society in the Roman Republican period, has been over the nature of Roman imperialism, especially (but not exclusively) Rome's extra-peninsular expansion. How were the Romans able to conquer the Mediterranean world so rapidly, and what motives drove this process? These are questions, as we have seen, that motivated Polybius' historical inquiry in the second century as well. Modern scholarship on these questions can be traced back to the emergence of ancient history as a specialist discipline, with Barthold Niebuhr and Theodor Mommsen both applying themselves to the task and addressing the issue, and the debate has remained a staple of most works on Roman history from that point on.[26] In his analysis, Mommsen interestingly did *not* follow Polybius directly, but instead relied more on Livy in promoting a view of "defensive imperialism" where Rome was drawn into conflicts, seemingly against her own will – a model expanded by Haverfield, Brown, and others in subsequent years.[27] For much of the nineteenth and twentieth centuries, this model, albeit with variations, was a cornerstone of Roman Republican history.[28]

Since the 1970s, however, the topic has been reinvigorated, driven in large part by the work of W.V. Harris, and especially with the publication of *War and Imperialism in Republican Rome, 327–70 B.C.* in 1979.[29] Harris challenged the more defensive and nationalistic approach to empire espoused by scholars like Mommsen and Badian, and offered a more aggressive and competitive model instead. In his own way, Harris, like Polybius, stressed Roman character – highlighting what he saw as their unusually militaristic characteristics and martial ethos which, he suggested, shaped their political structures and policy decisions. Indeed, Harris argued that Romans

---

25 On Roman defeats, see also Waller (2011); Rich (2012); Östenberg (2014b). For a wider consideration of defeat in the ancient world, see Clark and Turner (2018); Marco Simón, Pina Polo, and Remesal Rodríguez (2012).

26 See Terrenato (2019, 10–30) for an excellent summary of the *status quaestionis*.

27 Terrenato (2019, 20). The term "defensive imperialism" was coined derisively by Harris (1979, 163), its strongest critic, as a blanket designation covering a broad range of interpretations that all stress Rome's general unwillingness to annex territory or to set up formal imperial structures, among other considerations, as evidence of their lack of imperial designs.

28 For example, Badian (1958); Badian (1968).

29 Harris (1971) offered a glimpse of the more forceful arguments presented in Harris (1979). He has restated his position in numerous publications since then (for example, Harris 1990, 2016).

of every class consciously sought to expand Rome's empire, and that every class recognized and desired the economic benefits of war. This resulted in a conscious, and widely supported, policy of imperialism and expansion, with the goal of material extraction. His analysis assumes that the Roman martial ethos was distinct: that the Romans were something like a pathological war-making society, and that the entirety of Rome's aristocratic political culture was built on warfare – not only because of the value of *virtus* itself but also because the wealth generated by warfare was necessary in Rome's highly timocratic social and political structures.

Harris' study completely changed the nature of the debate, as virtually all subsequent scholarship on Roman imperialism has responded to his thesis in some way. His demolition of defensive imperialism is widely acknowledged, though scholars have challenged specific details of his formulation. North (1981), in an oft-cited review, and later Eckstein (1987) challenged the assumption that the Roman senate did – or even could – maintain any sort of conscious strategy. Alternatively, some have argued that Harris' analysis is too Rome-centric and does not give due weight to the allies and the peoples on the periphery in shaping international affairs.[30] Turning to Roman motives for conquest, several scholars have suggested that other factors, in addition to "greed," must also be considered. Sherwin-White (1980) noted that, by granting no room whatsoever for the Romans to act out of fear or self-defense, Harris at times forced the evidence to fit a monolithic thesis. Gruen (1984a) argued that material gain, while an important motive, was not the predominant factor driving Roman expansion: for example, the acquisition of glory was itself a major factor.

Much recent scholarship has explored wealth, state finances, and the economy in the Roman world and has served to further nuance the very concept of "greed" in this context. Nathan Rosenstein has revealed that the economic foundations of Roman warfare were much wider and more complex than Harris' model allows. As Rosenstein has demonstrated, the plunder brought in from most of Rome's wars did not usually cover the cost of those wars. Instead, warfare was largely funded by *tributum* paid by Roman citizens (at least before 167), while in the long run annual indemnities filled the treasury more than booty.[31] For individual Romans, especially the common soldiers, the prospects of enrichment through warfare were also uncertain, as recent studies have highlighted.[32] This economic turn forced a critical re-evaluation of some of the underlying assumptions of the links between wealth, warfare, and Roman imperialism.[33] Indeed, Rich (1993) argued that the causes of Rome's many wars are too complex, with motives

---

30 See particularly Gruen (1984b).
31 See Rosenstein (2016a, 2016b).
32 Coudry (2009); Gauthier (2016a, forthcoming).
33 Recent work on Roman warfare and finances in the Republic include: Bleckmann (2016); Serrati (2016); Tan (2017, 2017a).

fluctuating over time according to historical circumstances, to reduce the drive for empire to a single, simple formulation.

Lastly, scholars have further explored the apparent "Roman exceptionalism" in Harris' model, that is, the profoundly militarist nature of Roman society which he stressed. For example, Raaflaub (1996) accepted the basic premise that the Romans were distinctly militaristic, arguing that they were conditioned by highly hostile conditions in the early history of the Republic to be paranoid about their potential enemies and so tended toward highly aggressive foreign policy. More recently, Eckstein (2006) has rejected the notion of Roman militaristic exceptionalism. He granted that the Romans were certainly warlike, but no more so than other Mediterranean societies – and indeed he explicitly drew on Polybius, whose comparison between the Romans and Macedonians is revealing in this regard. Rather, according to Eckstein, Roman militarism is best understood through the interpretative framework of the "Realist" school of international relations theory. "IR Realism" posits that all interstate systems are anarchic (i.e., there is no effective law or force that governs the system), and so *all* states will attempt to maximize their own security and interests, with their behavior constrained only by the relative power of other competitor states within that system. In this view, Rome was no more or less militarist – or "greedy" – than any other peer state in the system. What set the Romans apart, then, according to Eckstein, was not their motives to expand, but rather their ability to manage resources and maintain alliances: in other words, the kind of statecraft that Polybius also highlights as being distinctly Roman.[34]

None of these various responses fundamentally invalidated Harris' core argument, although they have shifted the focus and trajectory of the discussion. Rome is now accepted to be much more aggressive than Mommsen suggested, but this does not, on its own, fully explain the "why" or the "how" of Rome's imperial success. The direction taken by Eckstein, Burton, and now most recently Terrenato (2019), which examines how the Romans shaped and managed the personal, family, and interstate relationships that formed the real foundation of their empire, is surely indicative of the future of the debate.

*   *   *

This volume sits, quite consciously, within this wider tradition of scholarship. It aims to bring together scholars whose expertise encompasses a diverse range of topics, approaches, and chronological periods pertaining to the modern study of the Roman Republican army, and who are actively

---

34 For example, Eckstein (2006); Eckstein (2008); critiqued by Smith and Yarrow (2012). Burton (2011) employs "IR constructivism" to propose an alternate model of ancient interstate relations. For Polybius' impression of the Romans' extraordinary ability to maintain allied loyalty: 2.90.13–14.

engaging with the questions and debates that define the current field. While many popular volumes continue the "nuts and bolts" approach to military history, this volume instead aims to reflect the wider focus of "war and society" (verging on "war and sociology") evident in the modern academic discipline. By emphasizing relationships, psychology, and social structures, scholars increasingly use warfare to explore the nature of Roman society as much as (if not more than) Rome's military systems themselves. In the present volume, this can be seen in the various ways that all of the chapters seek to explicate and define different aspects of Rome's social fabric through the lens of warfare.

Warfare is both one of the best attested activities in all of antiquity and one of the most ubiquitous and permeating. This is certainly true of ancient Rome. Indeed, as seen in the diversity of topics offered at the Celtic Conference in Classics panel in Montreal in 2017 that initiated the present volume, which ranged from poetry to history and from marriage to war-shouts, there is not an area of Roman society known to us that warfare did not impact or influence in some way. Since Polybius, if not earlier,[35] warfare has always loomed large in the history of the Roman Republic in particular, intersecting with virtually every social, political, religious, and economic institution or structure. A broad treatment that spans the full period of the Republic is, we think, justified, and even long overdue.[36]

---

35 The events of the First and (especially) Second Punic War certainly featured prominently in the histories of Q. Fabius Pictor and L. Cincius Alimentus, Rome's earliest historiographers, whose writing predated Polybius by at least a generation. The funerary elogium of L. Cornelius Scipio Barbatus (*CIL* 1^2.6–7), possibly composed c. 280 and perhaps the earliest surviving example of Roman historiography, is in large part a brief war narrative.

36 Diachronic surveys of the Roman military typically devote more pages to the late Republic and (especially) the early Principate. For example, Keppie (1984a), Le Bohec (1989, 1994), and now especially Goldsworthy (1996), all reflect this emphasis on the "high period" of the Roman army. This is not surprising given that this spans the era when literary, epigraphic, archeological, and even documentary evidence for the army becomes relatively abundant. It is a period for which Caesar, Tacitus, and Josephus (among other authors) can supplement Polybius, an entire volume of the *CIL* (vol. 16) can be devoted to military diplomas, and the Vindolanda Tablets can give an unparalleled glimpse into life in a frontier fort. Several important exceptions do exist. Rich and Shipley's (1993) volume devotes roughly equal space to the Republic and Empire. Michael Dobson's monograph, *The Army of the Roman Republic* (2008), is a rare example of a scholarly analysis devoted to the army of the Republic. Michael Sage's *The Republican Roman Army: A Sourcebook* (2008) contains enough exegesis to warrant being included as a study in its own right. His more recent *The Army of the Roman Republic: From the Regal Period to the Army of Julius Caesar* (2018) is aimed at a more general reader. There are also books dedicated to the Roman military systems of specific aspects and periods of the Republic, including by many of the contributors to this volume: Jeremiah McCall's *The Cavalry of the Roman Republic* (2001), Fred Drogula's *Commanders and Command in the Roman Republic and Early Empire* (2015), Jeremy Armstrong's *War and Society in Early Rome: From Warlords to Generals* (2016), and the many works of Nathan Rosenstein already cited, notably his 2004 *Rome at War: Farms, Families, and Death in the Middle Republic*.

This volume is not a "companion" *vel sim.*; it does not aspire to comprehensiveness. Rather, we knowingly attempt to give a snapshot that exemplifies the current state of the field as much as the limits of the medium (limited by word count, number and availability of contributors, etc.) allow. The chapters are organized roughly in chronological order, yet the volume does promote a coherent grand narrative or argue an overarching thesis. Each contributor's work stands on its own. Nevertheless, a number of thematic and topical threads weave through the volume and link individual chapters.

For example, in the chapters looking at the early Republic, there is a strong focus (seen particularly in the chapters of Drogula, Tan, Armstrong, and Helm) on the role of warfare in state formation and how this activity bound various segments of central Italian society together in a particular relationship. Similarly, Fronda expands on this theme through consideration of the military links between Romans and their Italian allies in the century or so before the Social War. The chapters by VanDerPuy, Tan, Roselaar, and Gauthier explore economic relationships and the role of warfare in defining and delineating economic structures. Serrati and Wells examine how myth and religion related to warfare and were deployed, often quite consciously, to help affirm and define social bonds. Milne, Fronda, and Clark's chapters consider different forms of commemoration and memory and their connection to war, whereas Barber analyzes the potentially disruptive impact of war on the relationships that regulated political competition in the Republic, despite the profound importance of martial success to elites competing in that system. Brice, McCall, and Gauthier all explore the power structures of command, with attention on the relationships between the general and his subordinates and between the general and his soldiers (a theme also touched on by Drogula, Armstrong, Helm, Milne, and Fronda).

The "nuts and bolts" approach of traditional military history, while not emphasized, is not entirely absent from this volume, and some more conventional topics typically associated with Roman military studies are dealt with. Thus, several chapters consider the organization and development of the legions and/or its increasing professionalization (Armstrong, Helm, and Gauthier), the nature of command (Drogula and McCall), and military discipline and indiscipline (Brice). Also, while no chapter stresses the mechanics of warfare nor presents an analysis of a particular battle, both Milne and Serrati address the psychological challenge of convincing soldiers to kill (especially in close quarters), and McCall considers several aspects of battlefield dynamics: all three gesture to the influence of Keegan's "face of battle" approach. And while no chapter is dedicated to the treatment of a specific war, another running theme throughout the volume is the impact of individual wars as pivotal drivers of historical change and development. Thus, the conquest of Veii, the Latin War, and the Samnite Wars figure prominently in the early chapters. Moving later, the Second Punic War is a critical fulcrum in the chapters by Fronda, Barber, and Roselaar, and the Social War is interpreted as a decisive turning point in chapters by Gauthier, Clark, and

Brice. Indeed, recognition of the profound importance of the Social War to the pace and course of events in the late Republic reflects a recent scholarly turn that increasingly emphasizes this conflict.[37]

Overall, this volume reflects recent trends in source criticism and analysis with regard to the Roman army. Scholars have long sought to move beyond the simple confines of what has been transmitted to us. However, in the field of Roman military history, with its traditional focus on the well attested army of the late Republic and early Empire, this work has typically involved "tinkering" with a fundamentally accepted tradition. In recent years, however, scholars have applied an increasingly critical approach to the evidence for Roman warfare and the Roman army with great effect, particularly for the Republican period. For instance, Nathan Rosenstein's study, "Phalanges in Rome?" offered the first detailed analysis of the literary evidence for what had been one of the key tenets of the early Roman army: the existence of a Roman hoplite phalanx.[38] Despite the fact that the existence of a phalanx in Rome is supported – both directly and indirectly – by numerous literary sources, and although the Romans of the late Republic seem to have been fully convinced of the tradition's authenticity, Rosenstein was able to demonstrate that this aspect of the Roman army's evolutionary narrative may have actually had its origins in the work of Hellenistic historians.

This is not "tinkering," but a fundamental shakeup of core beliefs. This type of work has picked up steam in recent years with the questioning of the "Marian Reforms"[39] and the reinterpreting of Roman military command.[40] It can be seen in the present volume, particularly in Drogula and Armstrong's contributions. The desire to push beyond the limitations of the explicit narrative is also seen in recent efforts at utilizing tools developed in other disciplines to break out of the limitations imposed by the "traditional" sources (see Barber in this volume). Other scholars examine aspects of warfare which the ancient sources arguably did not consider important (as with Helm and Milne's focus on the "common soldiers"), or even to use sources to explore collective memory rather than as tools for historical reconstruction (see Clark in this volume). These sorts of critical approaches are particularly relevant for the study of the army of the Roman Republic, for which so much of our extant source material originated from a much later period than the events described. Not only has there been a profound loss of evidence through the attrition of sources but also the evidence that does survive is subject to myriad biases and temporal distortions.

One area where the present volume is *not* indicative of the current scholarly situation though, is in its minimal explicit use of archaeological material, apart from scattered references to Roman camps in Spain. This is

37 For example, Mouritsen (1998); Keaveney (2005); Dart (2014); Kendall (2014).
38 Rosenstein (2010).
39 See particularly Gauthier (2016a, 2016b), and Cadiou (2018).
40 See particularly Vervaet (2014) and Drogula (2015).

an area which the editors would have certainly liked to expand if given the space, with the work of scholars like Fernando Quesada Sanz and Mike Bishop being some of the most important for the Republican period. There are other areas as well which could have been explored in more depth. The economy of the army and Rome's burgeoning Republican empire have been studied extensively by Nathan Rosenstein – and indeed this volume's title pays homage to a central study on the topic, his 2004 *Rome at War: Farms, Families, and Death in the Middle Republic*. While the contributions of Tan, Roselaar, and Gauthier each delve into this area, the volume could have certainly benefited from the addition of other voices – for instance, the recent work of Michael Taylor, among others. The Roman navy is mentioned only tangentially in a couple of chapters; this subject, too, merits its own chapter. And, interestingly, no contributor directly engages with the "imperialism debate," though the means and motives behind Roman imperialism are implicit in several chapters. The list of omissions is endless.

These caveats aside, the editors hope you enjoy the volume in front of you and that it does – albeit likely to a very small degree – both advance the field of Roman military history and pay respect to the tremendous career of Prof. Nathan Rosenstein to whom it is dedicated. And indeed, it might also be best here to return to Polybius (1.1), with whom we started (and indeed with whom the field of Roman military studies began), whose introduction might also serve as an appropriate introduction to the work before you:

> Had previous chroniclers neglected to speak in praise of History in general, it might perhaps have been necessary for me to recommend everyone to choose for study and welcome such treatises as the present, since men have no more ready corrective of conduct than knowledge of the past. But all historians, one may say without exception, and in no half-hearted manner, but making this the beginning and end of their labour, have impressed on us that the soundest education and training for a life of active politics is the study of History, and that surest and indeed the only method of learning how to bear bravely the vicissitudes of fortune, is to recall the calamities of others. Evidently therefore no one, and least of all myself, would think it his duty at this day to repeat what has been so well and so often said. For the very element of unexpectedness in the events I have chosen as my theme will be sufficient to challenge and incite everyone, young and old alike, to peruse my systematic history. For who is so worthless or indolent as not to wish to know by what means and under what system of polity the Romans in less than fifty-three years have succeeded in subjecting nearly the whole inhabited world to their sole government –a thing unique in history? Or who again is there so passionately devoted to other spectacles or studies as to regard anything as of greater moment than the acquisition of this knowledge?

2　The institutionalization of warfare in early Rome*

Fred K. Drogula

Introduction

Rome was already a large and powerful state when its citizens first began reconstructing the early history of their people and their city. By the time Rome's first historians began collecting their evidence and organizing it into written narratives in the third century, the city had already existed for several centuries, a vast stretch of time during which Rome had certainly experienced substantial (but mostly unrecorded) change and development. There was very little written evidence from this early period, so the first historians relied heavily on the popular legends, folklore, and family stories that had already created a powerful oral tradition and shaped how the Romans understood their past.[1] Raised in this tradition – and having little reason to question it –these historians allowed their preconceived notions to guide their work, even when they found discrepancies or inaccuracies in the evidence that appeared to contradict that tradition.[2] This tendency was supported by a powerful cultural dedication to traditional custom, which drove the Romans to believe they were following the example of their ancestors (the *mos maiorum*). Contemporary institutions were given simple origins and legendary creators: Romulus founded and built the city, religious practices were often attributed to Numa, organizational structures – such as the census and wealth classes – were credited to Servius, and the founders of the Republic created the consulship in 509 to assume most of the responsibilities and authorities of the exiled monarchy. In this way, the Romans tended to assume that their institutions were the deliberate and considered constructions of famous ancestors, overlooking the less remarkable possibility that some of their practices were slow developments that only gradually took shape.

---

* All dates are BC unless otherwise noted.

1　For discussion of sources on early Rome, see, Frier (1979); Oakley (1997, 24–41); Mellor (1999); Ungern-Sternberg (2000, 207–22); Bispham (2006a); Holloway (2008, 114); Feldherr (2009a); Raaflaub (2010, 127–35); Sandberg and Smith (2017).

2　Many ancient historians complained about the various inconsistencies, inaccuracies, and clear falsehoods in their records of early Rome: P. Clodius *FRHist* 16 F1 (= Plut. *Numa* 1.1); Cic. *Brut.* 62, *Leg.* 1.6 and 3.8; Dion. Hal. *Ant. Rom.* 11.62.3; Livy 2.21.4, 3.55.11–12, 6.1.1–2, 7.3.5, 8.40.3–5; Festus, *Gloss. Lat.* 249L.

DOI: 10.4324/9781351063500-2

This chapter has been made available under a CC-BY-NC-ND license.

While the Romans believed that the consulship and the republican system of government were established at the birth of the Republic by their revered ancestors – as if through some constitutional congress – modern scholars increasingly recognize that they probably arose instead from a long and slow process of development, experimentation, and compromise.[3] Flower summarized this point well when she emphasized that there was not one "Roman Republic," but rather several "Republics" reflecting different phases or stages in Rome's development, and Rosenstein has argued that the Roman military in particular went through a gradual process of transformation in the early Republic.[4] This paper seeks to contribute to this reconsideration of early Rome, and the early Roman army, by proposing an alternative explanation for how Roman thinking about military authority and command may have developed. Rather than being a singular regal prerogative transferred to the consulship at the founding of the Republic, military command underwent considerable variation and development in early Rome, and so the Roman understanding of military authority – the idea of *imperium* that was familiar to Rome's first historians – took shape only gradually. This not only provides a new lens through which to view the development of the Republic but it also underscores how even fundamental institutions were flexible and could evolve over time.

Military command in early Rome

The ancient narratives of early Rome have been familiar for over two millennia: the city was founded as a monarchy in which the *rex* held supreme civilian and military authority, and when the seventh and last *rex* was exiled, this authority was transferred to the consuls – new, annually elected magistrates specifically created to receive most of the exiled *rex*'s authority and responsibility. So, Rome was a traditional monarchy first, and then – suddenly and completely – it became a republic. While this narrative is attractive for its simplicity and for the tenacity with which the Romans clung to it, it is unlikely to be accurate. As noted above, modern scholars increasingly recognize that Rome's development was much more complex, and it stretched out over several centuries. There is no reason to doubt that there were men called *reges* in early Rome, as the word *rex* is attested in sixth-century inscriptions at least. Yet, there were also a number of wealthy and powerful oligarchic clans or *gentes* that used private armies or warbands to defend their land and seek profit by plundering hostile neighbors.[5] Indeed, it

---

3 For example, Beloch (1926, 231–32); Heuss (1944, 57–133); De Martino (1972, 1.406–15); Ungern-Sternberg (1990, 92–102); Urso (2011, 41–60); Drogula (2015, 8–45); Armstrong (2016c) *passim*.
4 Flower (2010); Rosenstein (2010, 289–303).
5 On the attestation of the word *rex* – which may refer to a "king" or to the priest, known as the *rex sacrorum* – see Cristofani (1990, 22–23). On the prevalence of *gentes* commanding

is possible that the government of the regal period (such as it was) was more an oligarchy than a monarchy, and that the men referred to as *reges* were simply the leaders of whichever clan was preeminent at that time – more chieftains than true kings. This is suggested by the facts that the *reges* came from several different clans (rather than the monarchy being the hereditary possession of one family), and that at least one *rex* – and perhaps others – seized power and demanded the loyalty of the people, much like a tyrant claiming ascendency over rival oligarchs.[6] The *rex* may have been the man who could summon the full levy of soldiers in Rome, and so command a "state army," but such major campaigns were probably less frequent than the private raids, battles, and skirmishes led by individual Roman *gentes* using their own resources. In earliest Rome, therefore, military command was not the sole prerogative of the *rex* or state, but rather was exercised by any *gens* with the resources to put together an armed force.

The removal of the monarchy does not seem to have changed Roman ideas about military command. According to the literary sources, the exile of the last *rex* triggered a wholesale restructuring of the government into a republic under consuls, but the reality was probably less dramatic. The *gentes* had enjoyed considerable independence in their private exercise of military force under the monarchs, and they would have continued to do so after the removal of the last *rex*.[7] It may even be that the *gentes* were responsible for removing the monarchy: the literary record suggests that it was a group of powerful noblemen who drove out the last *rex*, and afterwards the leading families prevented other ambitious individuals from seizing power in Rome.[8] The removal of the monarchy probably left Rome's *gentes* in control of the city as a traditional oligarchy, and they continued to conduct offensive and defensive military operations, which probably comprised the majority of war-making in the early Republic.[9] Examples of this survive in the traditional stories: the Fabian clan famously fought a private war against Veii; Attus Clausus (the future Appius Claudius) was said to have come to Rome with his retinue of 5,000 retainers; Cn. Marcius Coriolanus had an army of clients that followed him, and the Sabine Ap. Herdonius

---

private warbands in early Rome: Cornell (1988, 89–100); Smith (1996, 185–86); Rawlings (1999); MacKay (2004, 44); Forsythe (2005, 190, 198); Smith (2006a, 290–950); Rich (2007a, 15–16); Serrati (2011, 13–16); Drogula (2015, 8–33); Armstrong (2016c, 98–128).

6 Glinister (2006, 17–32).

7 An example of the oligarchic nature of the early government is found in both Livy (8.12.15) and Dionysius (9.41.2–5), who note that, until 339, all elections had to be ratified by the senate in order to be valid, demonstrating that ultimate authority lay with the elites. As I have argued elsewhere (for example, Drogula 2017, 101–23), it is likely that the civilian management of the city after the removal of the kings fell to the plebeian tribunes and priests.

8 Livy 1.57.6–60.4. On the attempts at tyranny in the early Republic, see Smith (2006b, 49–64).

9 Cornell (1995, 143–50); Smith (1996, 185–92); Rawlings (1999, 97–127); Rich (2011, 15); Armstrong (2016c, 69–72).

led a warband of several thousand men.[10] If one reads the ancient sources carefully, evidence of this kind of private warfare is plentiful: Livy describes Rome and Veii as being neither at war nor peace, but in a constant state of brigandage and freebooting,[11] Livy and Dionysius of Halicarnassus give numerous examples of private raiding by oligarchic warbands, and Dionysius even speaks of the Roman countryside as being dotted with fortresses each under a different commander.[12] While each of these stories must be treated with proper skepticism, together they reveal a clear picture of a time when individual Roman clans freely deployed their own private warbands to seek plunder and to defend their own possessions.[13] The *Lapis Satricanus* preserves a late sixth-century dedicatory inscription to (what is likely) a war god – Marmars – by the leader of a band of *sodales*, providing contemporary epigraphic evidence consistent with the literary evidence.[14] This has led some modern historians to suggest that the college of *fetiales* – the priests responsible for declaring war – were originally created to limit private wars among these powerful clans.[15] It is clear, therefore, that military command took many forms in the early Republic, and it was not the exclusive prerogative of the consuls or other heads of state as depicted in the literary sources.

While tradition insisted that the consuls were the commanders-in-chief of the early Republic, this is unlikely to have been true.[16] In spite of their clear and confident reckoning of consuls' names and actions going back to the founding of the Republic, several ancient authors knew that commanders in the early Republic were not called "consuls," but instead "praetors."[17] This is corroborated by epigraphical evidence, which suggests that the title

---

10 Fabii against Veii: Diod. Sic. 11.53.6; Livy 2.48.8, 49.5; Dion. Hal. *Ant. Rom.* 7.19, 9.15.2–3; Festus, *Gloss. Lat.* 451L; see Ogilvie (1965, 359–62); Richard (1990, 174–99); and Armstrong (2016c, 145–46). Attus Clausus: Livy 2.16.4–5; Dion. Hal. *Ant. Rom.* 5.40.3–5. Coriolanus: Livy 2.35.6. Herdonius: Livy 3.15.5; Dion. Hal. *Ant. Rom.* 10.14.1–17.7 (Livy calls Herdonius' men exiles and slaves, but Dionysius calls them clients). See: Cornell (1995, 174–82); Drogula (2015, 22–23); Armstrong (2016c, 131–37).

11 Livy 2.48.5–6.

12 Small scale raiding common: Livy 2.50.1, 3.1.8, 3.2.13, 3.26.2, 3.38.3, 5.16.3–4, 5.45.4, 6.5.4, 6.31.8, 7.15.8; Dion. Hal. *Ant. Rom.* 3.37.2, 3.39.2–3, 4.15.2, 5.50.3, 7.19.1. Roman territory full of fortresses: Dion. Hal. *Ant. Rom.* 4.15.2, 5.44.1–2, 5.45.3.

13 Garlan (1975, 31) notes that such private wars were common in archaic societies.

14 Stibbe, Colonna, de Simone, and Versnel (1980) 19 fig. 1. See also: Versnel (1980, 97–150); Bremmer (1982); Holloway (1994, 149–55); Forsythe (2005, 198–200); Armstrong (2016c, 142–45).

15 See Rawlings (1999, 113) and Armstrong (2016c, 71).

16 Livy's description of the creation of the consulship (2.1.7–11) displays these common beliefs, and he presents the office as being almost identical to how it existed in the late Republic.

17 Livy (3.55.11–12) and Festus (*Gloss. Lat.* 249L) state clearly that the Republic's first chief magistrate/commanders were called praetors, not consuls. Livy (7.3.5) also refers to the *praetor maximus* as the highest magistrate, while Cicero (*Leg.* 3.8) interchanges the terms consuls, praetors, and judges (*iudices*). Similarly, references to Rome's first law code – the Twelve Tables – mention only praetors, but no consuls (Gell. *NA* 11.18.8, 201.47; Plin. *HN* 18.12). See: Oakley (1998, 77–80); Smith (2011, 22); Urso (2011, 41–60).

consul was not introduced until the first half of the third century.[18] Thus, "praetor" was the older title for the Republic's military commanders.[19] This is usually presented only as an antiquarian curiosity by ancient writers, but it actually presents a serious problem because the same writers also knew that the praetorship – or at least the version with which they were familiar – was not created until 367, when it was established as a judicial office.[20] If the praetorship, as later writers knew it, was only created in the mid-fourth century, then who or what were the praetors before that? If the writers were correct that men called praetors exercised military command in the earliest Republic, these must have been something different from the familiar elected magistrates that were not created until 367. Since the *gentes* were responsible for the majority of military activity in the early Republic, it seems reasonable to believe that the praetors of the early Republic (the men referred to as consuls in the narrative sources) were actually the leaders of gentilician armies or warbands, as well as the occasional state army (see below). Thus, the removal of the monarchy did not produce a dramatic reconceiving of military command around the consulship; the *gentes* who had led private military ventures in the monarchy continued to do so in the early Republic.

The narrative account of military tribunes with consular power (henceforth: consular tribunes) similarly suggests that military command was largely an ad hoc activity in the early Republic, rather than a clear institution of the state. According to tradition, this new office was introduced in 444 as an alternative chief magistracy to the consulship, and was used with increasing frequency until 367, when the consulship was re-established as Rome's chief magistracy.[21] While this explanation seems clear, ancient authors were obviously perplexed by this new office, and they struggled to explain how it was different from the consulship, except that the number of consular tribunes elected each year was variable and not fixed, whereas

18 An inscription published by Cristofani (1986) suggests that Rome's chief magistrate was the praetor in the early third century. The earliest known use of the title consul appears on the surviving tombstone (*CIL* $1^2.7 = ILS$ 1) of Scipio Barbatus (cos. 298), but scholars have suggested it was carved in the second half of the third century. See Coarelli (1972, 36–106); Van Sickle (1987, 42–43); Wachter (1987, 301–42 and n. 9); Flower (1996, 171–75); Drogula (2015, 41–42). The first surviving unequivocal reference to a consul seems to be the statues set up by M. Fulvius Flaccus sometime after his consulship in 264. See Torelli (1968, 71–76).

19 The title "praetor" was closely connected to Rome's military terminology. It was probably derived from *praeire* (to precede) or perhaps from *praeesse* (to be preeminent), and the adjective form (*praetorius*) was widely used to describe military items, such as a general's tent (Cic. *Verr.* 2.4.65; Caes. *BC* 1.76; Sall. *Jug.* 8.2), the guards who protected that tent (Cic. *QFr.* 1.1.12; *Verr.* 2.1.36; *Fam.* 15.4.7), the main gate of a military camp (Festus, *Gloss. Lat.* 249L), and even a general's flagship (Livy 21.50.8).

20 Livy (6.42.11 and 7.1.2) emphasizes that the praetorship was created primarily to be a judicial office.

21 Livy 4.1.1–7.1; Dion. Hal. *Ant. Rom.* 11.53.1–61.3; Zon. 7.19.

there could be only two consuls each year. Livy assumed that a clear rationale must have existed for creating this new magistracy, and offered two different explanations: either it was a response to increasing demand for commanders, or perhaps it was a clever patrician maneuver to keep plebeians out of the consulship by creating an alternative chief magistracy that plebeians could hold instead of the consulship.[22] Both of these explanations seem to be nonsense – Livy himself records members of traditionally plebeian families holding the consulship before the creation of military tribunes, and Rome's military activity did not increase as rapidly as did the college of consular tribunes, undermining the idea that there was a correlation between them.[23] This confusion has led some modern scholars to suggest that the consular tribunate did not actually exist, and was merely a literary fiction created by Rome's earliest historians to explain why they usually (c. 70% of the time) found multiple men – rather than the expected two consuls –identified as annual military commanders between 444 and 367.[24] That is, the erroneous assumption that the consulship went back to the foundation of the Republic may have led early historians to fabricate the consular tribunate in order to explain why their research into early Roman military command did not fit their expectations – they postulated the creation of a new office to make their evidence fit their presuppositions. Whether this happened or not, the salient point is that these early researchers clearly found a large, but variable, number of men exercising military command down to 367, which strongly supports the reconstruction that military command was not yet a state monopoly located in a fixed college of elected magistrates, but rather that leaders of powerful *gentes* could and did engage in military ventures as they wanted.

Whether they were real or a literary fabrication, the title of these consular tribunes is also suggestive of early Roman thinking about the nature of military authority. The basis of the office seems to be the military tribunes (*tribuni militum*), although the frequent description of these tribunes

---

22 Livy 4.6.1–12. He later (6.1.2) includes the consular tribunes among the institutions of the early Republic that are: "obscure not only by reason of their great antiquity – like far-off objects which can hardly be described – but also because in those days there was but slight and scanty use of writing, the sole trustworthy guardian of the memory of past events...." (...*cum vetustate nimia obscuras, velut quae magno ex intervallo loci vix cernuntur, tum quod parvae et rarae per eadem tempora litterae fuere, una custodia fidelis memoriae rerum gestarum...*).

23 Holloway (2008, 107–25) gives a full discussion of why the explanations given by ancient authors for the creation of the consular tribunes are weak and should be dismissed.

24 See *MRR ad loc.* for references. No curule elections were said to have been held between 375 and 371. Pinsent (1975, 29–61) and Holloway (2008, 107–25) give the fullest discussions about the debates regarding consular tribunes, but see also Beloch (1926, 247–64); von Fritz (1950, 3–44); Staveley (1953); Adcock (1957); Boddington (1959); Sealey (1959); Ogilvie (1965, 539–50); Ridley (1986); Mitchell (1990, 137); Richard (1990); Cornell (1995, 334–39); Oakley (1997, 367–76); Stewart (1998, 54–95); Forsythe (2005, 234–39); Drogula (2015, 25–57).

as possessing consular *potestas* indicates that they were somehow different from ordinary military tribunes: military tribunes were important officers, but consular *potestas* was needed to be the supreme commander of
an army. Yet, this poses a problem: in the later Republic, *potestas* was the
regular civilian authority invested in all Roman magistrates, most of whom
did not exercise military command.[25] Since these consular tribunes were
clearly – and even primarily – military commanders, it seems strange that
the description of their office should emphasize their possession of consular
potestas (which did not confer military authority in the late Republic) instead of consular *imperium* (which did confer full military authority). This
emphasis on their possession of *potestas* and the fact that these consular
tribunes are not known to have celebrated triumphs for their victories have
led some scholars to hypothesize that consular tribunes did not possess *imperium*, and that this was the fundamental difference between their office
and the consulship.[26] Yet, this suggestion seems impossible if one is using the
understanding of *imperium* that was common in the late Republic: that *imperium* was absolutely necessary to exercise legitimate military command.[27]
According to this understanding, if the consular tribunes did not possess
imperium, then they could not have exercised military command, and yet
they most certainly did according to the ancient sources. This suggests either that *imperium* did not exist in the late fifth and early fourth centuries, or
that it existed but was not necessary to exercise military command. In both
cases, it seems clear that *imperium* in the early Republic was not the same it
was in the late Republic.

Shifting views

The classic thinking about *imperium* was established by Mommsen, who
accepted the Roman tradition that it represented the original monarchical
power of the kings that was transferred to the consuls at the foundation of
the Republic.[28] Heuss challenged this position, arguing that *imperium* was
first created in the Republican period to describe the authority of military
commanders.[29] Both positions have found support among scholars. In both
cases, Mommsen and Heuss assumed that *imperium* in early Rome was

---

25 See discussion in Drogula (2007, 419–52).
26 For recent discussion of this problem, see Armstrong (2016c) 194–95 and Armstrong
    (2017a, 124–48).
27 Cicero (*Phil.* 5.45; *Leg. agr.* 2.30; *ad Brut.* 1.15.7) makes this abundantly clear, and the
    necessity of *imperium* for the exercise of military command appears to have been unques-
    tioned until 54, when the consul Ap. Claudius Pulcher claimed that a *lex Cornelia* of 81
    made formal possession of *imperium* unnecessary, although many – including Cicero –
    questioned the legitimacy of this claim. See Drogula (2015, 107–9).
28 Mommsen (1887/88, 1.1–24, 116–36).
29 Heuss (1944, 57–133). For recent discussions of the debate on *imperium* with references, see
    Vervaet (2014, 17–53) and Drogula (2015, 81–130).

substantially the same as it was in later Rome, no doubt because this was how ancient writers spoke and seem to have thought about it.[30] Yet, this assumption increasingly seems wrong. The variety of modes of military leadership in early Rome makes it difficult to imagine a single, unified concept of military authority. Indeed, it is striking that no clear definition of *imperium* is found anywhere in Republican sources; our understanding of it is pieced together from a range of different authors and works.[31] Whatever *imperium* was, it seems unlikely that the leaders of private gentilician armies needed it. They led armies and warbands made up of relatives, clients, and retainers, who were bound to their commander through various types of reciprocal relationships. Such commanders could use ties of kinship and friendship, as well as the substantial economic power they held over their dependents, to summon and lead private armies.[32] And since the *gentes* generally sought profit and plunder in their ventures, the hopes of financial gain may have induced many to obey the orders of the commander voluntarily and without compulsion. Religious authority (*auspicium*) was surely necessary to consult the will of the gods before a battle, thereby securing divine favor and reassuring soldiers, but this need not have been the *auspicium publicum* that later magistrates held on behalf of the state, since every Roman possessed the right to consult the gods about his private affairs (*auspicium privatum*), and this was probably sufficient for a commander leading a private army.[33] The *gentes* may have carried out the majority of military campaigns and raids in the early Republic, but it is very unlikely that they possessed or needed *imperium*.

If *imperium* (or a similar concept) existed in the early Republic, it must have been held by those who led state armies made up of levies from the entire city, making it a public acknowledgment of a man's authority to give orders and compel obedience.[34] The *reges* had probably possessed this type

---

30 Cicero, for example, clearly states that the early *reges* possessed *imperium*, but gives no indication that their *imperium* was significantly different from *imperium* as it was known in his own day (*Rep.* 2.13.25, 2.17.31, 2.18.33, 2.20, 35, 2.21.38).

31 Augustus easily manipulated and redefined ideas of *imperium*, and although his power and position were unusual, his ability to reinterpret *imperium* suggests that such unwritten concepts were always open to some degree of development and change. See Drogula (2015, 345–73).

32 A famous example is the story of the Fabian clan, which is said to have assembled an army of 306 family members and retainers to fight a private war with Veii in 479 (Livy 2.49.4; Dion. Hal. *Ant. Rom.* 9.15.2–3). This story is questionable and seems to be influenced by the 300 Spartans who fought at Thermopylae, but it shows the general understanding that the largest *gentes* could recruit sizeable armies from among their retainers. Smith (1996, 189–98) discusses how wealth and land ownership may have created vertical divisions, and the society may have been organized around the territorial interests of the inhabitants.

33 Cic. *Div.* 1.28; Livy 4.2.5, 4.6.2–3, 10.8.9.

34 On this, see Armstrong (2013a). There is also considerable disagreement among scholars as to whether *imperium* was used for civilian governance, but all agree that it was a source of military authority. This paper accepts the position originally argued by Heuss (1944, 57–133), that *imperium* was solely military authority.

of authority by virtue of their positions as *rex*, so their status conferred the authority to command the citizens in war. It is possible that the Romans, in their *comitia curiata,* passed a formal law to confer *imperium* on the *reges* upon their accession as later tradition believed, or it may be that their military authority was less a legal principle and more a basic recognition of one's authority by one's soldiers, such as the military oaths of obedience that soldiers in the later Republic swore when joining the army.[35] This may be hinted at by the fact – still true in the late Republic – that the honorific title *imperator* could only be acquired by acclamation by one's soldiers rather than by a law. Whatever the nature of the *rex*'s military authority (the term *imperium* will be used), it was probably something he acquired for life when he assumed the office, and so *imperium* was closely identified with the *rex*. When the monarchy was removed, the oligarchy may have adapted the concept of *imperium* to enable one of them to summon and lead a large state army when needed. It is likely that the oligarchs selected one or two of their own members to lead the army and presented them to the citizens (probably in the *comitia curiata*) for confirmation, but not necessarily election. This is hinted at in the literary accounts of early Rome, where the people are regularly portrayed as having no choice of candidates in elections; the elites put forward only as many names for command as were needed, leaving the people with no choice but to accept those who had been preselected.[36] While the approval of the people remained constant, the conveyance of *imperium* in the early Republic must have caused some new thinking about the nature of military authority. Rather than conveying a lifelong grant of *imperium* to a king, the people began making short-term grants of *imperium* to individuals for specific campaigns. Thus, *imperium* was discussed and conferred far more regularly than merely once in a generation, and it could be conferred to multiple individuals at one time, thus distributed among several men rather than concentrated in one *rex*. In this way, the removal of the monarchy must have caused significant change and development in the way the Romans thought about military commanders and authority.

35 Tradition that *reges* received *imperium*: Cic. *Rep.* 2.13.25, 2.17.31, 2.18.33, 2.20.35, 2.21.38. Military oaths: Polyb. 6.21.1–3; Livy 2.32.1–2, 3.20.1–5, 22.11.8, 22.38.1–5, 28.27.4 and 12; Dion. Hal. *Ant. Rom.* 10.18.1, 11.43.2. Latte (1934/36) 59–77 suggested that early military authority may have been based upon an oath, and that the *lex curiata de imperio* was a survival of an earlier time before the nature of magisterial authority was fully defined. Magdelain (1964, 198–203) argued that *imperium* could not be a type of oath because the *lex curiata* that conferred *imperium* later in the Republic was clearly a legal text and not an oath, but this assumes that the Roman understanding of *imperium* had not changed over several centuries, which is possible but perhaps unlikely. Compare with the better documented kings of Macedon, whose ascension to the throne was only possible if they received a positive vote from the army by acclamation; see Hammond (1989, 271); Errington (1990, 7, 15–16).
36 For example: Livy 1.43.10–12; Dion. Hal. *Ant. Rom.* 4.21.1–3, 7.59.3–10, 8.82.6, 10.17.3, 11.10.4.

A passage probably from Lucius Cincius Alimentus (pr. c. 210), preserved by Festus, offers some hints about how Roman thinking on these subjects was advancing. Alimentus describes a religious ceremony used in the mid-fourth century to select a commander (a praetor) to assume command of Rome's full army, including Latin allies:

> In his book on consular *potestas*, Cincius said this was the custom of this office: "The Albans held authority up to the time of King Tullus, but Alba then being destroyed, up to the consulship of P. Decius Mus [in 340] the Latin people were accustomed to take counsel at the headwaters of the Ferentina, which is under the Alban Mount, and to administer *imperium* by common council. Yet in a year when it was appropriate for the Romans to send a general to the army by order of the Latin name, many of our men were accustomed to consult the auspices on the Capitol from the rising sun. When the birds were favorable, that army, which had been sent by the Latin community, was accustomed to salute as 'praetor' that one whom the birds had approved, who would undertake that *provincia* with the title 'praetor.'"[37]

This passage could simply record an archaic version of the religious ceremonies that all Roman commanders in later periods performed when preparing to exit the city and take up a military command, but there are some unusual details.[38] First, the passage indicates that a praetor with *imperium* was only created as needed ("in a year when it was appropriate..."), suggesting that there was no annual office of praetor as it existed in the late Republic – the praetor was an ad hoc commander. Second, the conferral of *imperium* is connected to the leadership of the whole Latin people (Romans and allies), suggesting that other types of military command may not have required *imperium*. Third, Ziółkowski noted that the description of many men (*conplures nostros*) conducting the auspices in this ceremony is unique in our

---

37 Festus. *Gloss. Lat.* 276L (trans. Drogula): *cuius rei morem ait fuisse Cincius in libro de consulum potestate talem: Albanos rerum potitos usque ad Tullum regem: Alba deinde diruta usque ad P. Decium Murem consulem populos Latinos ad caput Ferentinae, quod est sub monte Albano, consulere solitos, et imperium communi consilio administrare: itaque quo anno Romanos imperatores ad exercitum mittere oporteret iussu nominis Latini, conplures nostros in Capitolio a sole oriente auspicis operam dare solitos. Ubi aves addixissent, militem illum, qui a communi Latio missus esset, illum quem aves addixerant, praetorem salutare solitum, qui eam provinciam optineret praetoris nomine.* For commentary, see Ogilvie (1976, 103–4); Cornell (1995, 299–301); Oakley (1997, 340). It is not certain that the Cincius cited by Festus is the historian Cincius Alimentus or a later antiquarian author. Forsythe (2005, 187–88) assumes the latter, and this passage is not included among the fragments assigned to Cincius Alimentus in *FRHist*.

38 Even in the late Republic, a consul or praetor needed to perform certain religious ceremonies before exiting the city to take up a military command, and the failure to perform those ceremonies properly invalided his authority to command (Cic. *Att.* 4.13.2; *Pis.* 55; Caes. *BCiv.* 1.6; Livy 21.63.5–13, 36.3.14, 37.4.2, 41.10.5, 41.17.6, 42.49.1–2, 45.39.11; Varro, *Ling.* 7.37).

knowledge of Roman auspication practices, leading him to argue that these "many men" were not augurs or magistrates, but may have been candidates for selection as praetor.[39] He thus suggested that this ceremony reveals an early method for selecting (and not just confirming) a man as praetor. If true, this would echo the famous auspication ceremony said to have been conducted by Romulus and Remus, when they consulted the auspices to determine which of them was to rule the new city. Whether Ziółkowski is correct or not, this passage reflects a period when the concept of *imperium* was still developing into what it would become in the late Republic.

Evidence in the literary sources is also suggestive of how Roman thinking about *imperium* developed in early Rome. Although later writers interpreted the so-called "Conflict of the Orders" as a social revolution, the reforms the Romans struggled over during the fifth and fourth centuries focus heavily on the definition and control of the use of force, including *imperium*: in 494, the college of plebeian tribunes was created to protect the common folk from arbitrary abuse by the powerful; in 462, the urban population – led by the tribunes – began demanding a special commission to devise and codify laws defining and limiting the use of *imperium*; in 457, the college of plebeian tribunes was increased from five to ten to further limit abuses; in 454, the plebeians are said to have compelled the ruling elite to send an embassy to Greece to research law, which ultimately resulted in the publication of the Twelve Tables in 451 and 450; and a *lex Valeria* is attributed to 449, granting the people the right of appeal from arbitrary use of *imperium*.[40] Many of these reforms are associated in the literary sources with conflicts over military recruitment and the refusal of military service, emphasizing the tradition that the "Conflict of the Orders" focused heavily on reforms to the use of military force.[41] While one may fairly question how accurately these events are described by our sources, the early historians who gathered this material clearly believed that the Roman people worked strenuously to change how force – especially military force – was used in the fifth century. Step by step, the Romans were thinking about how the use of military force needed to be brought within public control.

While the concept of military authority was clearly evolving in the early Republic, at some point before the mid-fourth century it became the norm for all commanders – not just commanders of state armies – to receive *imperium*

---

39 Ziółkowski (2011, 465–71) notes, "Yet even if anything like this would be practicable in real life, this would not be an *auspicium* before the departure to a province, but the gods' deciding the assignment of the said province, i.e., something entirely different, which in truly historical times was managed through *sortitio*."

40 Cic. *Rep.* 2.54; Livy 3.9.1–13, 3.30.5–7, 3.31.7–8, 3.55.4–6 and 14; Dion. Hal. *Ant. Rom.* 10.1.1–2, 10.26.4–30.6.

41 For example: Livy 2.24.1–8, 2.27.10, 2.32.1–8, 6.27.10, 6.31.5, 6.32.2; Dion. Hal. *Ant. Rom.* 6.23.2, 6.25.1, 6.28.2–3, 8.81.3. See Armstrong (2008, 52, 2016c, 165, 179) for a full list of these refusals of service.

by a vote of the *comitia curiata*.[42] When this happened is not clear, but by 367 the independent command of private armies seems to have disappeared in Rome, and only men who received a *lex curiata* conferring *imperium* were entitled to hold command (see below). It is likely that the struggles to reform and define *imperium* in the fifth century led to this change, which was facilitated by the changing migratory and economic conditions in Latium (and Italy) that were driving the *gentes* to start prioritizing the conquest and acquisition of land rather than seeking moveable plunder.[43] This change meant the *gentes* needed larger armies that drew on the urban population, which would have enabled the *comitia curiata* to assert that their sanction and the conferral of *imperium* was necessary to authorize the leader of a gentilician army to recruit large numbers of soldiers from the city. This development may be reflected in the dramatic rise in the number of commanders recorded in the late fifth and early fourth centuries. Although this was later explained as the introduction of a new magistracy (the so-called "consular tribunes"), it may instead be the result of more men – mostly leaders of gentilician armies – seeking *imperium* from the *comitia curiata* so they could raise larger armies.[44] A man whose authority to command had been confirmed by a *lex curiata* was more likely to be recorded than a man who simply led a group of retainers on a private raid, so the proliferation of recorded commanders between 444 and 367 may have been caused by new thinking about *imperium* that led more commanders to seek a *lex curiata*.

While a large number of men were recorded as commanders in the late fifth and early fourth centuries, this does not mean they exercised command for the entire year. In all likelihood, most of the men who exercised military command in the fifth and fourth centuries only did so for a few days or weeks at a time, and so they only needed military authority for the duration of a campaign or raid. This was surely the case with leaders of gentilician armies, but the excerpt from Cincius Alimentus above suggests that even commanders of state armies were only appointed as needed, and only for particular campaigns. Some men might have sought and received *imperium* for spring campaigns, whereas others may not have sought a *lex curiata* until later in the year as events unfolded. In other words, the large number of men identified as consular tribunes probably did not all hold command at the same time, or for the entire year; each probably exercised command only for the short period necessary for his venture. The office of the dictator may

---

42 Coli (1951, 385) suggested that the *lex curiata* dated to the Republic, but Magdelain (1968, 32–33) argued for the regal origins. See further discussion in Palmer (1970, 67–79, 184–97); Smith (2006a, 184–234); Humm (2011); Mouritsen (2017, 15–21, 25–26). It is widely but not universally accepted that *imperium* was conferred by a vote of the *comitia curiata* (a *lex curiata*). See Drogula (2015, 71–78) for discussion and references.

43 See Armstrong (2016c, 129–232) on the changing military priorities of the *gentes*.

44 As Holloway (2008, 122) pointed out, "if we could consult the family records which formed the basis of early Roman history, we would surely find many individuals who had commanded in war but had never occupied a magistracy."

have preserved this ancient way of thinking, since this military commander could be created and authorized to command if and when needed, receiving *imperium* at any time in the year.[45] And even later in the Republic, although consuls entered office on a fixed date, they could seek a *lex curiata* at any point during their magistracy in order to receive their *imperium*.[46] This reinforces the ad hoc nature of military command in the early Republic: *imperium* could be granted at any time because military expeditions were usually extemporaneous in nature, in response to the unique opportunities and necessities that arose throughout the year.[47] Within the civilian sphere (*domi*), priests and elected officials – such as plebeian tribunes – certainly exercised year-round administrative and judicial authority over domestic affairs. The reconstruction above suggests, however, that the Romans did not yet think of their commanders as annual magistrates, but rather as men who exercised command for only a part of the year, and their system of conveying military authority developed to accommodate this.[48]

The fourth century

A major reorganization of Rome's system of military command occurred in the mid-fourth century, when the Roman citizens as a whole took firm control of the selection of military commanders as well as the conferral of *imperium*. According to tradition, this happened in 367, when two plebeian tribunes – C. Licinius Stolo and L. Sextius Lateranus – were said to have finally pushed through a series of social and political reforms that, among other things, ended the use of military tribunes with consular power, "restored" the consulship as the normal chief magistracy, opened the consulship to men of plebeian family, and created the praetorship.[49] Of course, it is unlikely that this was true as written. As discussed above, educated

45 On the Latin dictator, see Cato *FRHist* 5 F36; *CIL* 10.5655; Livy 1.23.4, 3.18.2, 6.26.4; Dion. Hal. *Ant. Rom.* 5.74.4. See also: Alföldi (1965, 42–56); Ridley (1979, 303–9); Cornell (1995, 227–30, 297–98).

46 This is demonstrated by Ap. Claudius Pulcher (cos. 54), who neglected to receive a *lex curiata* during his term as consul, and as his year was coming to a close he attempted to suborn some augurs to swear falsely that he had in fact received a *lex curiata* (Cic. *Att.* 4.17.24.18.4; *Fam.* 1.9.25; *QFr.* 3.2.3). This episode suggests that he could have sought a *lex curiata* at any point during his year in office, and that receiving a *lex curiata* and *imperium* was something different from holding office as consul in the late Republic.

47 This is further emphasized by the later Roman practice of assigning provinces or *provinciae* to their commanders. See Drogula (2015, 131–81).

48 As I have argued elsewhere, plebeian tribunes were probably the original chief magistrates in the civilian sphere of Rome, and – with the help of priests and other officials – exercised primary authority over domestic affairs: see Drogula (2017).

49 Discussed by Livy (6.35.5, 6.42.9–11) and Plutarch (*Cam.* 39.1, 42.1–5); see *MRR* 1.114 for all references. Roman tradition held that the reforms of 367 required that one consulship be reserved for a plebeian, but Billows (1989, 112–33) argues that this requirement was only established by the *lex Genucia* of 342, and that the reforms of 367 only opened the consulship to plebeians.

Romans knew that the praetorship had existed long before the consulship, and inscriptional evidence suggests that Rome's chief magistrates were called praetors – not consuls – as late as the early third century.[50] In addition to these inaccuracies of fact, the presentation of these developments as (primarily) social reforms is also likely to be an anachronistic interpretation by later historians who were influenced by events of their own times.[51] If one examines the essential details of these reforms, they were clearly military in nature: they gave structure to Rome's system of military command by reducing and fixing the number of men who would receive and exercise *imperium* each year, and they made military command an annual institution rather than an ad hoc measure.[52] Scholars have offered different suggestions about the magistracies created by these rogations, but it seems most likely that the reform created a college of three annually elected military commanders called praetors, two of which would eventually come to be known as consuls while the third remained the *praetor urbanus*.[53] These were very significant changes, and represented a dramatic reduction (roughly 50%) in the number of men authorized to exercise command each year, which was an unparalleled step for the Romans, who were better known for expanding their military capacities. Rome was actually becoming more militarily aggressive in the fourth century, so the reduction in the number of commanders makes little sense unless the establishment of three annual commanders provided as much (or more) military leadership as the previous system. That is, that three praetors who held *imperium* all year provided more (and perhaps better) leadership than the previous ad hoc appointment of commanders in response to events as they occurred. So while the total number of men who exercised command each year went down, Rome's capacity for command was actually improved by the introduction of regular, year-round commanders who were on-hand and ready to undertake any campaign. This was a critical change in Rome's thinking about what a commander was, since it made the commander a regular, annual magistrate. This probably required the final suppression of private warbands, which were already going out of widespread use as Rome was increasingly relying on large state armies (see below). In this way, the Licinio-Sextian Rogations of 367

50　See nn. 17 and 18 above.
51　For discussion of the sources, see von Fritz (1950, 3–44); Hölkeskamp (1987, 62–109); Cornell (1995, 334–37); Raaflaub (1996, 279–84); Brennan (2000, 59–69); Forsythe (2005, 262–67); Bergk (2011, 61–74).
52　von Fritz (1950, 3–44) emphasized the essentially military nature of these reforms.
53　The argument is that two of the praetors were regularly placed in command of armies and so gained greater prestige and honor and came to be known as consuls, whereas the third praetor was usually kept for the defense of the city. This explains why consuls and praetors were always understood to be members of the same college, albeit different in some way that was not entirely clear (for example, C. Sempronius Tuditanus *FRHist* 10 F2). See discussion in Hölkeskamp (1993, 26–31); Bunse (1998, 189); Stewart (1998); Beck (2005, 63–70); Beck (2011a, 82–91); Bergk (2011, 61–74); Drogula (2015, 183–93).

represented the end of an earlier military structure – one that was not well remembered by later writers – and the start of a new way of thinking about military command. As the new system took root, family traditions no doubt adjusted to reflect the change, and ancestors who had won glorious victories as leaders of private warbands came to be described in family histories as praetors and (later) as consuls – the newer titles were transferred to earlier generations of commanders.[54]

The introduction of regular, annual praetors also involved a change in Roman thinking about how commanders were selected. It is not clear how commanders of state armies were selected before 367, although the passage (quoted above) by Cincius Alimentus suggests that it was done on an ad hoc basis, and only when major campaigns were needed. After 367, however, the *comitia centuriata* – the assembly of citizens voting in military units – were responsible for electing commanders annually. Roman tradition held that this assembly dated back to the monarchy, but recent scholarship has argued that *comitia centuriata* did not take their familiar shape until the Republican period, in the fifth or even fourth centuries.[55] The decision to elect commanders in the *comitia centuriata* shows the citizens-as-soldiers using their increased influence to assert their right to select which men they would follow in war.[56] The shift to larger, state armies gave the soldiers a louder voice in the choice of commanders. Of course, the election in the *comitia centuriata* did not eliminate the older requirement that commanders receive their *imperium* from the *comitia curiata*, creating the familiar system in which two votes were needed to create a commander – one by the *comitia centuriata* selecting a man for command, and one by the *comitia curiata* to invest him with *imperium*. The cumbersome nature of this double election emphasizes that Rome's system for appointing commanders was not designed all at once, in which case one would expect a simpler process with only one vote. Rather, it resulted from a gradual process of development during which several ideas of military authority eventually coalesced into a single, somewhat awkward

54 An example of this may be the tombstone of Scipio Barbatus (*CIL* 1².7 = *ILS* 1). Barbatus was consul in 298, but scholars believe the tombstone we have was carved in the second half of the third century, decades after his death. It is possible that the original tombstone was simply damaged and needed to be replaced but it is also possible that the replacement was made to reflect the recent adoption of the title "consul" to describe Rome's chief magistrates. Indeed, an inscription published by Cristofani (1986) suggests that Rome's chief magistrates were still referred to as praetors as late as 276 or 270, after Barbatus held the office. The original tombstone may have listed his highest office as praetor, but a replacement was made decades later when the term "consul" came to be used to describe Rome's highest magistracy, thereby protecting the status of Barbatus and his clan, which was very powerful in the late third century. See Coarelli (1972); Wachter (1987, 301–42); Van Sickle (1987, 42–43 and n. 9); Flower (1996, 171–75); Drogula (2015, 41–42).

55 See Cornell (1995, 187); Forsythe (2007, 24–41); Rosenstein (2010, 299); Mouritsen (2017, 40); Armstrong (2016c, 74–86).

56 Compare with the right of Macedonian soldiers to approve their new king: see Hammond (1989, 271).

system. Cicero demonstrates this well: despite being highly educated, a great lawyer, and a senior statesman, he was perplexed by the need to have two different votes to create a commander, and came up with the rather flimsy explanation that the second vote gave the Romans a chance to "change their mind" if they regretted the first vote.[57] Clearly, he imagined that Rome's system for creating a commander was a single, carefully planned process established by the founders of the Republic, and therefore it must have a logical explanation and purpose to its cumbersome organization. He did not understand that the system was the product of a gradual, layered process, through which Rome's thinking about military authority developed in stages, many of which were preserved into the late Republic.

There were many factors in the fourth century that pushed the Romans to reform and centralize their military structure. The sack of the city by the Gauls (c. 390) was a decisive event that no doubt provided a catalyst to reorganize and put bigger and more effective armies into the field.[58] The Romans recovered from this setback and subdued many nearby states, and it was the reorganization of Rome's military command structure in or around 367 that seems to have enabled a rapid expansion of Roman territory. Shortly after the sack of the city, the Romans began mobilizing their considerable manpower for new military projects, including the construction of a large defensive wall around the city. Cornell notes that "the resumption of warfare in 362 BC opened a new phase in the history of Rome's external relations. A decade of vigorous and successful campaigning brought an unprecedented series of victories...and placed Roman power on a new footing."[59] By 338, Roman arms had conquered the whole of Latium, had incorporated the Latins into their military system, and had established military colonies.[60] All of this was possible because the reforms in the mid-fourth century enabled the Romans to project their military power more extensively, regularly, and effectively.[61] As Rome increasingly focused on expanding its territorial conquests, its military structure continued to change to deal with these challenges: pay for the soldiers was introduced, which enabled the Romans to fight in larger numbers and for much longer periods each year.[62] To pay for this, a new tax seems to have been established – the *tributum* – that

---

57 Cicero (*Leg. agr.* 2.26–7) would struggle to explain why the *comitia curiata* conferred *imperium* when it was the *comitia centuriata* that elected consuls.
58 See Armstrong (2016c, 241–42).
59 Cornell (1995, 423).
60 See Cornell (1995, 347–52) for discussion.
61 Armstrong (2016c, 251–54) discusses the new Roman focus on the conquest of land in this period, which he convincingly argues was a powerful motive for Rome's military reforms.
62 Livy 4.59.11; Diod. Sic.14.16.5. Rosenstein (2010, 293) points out that the date for the introduction of the *stipendium* (pay for soldiers) is uncertain, with some scholars arguing for the late fifth century in relation to the siege of Veii (where Livy and Diodorus place it), and others during the Second Samnite War (late fourth century), where Rosenstein agrees that it fits better.

citizens paid to fund the army, demonstrating the state's commitment to supporting these wars. The *lex Genucia*, which Livy places in 342, imposed a ten-year hiatus between the repetition of a magistracy, presumably to prevent individual elites from gaining too much influence and power through repeated tenure of command.[63] Sometime after this, the Romans were said to have abandoned their practice of fighting in Greek-style phalanxes, and instead began fighting in tactical units called maniples (see Armstrong and McCall in this volume), which may have required a greater degree of military skill, and seem to have involved new types of armor and equipment.[64] These were all dramatic changes to Rome's military structure, and nearly every aspect of the army was retooled and reorganized to put larger and better armies in the field.[65] Given this degree of restructuring, it is to be expected that the Romans also reconceived their system of military command, establishing regular praetors who were elected annually by the soldiers they commanded. While these reforms enabled further Roman expansion, the centralization and regulation of military authority and command started the transformation of the elite *gentes* from an oligarchy that relied on its own military resources to an aristocracy that derived its special claims to honor, status, and privilege from its service to the Roman people as military commanders. As the generations went by, the aristocracy claimed special status based on the inherited honors achieved by their ancestors through advancement to elected offices, and so the institutionalization of warfare also transformed Rome's archaic private warlords into the nobility of the late Republic.

Conclusions

The fact that the consulship was not created at the outset of the Republic, and that oligarchic *gentes* freely used their own resources to engage in private military operations in the monarchy and early Republic, strongly suggest that military authority and command in early Rome were not the same as they were in the later Republic, when Rome's first historians were writing. Although those early writers clearly imagined that military command had always been the sole prerogative of the heads of state, and had

---

63 Livy 7.42.1–2; see Rosenstein (2010, 301).
64 Livy 8.8.3–18. The date of this change is unknown, and it may have taken place over a period of time, but seems to have been in place by at least the mid-fourth century. See Oakley (1998, 455–57). Rosenstein (2010), Armstrong (2016c, 265–59), and others who argue that Livy is mistaken, and that the Romans never used phalanxes.
65 Rosenstein (2004, 26–31) gives a good description of how such innovations influenced Rome's war-making capacity. Raaflaub (1996, 273–314) discusses the number of changes in fifth- and fourth-century Rome that gave rise to its imperialist qualities and its sudden expansion starting in the late fourth century. The arguments made in this paper contribute to both sets of arguments and help explain how and why Rome became a more effective military power in the late fourth century.

remained largely unchanged since the days of Romulus, it seems more likely that Roman thinking about warfare and command went through considerable development, only gradually arriving at the system that was familiar to those early writers. In early Rome, private armies and warbands existed side-by-side with larger state armies for centuries, and in the early Republic the selection of commanders for state armies seems to have been an ad hoc affair, with men being chosen only when and as needed by circumstances. The *comitia curiata* may have been granting *imperium* (or something like it) to commanders of state armies since the days of the *reges*, but the existence of private warfare suggests that this "*imperium*" was not what *imperium* would become later in the Republic – absolutely necessary for command. In the fifth and fourth centuries, however, the Romans' thinking about military authority and command gradually developed, driven by the growth of the city and its increasing military aggressiveness. *Imperium* conferred by the *comitia curiata* became a necessity for all military command, and by 367 private military ventures by gentilician warbands had disappeared or had been banned, leaving the selection of commanders and the conferral of *imperium* in the hands of the citizens. This gradual development of Roman military thinking was eventually forgotten, but it left telltale signs, such as the stories of private warbands in the monarchy and early Republic, the inconsistencies in the origins of the consulship, and the strange "double election" of commanders that perplexed Cicero. The Romans preferred to imagine that they were following closely the practices of their ancestors, but, in reality, their thinking about military authority and command was dynamic and evolved over the centuries.

3 The price of expansion

Agriculture, debt-dependency, and warfare during the rise of the Republic, c. 450–287*

Peter VanDerPuy

Introduction

Debt has long been recognized as one of the major issues underpinning the social and political unrest that characterizes the literary narrative for Roman society in the fifth and fourth centuries.[1] Yet, despite this traditional importance, its influence is still arguably underappreciated. Problems with debt went beyond just social and political issues. This chapter will outline the ways in which the social, political, agricultural, environmental, and even the changing nature of debt interactions, were all inherently related to each other, and to warfare – particularly Rome's cycle of ever-expanding territorial conquests and land distributions that emerged in the fourth century. Political, agricultural, and military changes did not occur in separate spheres with their own impetuses, but rather were intimately connected with each other. And they were all connected, in one way or another, to the principle of debt.

The points in this chapter can be summarized as follows: (1) the Roman community transitioned to a more robust civic structure in the fourth century that established people on the landscape as individualized, purposely distinct farmers, whose contributions of taxation and manpower could be calculated to serve the military needs of expanding warfare; (2) this transition imposed new civic and agricultural burdens on farmers, new pressures that could only lead toward forms of agricultural intensification and potentially unsustainable uses of landscape and ecosystem; (3) the redefinition of forms of land use involving small farms, as well as the new pressures associated with them, transformed older, more feudalized modes of dependency and created new types of debt as the relational process of borrowing as a whole became less embedded in an ongoing process of survival, stability, and continuity, and became more externalized and fiscal; and (4) the Roman

---

* All dates are BC unless otherwise noted.
1 On the "Struggle of the Orders" and land and debt problems, see Brunt (1971b, 47*ff.*); Hölkeskamp (1987); Momigliano (1994, 226–48); Cornell (1995, 327–44); Oakley (1997, 365–74).

DOI: 10.4324/9781351063500-3
This chapter has been made available under a CC-BY-NC-ND license.

process of territorial conquest, annexation, and land distribution was secured by maintaining the presence of individualized farms and civic entities
on the land precisely at the same time as the pressures of this paradigm
contributed to forms of competition, intensification, and ecological destabilization, which then demanded the acquisition of new lands. Thus, the entire
paradigm became a self-perpetuating, yet unstable, foundation of Roman
imperialism. In a context where at least 90% of wealth in the ancient world
was in the form of land, and where 90% of the people were farmers (including elites), the fortunes of warfare, imperialism, and expansion must also
have been bound up with the fate of the landscape and environment.

From the outset, it should be noted that this chapter follows several general premises developed by other scholars. The first is an argument that
emerges from a volume co-edited by Rosenstein and Raaflaub, whereby a
society's warfare should be understood as an expression or product of its
political and economic bases (though it also, certainly, went beyond those
bases).[2] In that vein, I am arguing that Rome's fourth-century warfare was
a product of the Romans' evolving civic structure and, specifically, the
changes that correspondingly arose in its agricultural structure and land
use. We will see how the transformation of that agricultural basis from the
fifth to the fourth centuries contributed to a new debt problem in connection to the community's cycle of warfare. Second, this paper follows the
emphasis in Armstrong's recent book, that the Roman community of the
fifth century transitioned from a more gentilicial, warband-based military
model – in which military objectives were often short-term goals centered
around portable plunder – to a larger, centralized state-based mode of warfare, geared around the acquisition and defense of land.[3] While Armstrong
highlighted the transition from gentilicial to state-based structures in early
Roman society through the lens of warfare, the present chapter examines
the same process through consideration of forms of debt and dependency in
what was always an overwhelmingly agricultural lifeworld.

A record of debt

Let us begin with a pattern of notices regarding debt in Livy's narrative.[4]
There is a cluster of notices concerning debt beginning around 387. From
there, these notices crowd the narrative with a startling frequency and carry
on through most of the fourth century and down to 287, with the infamous
"Third Secession of the Plebs" over debt.[5] Though the modern scholarly

---

2 Raaflaub and Rosenstein (1999a, 1–2).
3 Armstrong (2016c, 152, 272–80).
4 Also highlighted by Bernard (2016, 317).
5 For debt notices: Livy 6.11–20 (in 385–84, sedition of Manlius Capitolinus), 6.27 (in 380),
   6.31–7 (in 378–68, spiraling debt), 7.16 (in 357), 7.19 (in 353), 7.20–1 (in 352, major initiative
   on debt including the creation of a five-man board of "bankers" [*mensarii*] for discharging

debate has been fraught with disagreement on the issue, there are no good reasons to doubt the authenticity of these kinds of bare notices, and they likely arise from basic archival materials that later historians could draw upon.[6] The real curiosity here is why debt notices suddenly spiked beginning in the fourth century, particularly as other agitations over land had continued unbroken in the annalistic record from the fifth to fourth centuries.

The answer is to be found in the fact that at least some segments of the Roman community had begun to live and interact with both each other and the community in very different ways. Starting slowly, from about 450 onward, the community seems to have transitioned from a structure of hierarchic dependency networks to a much more individualized civic structure. This structure reoriented farmers to an emergent state centrality, enmeshed them in calculative civic apparatuses like the census, and imposed newfound burdens upon them in the wake of widening horizons of ownership and participation.[7] The pattern of debt notices in the narrative forms a cache of evidence that testifies to the transformations that were going on socially, politically, agriculturally, militarily, and – naturally – economically.[8] This evidence suggests that there were unforeseen consequences and associated problems with this transition to a more individualized, civic structure for those portions of the Roman community that embraced it.

In the most basic equation, Rome's increased capacity for warfare in the fourth century depended on two things: money – that is, taxation through the *tributum* – and available manpower. Both of those variables required that a certain number of individual farmer-taxpayer-soldiers be maintained on the

---

debts), 7.27 (in 347–46), 7.28 (in 345–43, prosecution of *faeneratores* by the aediles), 7.38 (in 342), 7.42 (in 342, *lex Genucia*), 8.28 (in 326, *lex Poetelia* abolishing *nexum*), 8.34 (in 310); 10.23 (in 295, prosecutions of *faeneratores* by aediles), *Per.* 11 (in 287, "Third Secession of the Plebs" over debt); in total, at least 14 major notices spanning the whole of the fourth century and early third century.

6 Oakley (1997, 27, 72). See also Cornell (1995, 13) and Raaflaub (2005a, 5–6). Rawson (1971b) and Frier (1979) critiqued aspects of the *annales maximi*, but see now Forsythe (1994, 53–73).

7 On this widening participation, Armstrong (2016c, 163–64).

8 Bernard (2016) sees the transition of the early Republican economy as a shift from insufficient production (famines) to unequal distribution (fourth-century debt), where the definition of famine seems to rest on low food production levels, and the definition of debt rests mainly on the size of land holdings and amounts of labor. But note Sen (1993, 36–37): the causes of famines are more complex than insufficient production and a large decline in food. Famines also concern hindrances from broader "entitlements" and assets to which farmers can potentially lay claim. Additionally, while we generally see a shift from notices of famine in the fifth century to increasing notices about debt in the fourth century, the causes of this shift have to do as well with deeper structural changes to the nature of society, its agriculture and its farmers, in addition to questions of quantity of land and labor. Qualitative questions about the nature of peasant agriculture, the diversity of strategies, the *kind* of land peasants worked, how they interacted with each other, and how they interacted with an emerging state structure are indispensable parts of a complex understanding of agrarian debt.

landscapes of Rome's increasing territory at all costs.[9] Yet that cost potentially came at a high price for many citizens, farms, and landscapes. For the Roman state system that emerged in the fourth century was chiefly a military war-making apparatus and hegemon, almost corporate in the nature of its calculative relationship to its citizens, restlessly widening scope of acquisitions, and only limited attempts to ameliorate the growing problems of its farmers and lands.[10] But it is also important here to make some observations about the earlier fifth-century societal structure and the character of debt-dependency before looking at the problems of the fourth century in earnest.

The world of the Twelve Tables

While the Twelve Tables[11] – one of the few sources of evidence held to be a genuine product of the period – do speak of some institutional state structures, these were likely very rudimentary in the mid-fifth century.[12] Instead of mining the Twelve Tables for legal and constitutional history, it may be more productive to examine them for what they can tell us about the life-world and thought-world of the fifth century community – and for this there is actually quite good evidence in the mundane statutes about agrarian life. Elsewhere, I have more thoroughly engaged in an exegesis of this material, but I will give a basic summary of the way of life they seem to indicate.[13]

The various statutes that deal with the trimming of trees, boundary issues, the building of walls and fences, early servitudes for access and rights of way, procedures for recompense if animals cause some form of damage or loss, directions for containing rain-water and seasonal torrents, and the maintaining of simple forms of roads, provide a striking reminder of the sheer

---

9 On this point, see Tan in this volume.
10 North (1981) on the importance of Rome's alliance system as a collective apparatus in the production of its warfare and imperialism. See, also, Tan in this volume, on the nature of the Roman state as an extractive "turbo-charged war machine" mostly concerned with the acquisition of resources. Due to the theoretical nature of this piece, which conceptualizes the results of a war-producing state on a farming community, I have purposefully chosen in places to make use of language that seems quite modern and industrialist in tone, as one way of understanding the results. I am, however, aware that ancient farmers did not think about such things in the same manner as we do.
11 References to specific statutes in the Twelve Tables in this chapter follow the numbering scheme in Crawford (1996) (= *Roman Statutes*).
12 Cornell (1995), Smith (2011, esp. 339), and Raaflaub (2005, 15) are all fairly comfortable likening Rome of this period to a Greek *polis* structurally. However, Anderson (2005), Anderson (2009), and Capogrossi Colognesi (2014, xxii–xxiii) caution about how we conceptualize an ancient state or *polis*.
13 VanDerPuy (2017) Ch. 4. It should be noted that the sources for the Twelve Tables are fairly late. For discussion of the source tradition, see Crawford (1996, 2.556*ff*); Smith (2006a, 21). Cornell (2013, 67–68), however, argues that a majority of the antiquarian Festus' information on the Twelve Tables appears to have come from the writings of the third and early second-century historians.

complexity, interdependence, and fragility of mixed-farming agriculture.[14] The statutes indicate that the best model for sustainable agriculture is the mixed farm – mixed polyculture, underscored by a high degree of cooperation.[15] Further, the enormously detailed writings of the Roman agronomists (Cato, Varro, Columella, and even Virgil) provide thorough descriptions of the bewildering variety of interconnected strategies that farmers had to master, in tune with the rhythms of the agricultural year and, importantly, in balance with the microecological niches in which they found themselves.[16] These writings form a back-fill of supporting information for the statutes of the Twelve Tables, and it is clear from these various sources that agriculture (at least in these later periods) operated under the imperatives of occupational diversification and the need to make deposits back into the landscape toward its future resiliency; in this way, present production and future fertility merged conceptually in the mind of a good farmer.[17] It is also important to acknowledge the extent to which cooperation and mutual dependencies, both equal and asymmetrical (or abusive) forms, were inherently written

14 In particular, note the statutes in tables VI–VIII. On mixed farming: Cato, *Agr.* 1.3–7; Columella, *Rust.* 1.2–3; Varro, *Rust.* 1.6. See also: Frayn (1975); Evans (1980, 135*ff.*); Horden and Purcell (2000, 59).

15 In defense of the intelligence and productivity of small farmers: Evans (1980); Horden and Purcell (2000); Kron (2008); Halstead (2014); and Hughes (2014).

16 On Latium's microecological diversity and the peasant response to sustaining ecosystems, see Frank (1919); Platner-Ashby (1927, 24–25); Horden and Purcell (2000, 61*ff.*, 171, 181, and 273); Bolle (2003, 10–11); Hughes (2014, 120); Berry (2015, 91). See also Columella, *Rust.* 2.2, 2.6.4 and 2.9, whose classifications of regions and soil types speak well to the idea of microregions and microecologies. On the reliability of the agronomists for capturing the complexity of peasant farming, in particular, see Spurr (1986, 23): "any ideas that the Roman agronomists only wrote theoretically...must be dispelled. Their instructions acknowledge the varied reality." One may argue that the agronomists represent a later, theoretical and idealized portrait of farming; so Marzano (2007) 85. But the difference between Cato's large estate, for example, and that of a small farmer was a quantitative one, the difference being the number of *iugera* and use of slave labor; it was not a qualitative difference in the complexity and types of agriculture worked on either kind of farm.

17 Columella, *Rust.* 2.1.6–7, 2.2.13 on the need to make deposits back into the soil in order to keep it healthy. The following statutes of the Twelve Tables are borne out by parallel evidence from the agronomical writers: VII.8 on the containment of rain-water and runoff receives attention in: Cato, *Agr.* 155.1; Columella, *Rust.* 1.6.24, 2.2.9–10, 2.16.4–5; Verg. *Ecl.* 3.111, *Georg.* 1.106–10, 1.269–70. See Thommen (2012) 86 on flooding problems. Statutes VI.7 and VII.9 on the trimming of trees. See also: Cato, *De ag.* 2.3–4, 3.1, and 6; Columella, *Rust.* 1.3.7; Varro, *Rust.* 1.15; Verg. *Ecl.* 9.60–1. Many strategies involved in agricultural life had more than one use or area of impact. As such, these references on trimming and planting trees also dovetail well with the statutes in VII.2–5 on boundary marking between farms, and see also Varro, *Rust.* 1.16.6. For statutes VII.1, VII.6–7, and VIII.2–3, on road/track maintenance, and rights of access (early servitudes, including problems of damage caused by or to animals): Columella, *Rust.* 1.3.3, 2.10.11, 2.17.1–7; Varro, *Rust.* 1.2.17–19. On early servitudes see Grosso (1969, 14–19); and Bannon (2009) generally.

into the process of sustainable agriculture.[18] Group-based forms of dependency hierarchies therefore tended to be a normal part of the paradigm in the fifth-century community, and forms of debt must be understood as deeply embedded in the functionality and continuity of these networks.[19]

Not only do well-integrated, sustainable, mixed forms of agriculture demand a certain level of interdependence among farmers, the low level of institutional development in the fifth-century Roman state also created a fabric in which the lower socio-economic levels of Roman society would have perceived "the state" as effectively the localized collection of networks of local strongmen, dependent farmers, and neighbors.[20] For much of the Roman community, I suggest, the immediacy of their local agrarian lifeworld and its hierarchies *were* the form of state that they knew. The great military and civic institutions of the *rex*, praetors, and *comitia* may have been visible and known, but their importance likely often paled in comparison to the very present and tangible power of local clan leaders – particularly out in the rural hinterland. Further, in light of the serious problems of the fifth-century community – migrations, population movements, hostile and peaceful resettlement, famine, drought, and pestilence[21] – a societal structure of dependency networks, with a spectrum of hierarchic roles, was likely something of an imperative for survival, stability, and continuity, however abusive this might become.[22] Statutes attesting to forms of service and subordination, such as *clientela* and *nexum*, provide support for a spectrum of subordinated statuses within a dependency fabric, and these statutes are sanctioned with the binding force of *fides* ("trust").[23] All of this supports the existence of what Wiseman once called a "quasi-feudal" society – in which hierarchy, service, and dependency were the norms.[24] As Crone reminds us, hierarchy is an expression of order and stability in premodern communities: "equality meant chaos, conjuring up much the same image of disorder as does the idea of mixing up the different parts of a car engine to us."[25]

---

18 On cooperation between agriculturalists, see Horden and Purcell (2000, 84–86 and 180), as well as Bannon (2009, 24). Smith (1996, 190–92, 2006a, 247–50), on the *gens* as a dependency hierarchy and its potential control of lands.

19 See also Wallace-Hadrill (1989b, 69).

20 Johnson and Dandeker (1989, 235) suggest a patronage structure itself can be the dominant form that the state takes. Though I follow Armstrong's reconstruction of a gentilicial society for the fifth century in particular, I have purposefully left such terms as "strongmen" somewhat vague in order to connote generally powerful, landed, and likely aristocratic individuals whose power loomed quite large in a given locale.

21 See Cornell (1995, 304–9) and Rosenstein (1999, 195) on the migrations/invasions.

22 Cornell (1995, 291 and 330) suggests that forms of clientage might shade into more severe forms of servitude.

23 Cf. the language in statutes VI.1–2, dealing with *nexum*; Livy, 8.28.8, notes as well that with the abolition of *nexum* in 326 "a strong bond of trust was broken" (*victum...ingens vinculum fidei*); Crone (1989, 32 and 54), on trust in pre-industrial societies.

24 Wiseman (2004, 67).

25 Crone (1989, 106). Bernard (2016, 323) suggests *nexum* should be seen, not as something avoided, but as a strategy for exchanging labor.

The other significant point to consider is that early Roman society also featured a largely pre-coinage economy.[26] This has important ramifications for how we understand debt in this society, as the substance of loans must have been almost exclusively in forms of kind made for consumption and therefore survival.[27] Importantly, there could not have been much of a role in this system for the character of the moneylender-middleman, who financed loans with capital for the specific purpose of leaving the loans uncollected in order to profit from the interest.[28] This is a type of interaction that became more apparent only from the fourth century onward. The role of "moneylender" in this earlier socio-economic structure could only have been held by the local strongman and/or large-scale landowner with surplus crops. As has been fairly well understood by Finley and others, powerful elites made loans that secured the services of their local farmers as debtors in a relationship akin to vassalage.[29] What bound them together was trust (*fides*) and obligation, and this typically meant that the debtor became that powerful lender's man, since labor in premodern societies was not a neutral commodity distinct from the person.[30] And it is tempting here to identify sites like that of the "Auditorium Villa" as the centers of power for local strongmen/clan leaders and their dependency networks, in which debt relations of service and subordination were carried out through collection and redistribution of surpluses, where the villa served as a facility of processing goods into secondary products serving elite consumption and/or a market orientation.[31]

Supply and demand

It is also important to look at the dynamics of supply and demand for both food and labor, using one type of scenario that seems to feature quite commonly in our sources for the fifth century. Considering the notices of grain shortages and grain missions to places like Etruria, Campania, and Magna Graecia, the accounts seem to indicate that it was the elite who mediated and conducted trade on behalf of their community.[32] Crone noted that, in premodern societies, the first type of trade to develop is that of long-distance trade between elites, and not centralized, state-controlled trade.[33] Here then is a situation where local strongmen used their connections to situate

---

26 Crawford (1974); Nijboer (1998) 38.
27 Though our focus here is on agricultural economy, we ought to also acknowledge that there existed other possible reasons for loans, such as marriages, religious sacrifices, or other interactions.
28 On money-lending in the Republic, see Gruen (1995, 426–27).
29 Finley (1973); Finley (1981); Foxhall (1990).
30 Crone, (1989, 30–31).
31 For the Auditorium Villa: Terrenato (2001, 8–9, 2011, 238–39).
32 For example, Livy 2.34 (in 492–91), 4.12 (in 440), 4.25 (in 433); Dion. Hal. *Ant. Rom.* 10.54.1–2 (in 451), 12.1.2 (in 440).
33 Crone (1989, 22 and 33).

themselves in an advantageous position within an economic regime.[34] In times of pestilence, drought, and particularly famine, all would have felt the desperate need for grain at the same time, thereby skyrocketing the demand for, and value of, this commodity – such that these powerful individuals could create relations of dependency or shore up preexisting relationships by lending on extortionate terms, particularly as they stood as conduits of trade and were already more likely insulated by their surpluses.[35] Of course, these are loans made for consumption, so they are never meant to be repaid in one sense. Instead, they form part of an economic regime and continuous, functional order of survival. As most farmers experience of "the state" in this period was embedded in their immediate agricultural lifeworld and dependency network, so forms of borrowing and debt were also deeply embedded in that structure as mechanisms of continuity.

On the other hand, the poorer farmers' attempts to manipulate the exchange can be seen in moments where the logics of supply and demand for *labor* came to the forefront. Each year there are key moments within the natural rhythms of the agricultural cycle when large amounts of labor are needed, and they are needed right away.[36] At times of planting and harvest in particular, the demand for labor would have skyrocketed, and not all elite landowners would have been able to meet the needs of their landholdings on their own. This put poorer farmers – at least theoretically – in a more advantageous position, and it was likely the shrewdest among them used the advantage of surplus labor through adolescent sons to negotiate the terms of their debt-dependency more favorably. Those with adolescent sons might strike the best of terms, particularly remembering that large-scale, far-off wars had not yet begun to relieve farms of the burdens of adolescent males and their caloric needs at a crucial moment in the family life cycle when it could destabilize a farm.[37] The statute providing that a father may sell his son three times makes a good deal of sense here, within the logic of this economic regime.[38] It allowed those further down the socio-economic hierarchy to make labor payments toward their dependency at the same time as they relieved their farms of a caloric burden for a crucial period of years,

---

34 On "regime-building," see van der Vliet (2011, 129); Gallant (1991, 9–10). Sen (1993, 33) notes that elites faced incentives to take preemptive action if they seek to continue in power.

35 Elites also likely enjoyed insulation through the possession of different types of seed – a valuable form of diversification and insurance; see Columella, *Rust.* 2.9.7–8; see also Rickman (1980, 4). Spurr (1986, 41) notes the tendency of elites to benefit from the ability to engage in careful and discriminating seed selection year after year.

36 Ancient agricultural writers agree on the importance of timing and a deft response to seasonal demands: for example, Cato, *Agr.* 5.1; Columella, *Rust.* 2.9.15, 2.20.1; Hes. *Op.* 486–90; Verg. *Georg.* 1.43–70, 1.208–30.

37 On life cycles: Gallant (1991, 25–33). On labor surpluses: Rosenstein (2004, 66–69).

38 Statute IV.2: "if a father thrice sell his son, his son is to be liberated/emancipated from his father" (*si pater ter filium venumduit, filius a patre liber esto*). Crone (1989, 110): "children were expected rapidly to begin repaying the investment which had gone into their creation." See also Arist. [*Oec.*] 1343b, for this view.

after which the son(s) might return to marry and take up the farm as the parents neared the end of their own lives, thus restarting a stable life cycle of family and farm.[39] Importantly, such a rhythm would have also benefitted the powerful lender, who also depended on the continuity of the farm for future produce and potential forms of labor. It benefitted no one to destabilize the population of the lower classes, their farms, or indeed the landscapes, beyond the point of resiliency. Thus, a certain degree of flexibility between participants and their competing demands can be expected in this type of symbiotic system and societal structure.

What we need to understand here is that this political and economic mode of living would inevitably tend toward the blurring of distinctions between individuals and powerful lenders, as well as blending together properties and produce. This is not necessarily to argue that the social categories of identity – such as that of *nexus* or *senatus* – became indistinct between "peasant" and elite landowner. But both at the level of neighborly codependence, and at the unequal level between "peasants" and powerful strongmen, the tendency of this system is to make members less economically distinguishable as individuals, and more identifiable as parts of a functioning whole or organism.[40] This is not a utopian vision by any means; it was always prone to horrific abuses. But the point is that it constituted a functional regime and, more importantly, it helps us understand why it produced warfare that was characterized by short-term raiding, portable spoils, and only rare if any moments of large-scale participation in state-based armies.

The growth of the Roman state

The world of the Twelve Tables was not a static one, however. Around 450–40 there was a substantial reorganization and reformation of the nascent Republican state, as a code of law was created for the community, the new consular tribunes supposedly came into being, and the censorship – as a distinct and discrete magistracy – was supposedly created.[41] We begin to see what looks like a more rigid administrative structure, and the censorship in particular was likely designed to meet the slowly emerging need of calculating broadening layers of participation in the community's warfare.[42]

---

39 Rosenstein (2004, 89): marriage age for men in late twenties to early thirties. See also Hes. *Op.* 700–1; Pl. *Leg.* 721b, 785b, *Rep.* 460e; Arist. *Pol.* 1335a6: age of marriage around 30 years old. As well, in premodern societies in general, adolescent sons and their reproductive potential are also agents of social chaos; see Arlacchi (1983, 42–46).

40 Crone (1989, 99–100) suggests that hierarchy was the rule of the community, the image of preindustrial society "highlighted the importance of coordination and subordination... and they were not perceived as contradictory."

41 On the law code: Livy 3.31–4. Creation of the consular tribunes: Livy 4.6; Dion. Hal. *Ant. Rom.* 11.62; see also Drogula in this volume. Creation of the censorship: Livy 4.8; Dion. Hal. *Ant. Rom.* 11.63 is fragmentary.

42 On the emerging importance of the censorship at this time: Cornell (1995, 188); Torelli (2012, 9); Armstrong (2016c, 231 and 241).

The administrative task of categorizing people on a more individualized basis for the purposes of calculating the state's needs would only become increasingly important throughout the fourth century.[43] Significantly, the basic logic underpinning the system changes now toward an increasing focus on individualization and distinctions made between farms, and it is important to remember that the emerging administrative apparatus possibly held less room for negotiation for members of the lower classes accustomed to older forms of dependency.

The watershed moment that would accelerate this process was the conquest of Veii in the opening decade of the fourth century.[44] This conquest installed a new paradigm that would characterize the community's warfare from the fourth century onward: a cycle of conquest, annexation, and land distribution emerged, whose logics eventually became self-perpetuating. In the wake of the conquest,[45] the territory of Veii was distributed to a large number of people as more individualized farmers, landholders, and civic entities – organized, ultimately, through tribes.[46] As a measure of this new paradigm, in total, the years from 387–58 alone saw the creation of six new tribal installments, representing a major expansion of farmers on the landscape and, correspondingly, a major increase in the role of the census as an institutional apparatus [see Map 3].[47] It is worth stressing here that the change to a civic mode of living – the trend toward individualization – is a radical ideological development for premodern communities, and nowhere near as intuitive as a modern reader may naturally think. Accordingly, we must consider some of the problems and difficulties associated with receiving a new individual allotment.

Though the Veientine settlement scheme in particular may have made use of preexisting established farms, we must consider that some settlers in the various allotment schemes of the fourth to early third centuries would

---

43 Armstrong (2016c, 184).

44 Bernard (2016, 324–26).

45 Beloch (1926, 620) estimated that Roman territory grew from 822 to 1510 km² after the conquest. Followed by Oakley (1997, 345) and Briquel (2000, 206) with a slightly different estimate.

46 For the Veientine land allotments: Livy 5.30.8, 6.4.4; Dion. Hal. *Ant. Rom.* 14.102.4. The archetypal allotment for many of our sources was two *iugera*. See: Varro, *Rust.* 1.10. 2; Plin. *HN* 18.7; Cic., *Rep.* 2.14.26. A number of Livy's notices of land allotments tend to be in that range – two *iugera* at Labici in 417 (4.47.7); two and a half *iugera* at Satricum in 385 (6.16.6); and two *iugera* at Tarracina in 329 (8.21.11). Diodorus, (14.102.4) suggested four *iugera* as the norm. But Columella (*Rust.* 1.3.10) indicates seven *iugera*; Valerius Maximus (4.4.6) highlights the seven-*iugera* farm of M. Atilius Regulus. For modern scholars, it is often felt that we are on better footing with the seven-*iugera* allotments of Veii; for example, Brunt (1971b) 35; Rathbone (2008) 307. Rosenstein (1999, 196) estimates plots of land between one and a half to two and a half hectares, somewhere between six to ten *iugera*. On the imposition of *tributum* on the Veientine land allotments, see Tan in this volume.

47 Livy 6.5.8 and Taylor (2013, 9 and 47–49), on the creation of the *tribus* Arniensis, Sabatina, Stellatina, and Tromentina. On the Pomptine and Publilian tribes: Livy 6.6.1–2, 6.21.4–5, 7.15.12.

have found themselves facing potentially devastated land, virgin territory, diverse ecologies (not all land is equal), and what was an overall new beginning on a piece of land.[48] To make a start, a farmer needed several key things: (1) seed for planting (which many, particularly if they came from the urban poor, were unlikely to have stored from the previous year); (2) building materials for dwellings and storage, building tools, and farm equipment, possibly some animals[49]; and (3), most crucially, food in the form of grain and other types to make it through to the first harvest.[50] Additionally, and no less important, the sustainable mixed farm, with its plethora of occupational strategies (involving animals, diverse ecosystems, and, of course, cooperative agreements for servitudes), took *time* to establish – generations even – and this was a variable or entitlement that few could lay claim to within the immediacy of survival on a new farm.

Much of this could have represented a severe financial outlay up-front, and while the Roman treasury seems to have been able to finance some forms of resettlement in later periods, the treasury of the early fourth century could not have made this kind of enormous outlay for its citizens.[51] This meant there was a high likelihood that many members of the community, who had been granted land, had to turn to forms of debt simply start their farm on a new allotment. Therefore, the whole equation which settlers faced in land allotment schemes was likely a very difficult one, in which debt figured very heavily from the beginning. It is no wonder then that debt notices suddenly reappear with a crescendo of frequency in the first half of the fourth century.[52] While many scholars have highlighted the defensive building projects of this era as a major cause of debt, this is likely only part of the reason; the point here is to identify the deeper, structural-level changes to society and

---

48 The region of Veii, and Etruria in general, was likely already well developed with, as Bradley points out, a previous road system. Thus, we may expect that a decent number of settlers may have received good, pre-developed land. Elsewhere in Latium, in the Pontine region of the later *tribus* Pomptina and Oufentina, land allotment schemes very well took the form of virgin land and/or land reclamation projects. See Bradley (2014, 68) on Etruria and Attema, de Hass, and Termeer (2014) on the archaeology of settlement and land distribution in the lower Pontine region during this period. The results do not seem to show as much new settlement in places closer to Rome, like Satricum and Antium, while expansion further to the south in the Pontine region does tend to indicate an increase in settlement in the late fourth century. Torelli (2012, 11) reminds us that the archaeological evidence for farms, villas, and such is unsecure for anything before the third century.

49 On the importance of storage and building materials: Cato, *Agr.* Chs. 23, 31, 37, and 39; Columella, *Rust.* 1.6.9–20, 2.10.2–3, 2.10.16, 2.14, 2.18; Hes. *Op.* 600–1; Verg. *Georg.* I.259–75.

50 These considerations are rather theoretical and heuristic, and it is also important to consider that many settlers might have set out at a moment in the year when foraging strategies might tide them over until their first harvest.

51 Livy (40.38.6) notes the senate's use of funds to stock the farms of the Ligurian Apulani deported to Samnium in 180. As well, Tiberius Gracchus had intended to use the funds from the Attalid bequest – the wealth of a Hellenistic kingdom – to help stock and equip the settlers under his land law in the later second century, Plut. *Ti. Gracch.* 14.1.

52 See above, n. 5.

agriculture that caused debt to become such a problematic issue.[53] Rome's territorial expansion during this period certainly brought in wealth, but it was not delivered equally or contemporaneously for all involved. Indeed, for those who were not at the very top of Rome's socio-economic hierarchy, even if they were awarded land won in conquest, the immediate economic impact may have been significant debt which likely took generations to repay.

What is crucial to understand about the loans and debt that many must have had to enter into in order to start a farm is that (1) these were now increasingly made as an up-front, initial outlay, not as a part of an established feudal regime of consumption, dependence, and subordination that blurred distinctions and embedded individuals within a network; and (2) many of these forms of material loan were now no longer in kind, and therefore demanded new forms of capital and financial processes. As such, what we are seeing here is the *structure* of debt changing because the structure and circumstances of *loan* were changing. Additionally, this period has also been highlighted as an era of the "democratization of warfare," in which the new agricultural individualist paradigm allowed many more individuals to enter the community's armies.[54] This is the twin of the change toward individuality that we see in the establishment of the distinct civic farmer; the ideology extended to produce the concept of the individual farmer-legionary, who contributed to an expanded capacity of war-making. Yet, democracy based on citizen-soldiers is expensive. There was also a rather large financial outlay for military equipment that an individual had to be able to afford to participate – particularly as the manipular reforms may have begun to standardize equipment.[55] This represents yet another instance that was ripe for debt, and many must have had to borrow to afford their equipment and join the growing, community-based armies. Again, the type of loan here is of a very different kind; it represents an up-front financial outlay, and it is therefore a more externalized form of loan, less embedded in the rhythms of yearly survival than earlier, feudalized forms of dependency. As well, these kinds of loans were probably conditioned through newer structures involving forms of capital outlay, where the loan become increasingly nominalized through a balance sheet, a principle sum, and rates of interest.

It is also important to consider that many of the individuals who made themselves available for land allotments may not have come simply from the landless poor, but they may have been members of older dependency networks who were looking to "buy out" of their dependency. Powerful patrons or local strongmen may have been willing to discharge these individuals at

---

53 Livy 6.11, connecting building projects with debt; see also Cornell (1995) 331–32.
54 Armstrong (2016c, 256–65).
55 On equipment costs, see Foxhall (1997); van Wees (2004) Ch. 4 *passim*, and 55 in particular. The evidence seems to indicate that one needed to be in the range of a well-to-do, middling farmer. See also Armstrong in this volume for a renewed look at the development of the manipular legion in this period.

the cost of nominalizing what was their formerly embedded dependent status into a more calculated debt, now involving a principle sum and interest rates. Indeed, they may have even offered to make new loans to finance a new individual allotment and military equipment. The possibilities of how individuals contracted new forms of debt could be manifold, but individualization in general seems likely to have come with more punishing costs and a newer structure of debt. The ontology of debt changed from a feudal relationship, designed to deepen and continue over time, to a fiscal institution, theoretically dissolvable and terminable through payment, and therefore more punishing, as Livy indicates.[56] The willingness to be represented by money, as a kind of proxy, potentially divorced individuals from sustainable agricultural practices as they looked to intensify in the face of an increasing burdensome financial situation.[57]

It is not a coincidence that the debt notices in the narrative for the fourth century indicate a more fiscal type of loan and debt structure.[58] The basic notices that define debt through modifications of principle sums and interest rates are likely part of a reliable core of material. Though Roman silver coinage does not appear until the early third century, other forms of currency were certainly developing during this period, and it is not unreasonable to suppose that forms of money and coinage circulated as the *urbs* began to boast a more robust commercial sector in connection with commercially important regions such as Campania.[59] The debt notices paint a picture of a mounting problem, as newer forms of loan and debt produced a more complex set of interactions, with monetized interest that only compounded an individual's problems and further stacked the agricultural pressures and odds against them in making a new farm work.[60]

Newer, increasingly monetized, kinds of loan interactions widened the structure to now include the possibility of new agencies in the process. It is unsurprising, then, that the middle of the fourth century also sees the first

---

56 Livy 6.34.1–2: *cum eo ipso, quod necesse erat* **solvi**, *facultas* **solvendi** *impediretur* ("the fact itself that paying/discharging the debt was compulsory impeded the ability to pay"). [emphasis added].

57 Berry (2015) 25. Hughes (2014, 32) also notes that increasing urbanization in a region can have the effect of motivating farmers to substitute a posture of confrontation for the earlier feeling of cooperation with the land.

58 Notices including greater amounts of calculative language: Livy 6.14 (in 385, *aes et libram*), 6.17 (in 385, *faenus*), 6.18 (in 384, *faenus* and court procedures [*ius de pecuniis*]), 6.27 (in 380, *faenus*), 6.35–42 (in 375–67), 7.16 (in 357, *faenus* fixed at an *unciarius*), 7.19 (in 353, crushing weight of *faenus*), 7.21 (in 352, *faenus*), 7.27 (in 347–6, *faenus* reduced again), 7.28 (in 345–3, prosecution of *faeneratores*), 7.38 (in 342, *faenus*), 7.42 (in 342, passage of the *lex Genucia* making it unlawful to lend with interest).

59 On the appearance of forms of monetization by the fourth century, such as *aes signatum* and *aes grave*, see Forsythe (2005) 336, 339, and 358. See also Hughes (2014) 180 on commerce, road-building, erosion and agricultural destabilization.

60 Livy, 6.31–7 (in 378–68), 7.16 (in 357), 7.19 (in 353), 7.21.5–8 (in 352, major initiative of a five-man board of "bankers" [*mensarii*] for discharging debts).

notices in Livy that attest to the arrival of the *faenerator*, or moneylender.[61] Loaning in ways that contributed to balance sheets, with principle sums and interest rates, is just the kind of paradigm in which the classic role of the moneylender can flourish, and the role may have been assumed by both powerful local strongmen as well as a burgeoning new class of merchants with more readily liquid forms of capital, who saw the profits associated with financing these loans. While the narrative core reports instances of prosecutions of *faeneratores*, in what were likely attempts to restrain their abuses, by only modifying the rates of interest for debt, the authorities were not solving the problem. Rather, they maintained it at a simmering level, in order not to lose these civic entities from the landscape as both taxable farms and a pool of manpower recruits.[62]

Taxation, of course, formed another burden which the developing state imposed on its citizens – a point also explored by Tan in this volume. The *stipendium*, inaugurated at least on an ad hoc basis around 406, necessitated the *tributum* or war-tax, since all wars needed to be financed up-front.[63] It is not difficult to imagine that new individual farmers, facing the costs of starting a farm could have been even more destabilized through forms of taxation that financed Rome's armies, where the demands of production interfered with those of sustainability and continuity. It is also not unreasonable to suggest that it was local, powerful landowners as patrons who began to serve now as a local intermediary possessing the ability to marshal the resources for the *tributum* and large portions of their locality's tax burden in kind.[64] These men (some of whom may also have represented part

---

61 Livy 7.28.9 (in 344).
62 Though it is important to remember that a farmer would not likely have been called upon to supply manpower through the *dilectus* and pay the *tributum* at the same time. See Tan, in this volume, as well as Rosenstein (2016a).
63 On the dating of *stipendium* to *c.* 400: Cornell (1995, 187); Oakley (1997, 631); Rosenstein (1999, 212 n. 24). On *tributum*, see Nicolet (1976, 19–26).
64 How the census was carried out in our period is very difficult to know. Nicolet (1976, 46–55) indicated that the state eventually relied upon the *tribuni aerarii* who advanced the capital required. In the fifth to early fourth centuries, it makes sense to see local powerful elites as the middlemen. As a comparison, Fernandez (2017) notes that the early medieval Visigothic kingdom made use of local powerful landowners in the collection of taxes in local regions, and thus the despotic central power of the state used "non-bureaucratic" agents as infrastructural extensions of its power to mobilize resources. Ando (2017, 7–9) seeks to redefine state identity as a more discursive relationship where "the metropolitan, imperial, and local structures of social and institutional authority [exist] in a continuum rather than in oppositional relationship to each other." Such a situation I think makes sense of the messiness of what "the state" may look like in the fourth-century community: it was both a central, specialized, war-making aristocracy as magistrates, as well as a range of powerful individuals working outward to the local level, whose activities of collection on behalf of the state may have overlapped with their private functions as strongmen and lenders. This may as well have meant that there were a range of farmer-citizen types, from older preexisting feudal dependents to the more individualized members of the paradigm we have been outlining in our arguments.

of Rome's emerging war-making elite) may have fronted a good amount of the *tributum*, and then turned around to the business of squeezing the local independent, but now financially obligated, citizen-farmers for all they were worth – a bit like later forms of privatized tax-collecting we find with the infamous *publicani*.[65] Yet, collecting the *tributum*, designed to meet the grain payments of *stipendium* in kind, very likely led to increasing amounts of individual farmers turning to forms of intensification, closer to monoculture, in order to meet the particularly grain-based needs, at a dangerous destabilization of both farm and ecology, contributing to processes of erosion.[66] Diversification on small farms led to sustainability, and small farmers knew better than most how to keep the forces of growth and decay in balance. But this pursuit was disrupted the more that hindrances and pressures were placed upon a farmer,[67] and the natural result of the rise of a state structure in the fourth century was only to increase the number of new agents of whom a small farmer must beware.[68]

Thus, all of these new state burdens, farming and military costs, as well as the new forms of loan and debt interactions, combined to put immense pressure on individual farmers that tended toward two outcomes: (1) the acceleration of competition among farmers of all sorts, likely seen in the conflicts and legislation over *ager publicus*[69]; and (2) the turn toward forms of agricultural intensification that only contribute to ecological destabilization, erosion of landscapes, and an overall loss of resiliency, as small-scale farmers had to cope with all these new demands.[70] Arlacchi notes that peasantries do not always keep up with new modes of living: for modern Italy, industrialization ruptured the customary networks of peasant communities, it fractured the old *usi civici* (generations-old servitudes), and it atomized the peasantry, accelerating the process of individual competition.[71] I would suggest that a similar process of atomization affected many (but not all) Roman settlers and farmers across Latium's microregions, as the new

65 Erdkamp (1998, 16–17).
66 Hughes (2014, 124, 128, and 226–27), on the pressures of a tax system on sustainable agriculture and ecosystems.
67 Hindrances are not just financial but may also include lack of diverse seed, inability to sow at the right time, an inability to farm and cultivate through the most careful, responsible, and conservational strategies, as well as the need to over-intensify leading to soil erosion; see Spurr (1986, 41–45).
68 Berry (2015, 26). Peasantries showed a natural suspicion toward the larger world of government. Note this tone throughout Hesiod's *Works and Days*. See also Shanin (1972, 38); Gallant (1991, 58 and 98).
69 See Roselaar (2010, 25–26), on denial of access to *ager publicus* as a serious problem for smallholders.
70 In particular, see Hughes (2014, 30–31, 127, 140, 168, 227, 232–23), on the dire results of chronic agricultural decline. Many of these problems for Latium were highlighted a century ago by Frank (1919), yet studies on Roman expansion in connection with environmental factors are still somewhat rare.
71 Arlacchi (1983, 20–21).

civic organization of lands and people was increasingly applied by Roman leadership throughout the fourth century. Not all forms of progress lead to an uncomplicated or unbridled prosperity, and if the fifth century can be characterized as one of shrinking pains, then the fourth century was one of growing pains.

Conclusions

As Rosenstein has pointed out, land distribution schemes tend to work merely as a palliative for the problems of landlessness and poverty they aim to address.[72] In particular, environmental historians have noted that land distribution schemes often contribute to future rounds of debt and land-lessness, particularly as these tend to direct people into simplified, unsustainable forms of agricultural intensification that lead to erosion, and new waves of farmers look to move on to new land.[73] Though the cost came in the potential loss of ecosystems on which the state's war-making depended, the problem was never solved – it was nursed by the pursuit of ever-expanding territory, and the solution protected the problem. Population increase, small allotments, the simple preconditions of making a farm viable, and various other debt burdens placed an enormous pressure on farmers to compete with each other in ways that destroyed the cooperation on which mixed-farming depended and likely led them to over-intensify in their microecological niches. Forms of over-intensification, deforestation, and bad farming are immensely destructive to fragile microregions; they destroy topsoils and plant species which do not return, the loss of which leaves landscapes less retentive of water and nutrients.[74] The result is a landscape that slides relentlessly down the scale of ecosystem regression, for agriculture and humans cannot thrive at the expense of the soil that is the source of their future.[75]

In the Roman authorities' modifications of debt and interest rates during this period, we see a tendency to maintain civic entities on the landscape at what was probably something of a slow boiling point. Indeed, we see these

---

72 Rosenstein (1999, 198–99).
73 In particular, Hughes (2014), but cf. Rathbone (2008, 309–10). Attema, de Haas, and Termeer (2014, 221–2) have noted that the region of Antium and Satricum shows sign of population contraction in the early fourth century, after the foundation of a colony at Satricum in 385. Continuous conflict and warfare in the region may help to explain the decrease in population, but the gradual expansion of settlement southward in the Pontine region, and eventually into Campania in the late fourth and early third centuries, may also indicate a slow, stadial migration of farmers and settlers.
74 Frank (1919) on the soil exhaustion of Latium, desperation methods of farming, and the need to move on to new lands; he notes as well, at 272, that most small farmers could not command the resources to engage in many of the most effective land management techniques for mitigating chronic erosion.
75 Hughes (2014, 72–73, 81–82, and 140).

state authorities – aediles for example – restrict and prosecute *faeneratores* and *pecuarii* (grazing operations that are immensely destructive to ecosystems), likely only when they absolutely had to intervene in situations that threatened complete collapse of large numbers of citizen-farmers on the landscape.[76]

Overall, the mixed farm – an intricate, symbiotic web of strategies geared around mixed polyculture – was the sustainable ideal, and one that Latium's microecological diversity demanded of farmers. Yet, the new organizational mode of individualized allotments and civic farmers which appeared on the landscape in the fourth century produced a capacity for war and territorial expansion, with corresponding new demands, that likely only promoted unsustainable methods of land exploitation, resulting in eroded landscapes and thus the need for yet more land and war. This was the paradigm under which a new specialist war-making aristocracy gradually developed their own ethos in the fourth century.[77] Ultimately, we should understand the *secessio militum* of 342 and the "Third Secession of the Plebs" in 287 as two major moments in which the paradigm reached explosive points, fitting as these do with the logical outcome of the unstable political, economic, and agricultural foundations of land use and settlement on which fourth-century Roman warfare and imperialism was based.

The modern novelist, poet, and agrarian activist, Wendell Berry, has argued that the way a society treats its place produces that society's culture.[78] It is no wonder, therefore, that the Roman Republic's primary cultural export was its warfare, for its processes of distributing and overusing land produced a relentless, acquisitive need for new lands that could only be acquired through relentless conquests. As such, I would suggest that there was a deeper, structural-level motor pertaining to agriculture, forms of unsustainable farming and individualization that drove the logics of Roman imperialism over the long term.

76 On prosecutions of grazers, see Livy 10.13.14, 10.23.13, and 10.47.4.
77 See Hölkeskamp (1993), for a classic statement of the conquest-based outlook of the patricio-plebeian oligarchy.
78 Berry (2015, 24).

4 The *dilectus-tributum* system and the settlement of fourth century Italy*

James Tan

Introduction

For years I have passed on to my students what my teachers told me: the Romans did not tax their conquered Italian foes in money, but instead extracted manpower.[1] That was never entirely true, since allies were to some degree responsible for funding the contingents of troops they provided, but this seemed to be more a qualifier of the prior statement than a rejection of it.[2] It has become clear, however, that this view of the Romans as overwhelmingly extracting from their vanquished opponents in manpower rather than money is false.

The flaw in the traditional premise is that, as the Romans progressively extended Roman citizenship across much of Italy, every newly incorporated *assiduus* represented a new payer of *tributum* (however one imagines the *tributum* in the fourth century). This process of extending citizenship has traditionally been ascribed to a desire to increase the pool of available infantrymen, but as Rosenstein has pointed out, "a more immediate goal was undoubtedly the lightening of *tributum* on individual *assidui* by spreading the burden among more payers."[3] Thus, to the extent that Rome's settlement of Italy created more citizens, it increased the number of people contributing to public revenues.

In what follows, I will argue that the major (though not exclusive) consideration for the Romans when organizing newly conquered parts of Italy in the fourth century was not "Romanization," cultural affinity, or the

---

* All dates are BC unless otherwise noted. Much of the research for this paper was conducted during a generous leave provided by Hofstra University. It benefitted from feedback during the 2017 Laurence Seminar at Cambridge University, and by invaluable critiques from Jeremy Armstrong.

1 The notion has pedigree. See Momigliano (1975, 45): "Just as the organization of the Athenian empire had its own logic – more tribute and less military partnership – so the organization of the Italian alliance had its own logic – no tribute and therefore maximum military partnership."
2 Brunt (1988, 70); Nicolet (2000, 93–103).
3 Rosenstein (2016a, 96); see also Rosenstein (2012b, 108).

DOI: 10.4324/9781351063500-4

This chapter has been made available under a CC-BY-NC-ND license.

creation of juridical categories through the bestowal of civic rights on foreigners.[4] It was not to create a legal or administrative state across Italy. It was not even a myopic focus on manpower. Instead, it was the optimal extraction of *both* soldiers *and* money for the Roman war machine. Some communities were best-suited to the delivery of taxes, some to manpower, and some to a combination of the two. What drove those determinations, more than anything else, was the suitability of a community to the Romans' existing extraction apparatus: the *dilectus-tributum* system. By the 330s, the preferred method of settlement seems to have been enfranchisement and the creation of new citizens – to be enlisted through the *dilectus* and to pay taxes through *tributum*. It became clear, however, that not all communities were compatible with such forms of extraction. New categories had to be devised. The extension of Latin status, the imposition of bilateral alliances, and the rise of the elusive *civitas sine suffragio* were all responses to the failure of the *dilectus-tributum* system to extract from different peregrine communities the resources the Romans demanded.

One final note before progressing. This chapter seeks to explain the settlement of Italy in light of two institutional requirements of the Roman Republic – manpower and revenue – and thus has to make two claims about how resource extraction operated in the fourth century. The first claim is that Polybius' version of the *dilectus* was close enough to a fourth century reality to be taken as the best available model. There remain controversies surrounding army enlistment in this period (see Armstrong and Helm in this volume), but lest speculation be piled upon speculation, I have opted for the most conservative position possible as a control. The approach adopted here will of course have to be adjusted according to each reader's reconstruction of the *dilectus*.

The second claim is that *stipendium* and *tributum* were instituted in or around 406.[5] That has not always been accepted, and dissent has tended to focus on the difficulty of imagining *stipendium* and *tributum* functioning in the way described by the sources in a Rome before coinage.[6] Such fiscal transfers did not, however, require coinage. There are two main reasons for thinking this. The first is that the Romans were fully capable of effecting payments without coins.[7] The Twelve Tables presuppose the use of bronze to pay fines and pay debts, and the ritual of *mancipatio*, as Mersing has pointed out, required the touching of uncoined bronze on the scale in memory of a

---

4 The bibliography on these other approaches is vast. The most influential include: Sherwin-White (1973); Humbert (1978); Salmon (1982); Frederiksen (1984, 193–98).

5 See esp. Crawford (1985, 22–23) and Mersing (2007), with references to earlier scholarship.

6 Nicolet (1966, 35–42); Brunt (1971a, 641); Marchetti (1977); Harris (1990, 507, 2016, 32); Raaflaub (1996, 290); Rich (2007, 18); Ñaco del Hoyo (2011, 382); Serrati (2016, 107).

7 On Roman money before coinage, see Savunen (1993, 158); Morel (2008, 496–97); and Burnett (2012 esp. 298) on early Romans' comfort with money not only to effect payments but also to express value and define social status.

time when such payments were the norm.[8] The very word *stipendium* seems to imply the weighing of payments rather than the enumeration of coin.[9] There are, moreover, instances that suggest how effective such bronze payments could be in fourth century public finance: the lauded debt board of 352, for example, apparently deployed uncoined bronze to aid people in clearing their debts.[10] I take it for granted here that, if the Romans were capable of engaging in the rest of their economic pursuits without coinage, then they were capable of making these payments, too. The second reason to dismiss the notion that coinage was necessary for military pay is that there was no centralized mechanism for paying *tributum* and *stipendium*. The *tribuni aerarii*, as VanDerPuy also notes elsewhere in this volume, appear to have collected *tributum* and paid *stipendium* in a vast network of interpersonal exchanges. Without a centralized treasury process, there is no reason to assume that there was any fixed medium – even uncoined bronze – for any given payment. The medium would be negotiated between the *assiduus* and the *tribunus aerarius* on the one hand, and the *miles* and the *tribunus aerarius* on the other. Bronze might have been the ubiquitous medium of payment, but, so long as each party agreed otherwise, it need not have been. There was no procedural requirement for any particular form of payment. Of course, the treasury could receive large payments in tribute or booty – neither necessarily coined – and could distribute it through the army just as plunder was allotted, and I will argue below that *cives sine suffragio* probably made just such payments. But the *aerarium* was not involved in *stipendium* for most soldiers in most years. Thus, even in the absence of coinage, Rome was sufficiently monetized in this period to implement military pay, and the system they created was sufficiently decentralized and flexible to cope with what monetary limitations affected them.

The creation of the Italian categories

There is good reason for believing that Romans frequently found the burden of *tributum* to be oppressive in the first half of the fourth century. A refusal to wage war in 378, for example, was only ended once it was agreed that nobody would be hauled before the courts for debts and that nobody had to pay *tributum* for a year.[11] Similarly, 347 (a year without war) was remembered as blessed in part because debts were restructured and interest rates lowered, but mostly because nobody had to pay *tributum*: "Most

---

8   Mersing (2007, 231–32); see also Forysthe (2005, 336). For the use of bronze payments in the Twelve Tables (numbering follows Riccobono, *FIRA*): 2.1a (judicial *sacramenta*), 7.12 (inheritances and manumission), and 8.3, 8.4 and 8.11 (fines and compensation).

9   Boren (1983, 428); Crawford (1985, 22–23); Oakley (1997, 631); Mersing (2007, 231–32); Rawlings (2007, 49); Rich (2007, 18).

10 Livy 7.21.5–8. See: Hölkeskamp (1987, 83–84, 100–1); Savunen (1993, 152–53); Oakley (1998) ad loc. on the episode's historicity.

11 Livy 6.31.4.

of all, affairs were lightened because of the cessation of *tributum* and the *dilectus*."[12] Nor was warfare itself necessary for this problem of *tributum* to appear. The enterprise of building the great stone fortifications of 377 demanded financing, and Livy's sources believed that the resulting *tributum* exacerbated an already problematic debt situation.[13]

If they were having trouble meeting the demand for public revenue in the early fourth century, it is worth considering how the Romans went about repairing their abject fiscal condition. One answer was to create a sustainable fiscal base through conquest. War to the north brought the subjection of Veii in 396. A series of wars in the south then saw Rome's control extend through Latium, across the highlands to the east, down to the land of the Volsci and, by 342, even into Campania. By 338, the Romans, triumphant in the Latin War, had most other Latin communities at their mercy. As each entity was conquered along the way, the Romans had to decide what the ensuing peace and relationship would look like.[14] Traditionally, although they had absorbed some local areas, in the fifth century, the Romans had tended to embrace a pattern of "capture, plunder, and withdraw," which was an effective source of windfalls, but did not provide long-term revenues.[15] That method changed after the institution of military pay c. 400.

The fall of Veii presented the Romans with their first opportunity in this period. It can have been lost on no statesman that the conquest of a large neighboring city might prove lucrative, yet knowing that there was a potential revenue source was not the same as knowing how to tap it.[16] The obvious solution might have been to create a tributary. That is, the Romans could have preserved the Veientines as a peregrine community while demanding that some form of economic payment be delivered to Rome each year as the price of defeat. This might have resembled one of the myriad forms of tribute as they existed in the East before the fourth century, or perhaps a pre-coinage version of the indemnities imposed on various vanquished enemies after the Punic and Macedonian Wars.[17] Yet, for whatever unknown reason, this was not the road taken.[18] Instead of importing tribute from a foreign people, the Romans ultimately chose to export the *tributum* system into the *ager Veientanus*.[19] Because *tributum* was a tax on property instead

---

12 Livy 7.27.4: *Levatae maxime res, quia tributo ac dilectu supersessum.*

13 Livy 6.32.1, where the tax is referred to as *tributum*, even in the absence of a military campaign. For the oppressive demands of the wall, see Bernard (2016) ch. 4, focusing on labor instead of finances.

14 For a clear overview, see Cornell (1995, 309–26, 345–51).

15 Armstrong (2016c, 218–19) with examples. On a lesser but very real tradition of enfranchisement, see Humbert (1978, 78).

16 Neesen (1980, 6) on the range of extraction possibilities.

17 For a list of such indemnities, see Frank (1933, 127–35) and now Taylor (2017, esp. 173).

18 On the Romans' abiding focus on the city and its institutions over a more Italian focus, see Gargola (2017).

19 Livy 6.4.4, with Humbert (1978, 79) and Bernard (2016, 324–26). This should be seen as part of VanDerPuy's "agricultural individualist paradigm" elsewhere in this volume.

of people, the distribution of Veii's lands among *assidui* – or perhaps the enfranchisement of preexisting Veientines – was necessarily a fiscal act that expanded the tax base by increasing the assessable property of the Roman population.[20] Each new *iugerum* held by a Roman *assiduus* was a source of revenue through *tributum*.

That enfranchisement was the first of many in the fourth century. The Capenates and Falisci were given citizenship in 388, and after a defeat in 381, the Tusculans were "made Roman" as a consequence of their defeat in war.[21] This was no act of good will. As Cornell has noted, a hypothetical German enfranchisement of France in 1941 would leave us with no illusions of benevolence.[22] Furthermore, extending that analogy, so that one imagines the French each year having to pay a German tax collector to fuel the Wehrmacht, illustrates how gut-wrenching *tributum* must have been for the Tusculans. No wonder they joined the Great Latin War of 340–338. It is also possible that Velitrae, further south, was refounded as a citizen colony in 401 or after its capture in 380.[23] Though he mentions no such enfranchisement in his account for either year, Livy refers to the Veliterni as *cives* in 385 and 383, and later implies that by 338 they had long been citizens: "old

20 Bradley (2011, 244–45) and Bernard (2016, 325–26) make the case for enslavement. For the enfranchisement case, see Cornell (1995) 320; Armstrong (2016c, 246); Scopacasa (2016, 37).

21 Livy 6.4.4, 6.26.8; Oakley (1997, 357) on the strategic benefits of Tusculum's absorption. Despite the later date for the enfranchisement suggested by Armstrong (2016c, 249 n. 98), there are no obvious problems with a date around 381. The seizure of the *ager Pomptinus* in 387 was also a way of incorporating more land into the Roman *tributum* base. It is true that no new tribes were created until 358, but war and domestic turmoil – including the supposed anarchy years – probably explain that. Oakley (1997, 657) speculates intriguingly that there were not tribes because the rich few had snapped up the land, thus explaining in part the *lex Licinia* of 367. See Livy 6.5.2; Humbert (1978, 152–54); Roselaar (2010, 300 n. 11); Taylor (2013, 52), including further bibliography.

22 Cornell (1995, 323); on Tusculum: Humbert (1978, 154–61). Sherwin-White's (1972, 15) claim that Tusculans could be absorbed because they lacked their own political self-consciousness defies both plausibility and the evidence. Galsterer (1976, 65) argues that Rome's enfranchisement of Tusculum was the result of an alliance between the aristocratic families of the two cities, but the best evidence he has for this is that a Fulvius reached the consulship some six decades later and a Mamilius six decades after that. Even after that long delay, the consulship of Fulvius Curvus was apparently controversial among his compatriots: see Pliny's bizarre claim (*HN* 7.136) that Fulvius was consul of a rebelling Tusculum in the same year he triumphed over his home city as consul at Rome. There was then a motion to execute Tusculum's men and enslave the women and children (Livy 8.37.8–12; Val. Max. 9.10.1): see Oakley (1998, 755–57) and Helm (2017, 210) for discussion and further bibliography. Frederiksen (1984, 195–96) considers it "unlikely" given the volatility of Latium that Rome would enfranchise Tusculum as an act of "naked aggression." But Latium did indeed explode after this, with a war for Tusculum in 377: Livy 6.32.4–33.12; Toynbee (1965, 126); Sherwin-White (1972, 30); Humbert (1978, 159–61). These subsequent events make more sense if Roman policy was indeed menacing.

23 Year 401: Diod. Sic. 14.34.7–8, and the attestation of *Veliterni coloni* at Livy 6.36.1. Year 380: Livy 6.29.6. Cornell (1995, 349 n. 15) acknowledges the implication of citizenship in Livy's text, but dismisses it.

Roman citizens, who had so often rebelled."[24] This extra, recurring betrayal by citizens would also make sense of the unusually harsh treatment they received, which saw the destruction of their walls and the deportation of their leaders across the Tiber.[25] Moreover, the tales of Tusculum and Velitrae are entwined on more than one occasion. Tusculum was defeated and enfranchised in 381, whereas Velitrae was captured just one year later in 380, and each of these events occurred within the context of a financial crisis at Rome, which might reinforce the explanation of creating more taxpayers.[26] Livy claims that, as part of a wider war, the Latins then captured Tusculum in 377 because the latter had received Roman citizenship, and it is highly likely, given the involvement of the nearby Volsci, that the Veliterni were involved.[27] Velitrae again apparently marched on Tusculum in 370.[28] Finally, in his account of 338, Livy discusses how the Tusculani kept their citizenship, and proceeds immediately to the Veliterni without any mention of the status they received.[29] The simplest interpretation is that they were treated in the exact same way as the Tusculans: they retained a citizenship they already had.

With or without Velitrae, the record shows that the Romans at the very least understood that enfranchisement was a way to deal with a defeated enemy that did not require any new institutions. Furthermore, while it has long been understood that enfranchisement added to the reserves of military manpower, the Romans must have known that increasing the total number of taxpayers alleviated the fiscal burden on each individual citizen.[30] The

---

24 Livy 6.17.7 (where Velitrae is paired with Circeii), 6.21.3, and 8.14.5: *veteres cives Romanos, quod totiens rebellassent.* This may refer to the archaic status of Velitrae as a colony, though Livy is more insistent on the issue of Velitrae's citizenship than on other colonies. Humbert (1978, 185–86) assumes that Velitrae received only *civitas sine suffragio* in 338, but there is no evidence for this.

25 It is true, however, that hostile colonies were not uncommon in this period: see Roselaar (2010, 75). It is also true that the leaders in Privernum were similarly deported in 329, and they certainly were not citizens (Livy 8.20.9).

26 Livy 6.26.8 (Tusculum), 6.27.6–9 (finance), 6.29.6 (Velitrae). Velitrae's tribal status might be a problem here. Taylor (2013, 54–55) argued that Velitrae belonged to the *tribus Scaptia*, on the grounds that C. Octavius, the future Augustus and a native of that city, was a member of the tribe. The problem is that the Scaptian tribe was not established until 332, and hence could not have been the city's tribe prior to that date. It is possible, however, that the Scaptian tribe included the Roman colonists who were sent to occupy the old aristocracy's lands after their expulsion following the Latin War (*pace* Taylor 2013, 55), and that Velitrae had belonged to a different tribe prior to this, or had simply never been pacified enough to receive any tribal assignment.

27 Livy 6.33.6, with 6.32.4 for combined Latin-Volscian forces. The Veliterni were forever caught up in the hostility between the Volsci and Rome: see Armstrong (2016c, 225).

28 Livy 6.36.2.

29 Livy 8.14.4–8.

30 On manpower, see most recently Armstrong (2016c, 247, 255). Various scholars have noted that new citizens were obliged to pay *tributum* as well as serve in the legions; for example, Staveley (1989, 428); Cornell (1995, 322); Oakley (1998, 548); Eckstein (2006, 251). Yet few pursue that observation. Rosenstein (2016a, 96) and Harris (2016, 20) are rare in noting that conquest and enfranchisement represented an increase in revenue.

dilectus-tributum system thus meant that every act of enfranchisement was also a fiscal policy. There must be a full realization that the creation of these new citizens was in no small way the creation of new taxpayers. Despite the sharing of certain civic rights attached to citizenship, the benefits of such rights in the fourth century for those living at a distance from Rome were probably negligible. Enfranchisement should therefore be seen less as a sharing of benefits and more as a sharing of obligations. It constituted a peculiar tribute system. Ando has noted that the nature of enfranchisement depended very much on whether one adopts the perspective of the "metropolitans" or the "subalterns," and he is no doubt right to state that citizenship was being imposed *pro poena*.[31] Thus, to those Tusculans who now paid *tributum*, the modern notion that the Romans did not extract money from others would have been baffling and infuriating. Those new citizens must have understood – with crystal clarity – that the Romans, as they expanded through Italy and created more and more citizens, were also creating a larger and stronger base of taxpayers.

That process accelerated after the defeat of the Latins and the reorganization of the territory between Rome and Campania in and after 338.[32] Much as had been done to Tusculum, citizenship was forced upon several old Latin cities, including Aricia, Lanuvium, Pedum, and Nomentum.[33] Yet, other cities were assigned to other categories. There were Latins, especially in colonies, as well as independent allies (*socii*) providing troops. Finally, there were citizens without the vote (*cives sine suffragio*). But if the Romans had come to perceive enfranchisement as such an effective solution to diplomatic and fiscal problems, why not extend it to all communities? An examination of the fiscal structures of each status provides one part of the answer (see Table 4.1).

The existing category of the full Roman citizen (the *civis optimo iure*) provided the extractive cornerstone of the Roman war machine: the *dilectus-tributum* system. Part of the population of *assidui* (the citizens of property) provided military manpower through the *dilectus*, a draft mechanism which selected young men and assigned them to the legions to fight, where they were paid *stipendium*.[34] That *stipendium* was, in turn, paid for by the remainder of the citizenry through the financial levy known as *tributum*.[35]

---

31 Ando (2016, 175–79). On the punitive nature of citizenship, see also Humbert (1978, 151–207); Cornell (1995, 323); Oakley (1998, 547–51). Eckstein (2006, 251) acknowledges that the enfranchisement was burdensome, but holds that it still "seems a typical Roman attempt at political compromise."

32 For excellent overviews of the settlement, see Humbert (1978, 176–207); Oakley (1998, 538–59); Bourdin (2012, 295–96) (more briefly).

33 Livy 8.14.2–8; Vell. Pat. 1.14.3. On the likelihood that the settlement extended across several years in the 330s and 320s (per Velleius): Oakley (1998, 539).

34 Most of the evidence for the *dilectus* comes from Polyb. 6.19–20; for an intelligent discussion, see still Brunt (1971a, app. 19).

35 On *tributum*: Nicolet (1976, 1980, 2000); Rosenstein (2016a).

Table 4.1 Civic Status and Contribution in the *Dilectus-Tributum* System

Status	Method of Manpower Contribution	Method of Financial Contribution
Cives Optimo Iure	Dilectus	Tributum
Non-Latin Peregrine *Socii*	Mandated contribution to alae	None
Latini	Mandated contribution to alae	None
Cives Sine Suffragio	Mandated contribution to legions?	Tributum vel sim?

This nexus of manpower extraction and financial extraction relied on a balance between the two components: too many men ushered through the *dilectus* would leave too few to pay the bills through *tributum*; leaving too many on their farms to maximize *tributum* would provide too few to do the fighting. The *dilectus-tributum* system therefore required a demographic proportioning that kept both sides sustainable.[36]

That *dilectus-tributum* system was not, however, imposed on all Italians. Several peoples in central Italy avoided citizenship in the 330s. Trusted partners like the Hernici, for example, remained independent allies so long as they were loyal, and many other communities would be treated likewise in subsequent decades.[37] At least at a later date, these *socii* or *foederati* were required to contribute troops in numbers determined by the *formula togatorum*, though it remains uncertain whether this was the mechanism used in the fourth century.[38] In any case, it is all but certain that these allied communities had to contribute troops to Rome according to some arrangement from an early date; by the battle of Sentinum in 295, for instance, allies may already have outnumbered citizens in the army.[39] They did not, however, contribute money directly to the Roman state. Nor, importantly, were their troop contributions paid by the Romans. Instead, it seems that each allied community had to fund its own troops, if there was pay at all.[40]

36 See VanDerPuy in this volume.
37 Hernici: Livy 9.43.23–4 implies that the Hernici were not *cives* before 306.
38 Brunt (1971a, 545–49); Salmon (1982, 169–70); Baronowski (1984); de Ligt (2007, 116–17); Erdkamp (2007a, 47–48); Kent (2012a, 71–73). Lo Cascio (1991–4) offers a useful review of the history of scholarship on the term, though his conclusions – that the *formula* listed adult males eligible for conscription – are not adopted here. Treaties have conventionally been credited with determining a community's contributions, but Rich (2008) and Kent (2012a) have emphasized the role that other mechanisms likely played.
39 Cornell (1995, 361).
40 Pfeilschifter (2007, 31); Ñaco del Hoyo (2011, 383). The need to pay home troops may explain the proliferation of local bronze coinages in the third century: Burnett (2012, 308) (summarizing a wide bibliography). This interpretation does not explain why those coinages disappeared after 200.

From a fiscal perspective, there was no real difference between these allies and those in the third category: the *nomen Latinum*, or the Latins. The archaic set of rights and duties which came with Latin status (*ius Latini*) was, after 338, held almost exclusively by the Latin colonies.[41] Like other allies, these colonies would come to contribute men though the *formula togatorum* presumably in the same way as other allies, even in the late fourth century.[42] Also like the other allies, they paid no monetary taxes and received no pay from Rome for military service. Like the allies, they presumably had to fund their own contingents if they wanted their soldiers to be paid.[43] The fiscal interchangeability of Latin and non-Latin allies is expressed in the term *socii nominisve Latini* ("allies or those of the Latin name") on the agrarian law of 111.[44] The difference between the two lay in a few rights of interaction with Romans.

Comparing the contributions of allies and Latins with full Roman citizens, there is one obvious point to be made. Allies were cheap, because they received no *stipendium*, but they were also worth very little – at least in fiscal terms – because they contributed no revenue through *tributum*.[45] Citizens, on the other hand, were expensive because they required pay (*stipendium*) when serving as soldiers, but were valuable because they contributed revenue through *tributum*. Allies were low cost and low reward; citizens were high cost and high reward.

The final category comprised the citizens without the vote (*cives sine suffragio*) and appear to have occupied the worst of both worlds according to their contributions. This requires some more detailed argumentation, since there is little consensus about how the *civitas sine suffragio* operated. There is thus a need to establish how the *cives sine suffragio* contributed manpower and money.

It seems clear enough that the *cives sine suffragio* contributed men in the same way that the Latins and allies did, but that these men were *stipendium*-earning members of the citizen legions instead of the allied *alae*. Livy describes the Cumani, Acerrani, and Atellani as *cives sine suffragio*, and Paulus tells us that they earned *stipendium* in the legions from the start (*initio... in legiones merebant*).[46] At the same time, however, they seem to have done so in corporate groups that could be distinguished from other (Roman or allied) units in ways that made plain their place of origin. This is because, when they do appear fighting in the sources, they

---

41 On the nature and limitation of those rights, especially later in the Republic, see Coşkun (2009a).

42 Bourdin (2012, 289–96).

43 Livy (27.9.7) has colonists unable to provide *milites pecuniamque*, implying that they needed to raise money to pay their troops.

44 *Lex Agraria* (*CIL* 1^2.585) l. 21: *Romanus sociumve nominisve Latini, quibus ex formula togatorum [milites in terra Italia imperare solent]*. Cf. *lex Repetundarum* (*CIL* 1^2.583) l. 1.

45 Cornell (1995, 366).

46 Paul. 117L, 126L; Grieve (1983, 37); Oakley (1998, 545–46); Bispham (2007, 24).

are attested as deployed in precisely these kinds of recognizable detachments.[47] Campanian troops, for example, are listed among other Italian groups by Dionysius in his account of the battle of Asculum.[48] A garrison, probably comprising Campanians, is also attested at Rhegium at the end of the Pyrrhic War.[49] However one imagines the recruitment and deployment of full citizens in this period – see chapters by Armstrong and Helm in this volume – it is striking that the origin of allied and half-citizen troops remains in the evidence in a way that the origin of full citizen troops do not; there are no equivalent references to the glorious stand of the maniple from this *pagus* or the members of that *gens* (at least for this period).[50] The tradition did not know of such corporate categories for *cives optimo iure*, but it did for *socii* and *cives sine suffragio*. The citizenship of the *cives sine suffragio*, then, was reflected less in where they fought as members of the citizen legions, than in their drawing of *stipendium* like citizens and perhaps in their sleeping in the parts of the camp assigned to Romans.[51] It was not their position in the line of battle that determined their status, but their position in the fiscal system.

Much of this may have boiled down to the exclusion of *cives sine suffragio* from the tribes. As Humm and Mouritsen have noted, the *dilectus*, as Polybius describes it, had no way of processing troops without tribal affiliations, and *cives sine suffragio* were not members of the old or the new tribes.[52] Their exclusion likely had nothing to do with warfare; it is true that old Roman allies in Latium Vetus were fully enfranchised into the tribes, but this cannot have been due to a military relationship, since the Hernici – Rome's comrades in arms for over a century and a half – were excluded.[53] Whatever criteria determined this division, it would have prevented the use of the *dilectus* to raise troops from among the *cives sine suffragio*.

---

47 Sherwin-White (1973, 42); Grieve (1983, 37); Oakley (1998, 556). Lo Cascio (2001, 582) suggests that the *cives sine suffragio* originally fought in the legions like other citizens, but later fought in their own contingents after census reform. This seems unnecessary. Especially after the adoption of maniples, the legion was probably flexible enough to incorporate autonomous units who *in legiones merebant* while not being enrolled through the *dilectus*. On manipular flexibility: Armstrong (2016c, 268).

48 Dion. Hal. *Ant. Rom.* 20.1.5. For caution on the veracity of such lists of ethnic units, see Erdkamp (2007a, 67–74).

49 Dion. Hal. *Ant. Rom.* 20.4; Polyb. 1.7.6; cf. Val. Max. 2.7.15f, giving different names. It is possible that this garrison was made up of mercenaries or some other type of *auxilia* removed from regular contributions: Frederiksen (1984, 224); Kent (2012a, 80).

50 Contrast the district units of Alexander's army, which are described in the sources by locality: Diod. Sic. 17.57.2; Arr. *Anab.*1.2.5; Hatzopoulos (1996, 243); Rzepka (2008). Jehne (2006, 255–56) analyses the Polybian *dilectus* as a ritual negating local identity in favor of allegiance to Rome, though how early this began is uncertain.

51 On the camp as an expression of civic categories: Pfeilschifter (2007); Rosenstein (2012a, 93–103).

52 Humm (2006, 39, 50); Mouritsen (2007, 156). See also Armstrong in this volume.

53 On the *Latium Vetus* boundary (aside from the annexed *ager Falernus*), see Cornell (1995, 350–51) with map.

If exclusion from the tribes prevented the *cives sine suffragio* from being enlisted like the full citizens, did this mean that they were similarly unable to participate in the payment of *tributum* and the receiving of *stipendium*? This part of the fiscal system was, after all, similarly reliant on the tribe as the functional unit.[54] Since Beloch, however, various scholars have held that the *cives sine suffragio* paid *tributum* or some kind of revenue, and this is surely correct.[55]

The Romans knew how to move money in a variety of ways. They may have had fewer ways to raise citizen soldiers, but they were constantly handling booty, paying out contracts, collecting fines, establishing colonies, funding cults, managing the manumission tax, etc. It is absurd to claim that they could not manage public finance without the tribes. Despite the Campanians not being in tribes, moreover, we know that the censors of the third century were well-informed about the population details in Campania because Romans and Campanians were listed together in the census data provided by Polybius for 225.[56] The Campanians presumably ran their own census and forwarded the data to Rome, and it cannot have been difficult to calculate what they owed, have them collect it locally, and wait for it to be delivered in whichever way was mandated.[57] Thus, the collection and payment might not have been the same as *tributum*, but the cost and frequency were probably identical. Salmon suggested that a *praefectus iure dicundo* could have managed the census, though this seems unlikely given the small number and limited capacity of these officials.[58] A more plausible possibility is that the *cives sine suffragio* were listed on the *tabulae Caeritum*.[59] Strabo tells us that the *tabulae* were used to record citizens without *isonomia*, while Gellius and Pseudo-Acro believed they contained the names of citizens who could not vote; Pseudo-Asconius, moreover, noted that the only civic role for at least some of those entered into them was that they had

---

54 Varr. *Ling.* 5.181; Mouritsen (2007, 156); Taylor (2013, 8). See the more detailed discussion below.

55 Beloch (1886, 319); Cornell (1995, 351); Oakley (1998, 548), Humm (2006, 55); Cecchet (2017, 10–11).

56 Polyb. 2.24.14; Brunt (1971a, 20). Brunt notes that there was a *censor perpetuus* at Caere (*ILS* 6577–8), but linking that post with specific obligations to Rome is speculative in the extreme.

57 On the census in Campania: Galsterer (1976, 72); local registering of *cives sine suffragio*: Lo Cascio (2001, 585).

58 Salmon (1982, 69).

59 Bispham (2007, 23 n. 94); Nicolet (1980, 27, 86). Brunt (1971a, 515) argued that there were *tabulae* for each community, and not just one register named after Caere. Brunt's objection, however, that Fundani, Formiani, and others would not be listed on a table named after Caeritans carries little weight, given Gellius' information (16.13.7) that the *Tabulae Caeritum* contained the names of citizens stripped of the vote; though on this point see Grieve (1983, 28–29). This proves that the list was not limited to Caere, and thus it could have registered citizens from other communities. In any case, anachronistic terminology seems never to have been a problem to the Roman mind.

to pay *tributum*.[60] The tables were thus perfect for the fiscal registering of citizens without voting rights. There are enough alternatives to reject the notion that only those enrolled in the tribes could have paid some version of *tributum*.

It may be argued that, since Capua's demand that Hannibal levy no troops was not paired with a refusal to provide money, they must not have been paying *tributum*. There are, however, reasonable explanations for this: perhaps Roman sources omitted it so as not to make the Capuans' cause more sympathetic; perhaps, since it was clearly not in their interests for him to lose, they did not mind supporting Hannibal with money; perhaps Hannibal never asked. Capua's terms are suggestive at best, yet prove nothing.

There is little reason, then, to reject the idea that the *cives sine suffragio* could have paid *tributum*. It cannot have involved the same role for tribes and *tribuni aerarii*, but regardless of how it was collected and delivered, the nomenclature of *civitas* likely meant that this was a revenue source paid at the same level as citizen *tributum* and on the same occasions. It would have been a welcome way to offset the *tributum* obligations of the full citizens; if more fungible than the old *tributum* from the tribes, it might even have granted leaders greater discretion in spending it on public expenses beyond *stipendium*. All this, however, is a long way from what our meagre evidence can sustain. What matters is that, by framing contributions as the obligations of citizens, the system only demanded as much as the old full citizens themselves were paying, and only in the years that those citizens too were paying it. They thus asked nothing of others that they did not ask of themselves. The Romans could raise revenue while claiming not to be collecting tribute in the oppressive Assyrian or Athenian sense.

There are, moreover, good reasons for thinking that monetary contributions must have been part of the *civitas sine suffragio*. The first is that war became more expensive in the later second century. Fighting was further afield, and at some point in or before 311, the number of legions was doubled from two to four. By Sentinum in 295, there were at least six legions, and possibly eight or more in the field.[61] Twice as many legions required twice as many men, who required twice as much pay, which required twice as much *tributum*.[62] There had been an expansion of *ager Romanus* to strengthen the *tributum* base – four new tribes were added in 387, two in 358, and two more pairs in 332 and 318– to offset some of these expenses, but to assume that the *cives sine suffragio* did not pay taxes is to assume that the Romans, while

---

60 Strab. 5.2.3; Gell, 16.13.7; Ps. Acro, *ad Hor. Epist.* 1.6.62; Ps Asconius 189 Stangl; see Grieve (1983) for convenient summary of evidence.

61 Oakley (2005, 282–83).

62 That change need not have happened instantaneously in 311, but preceding years may have seen each of the two legions swell in numbers until the increase was ratified in a doubling of legions: see Sumner (1970, 71–72). Even this scenario would have represented a doubling of *stipendium/tributum*.

watching their costs rise, never required *cives* who *in legiones merebant* to contribute.[63] This would be especially odd given the wealth in Campania. It is true that they never asked peregrine allies to pay money either, but such allies were not drawing *stipendium*. These other allies might not have represented revenue, but nor were they costly. It strains credulity that the Romans would offer military pay to former enemies and watch their own public expenses rise, but then choose to pay higher taxes themselves so as to spare their new, often wealthy fellow-citizens the burden of the military pay they were receiving.

The final and best reason to accept that *cives sine suffragio* paid *tributum* is that it explains why the category of "half citizenship" was created in the first place. Sherwin-White noted the equivalence of Latins and *cives sine suffragio*, and, blind to any financial considerations, struggled to establish some differentiating mechanism:

> It is difficult to see any difference between the status of *municipes* enjoying the original form of *civitas sine suffragio* and the status of Latins enjoying *conubium, commercium*, and *ius civitatis mutandae*, except that the line of demarcation between Latin and *civis Romanus* was more clearly drawn.[64]

The inexplicability of the *civitas sine suffragio* has also struck Mouritsen, who in a counsel of despair has rejected its fourth century historicity altogether.[65]

Such a stance is excessive and, with no superior alternative, requires the rejection of too much ancient testimony. Festus confirmed that Cumani, Atellani, and Acerrani were indeed *cives* from the start, despite not being able to hold office.[66] Ennius' claim that "the Campani were then made Roman citizens" is explicit confirmation that the *civitas* (*sine suffragio*) was bestowed upon Capua, too, at some point prior to Ennius' day.[67] Mouritsen

---

63 See above n. 46.

64 Sherwin-White (1973, 46); *pace* Cornell (1995, 351). Sherwin-White's view that the difference might be explained by ascribing the categories to different periods does not solve the problem, since it is unclear why, if the two were so similar, they would switch from one technique of integration to another.

65 Mouritsen (2007); see also Pfeilschifter (2007, 27 n. 2).

66 Festus, *Gloss. Lat.* 117L, 126L.

67 *Ann.* 169 Vahlen: *cives Romani tunc facti sunt Campani*. Mouritsen is correct that the reference to Campani cannot be dated within Ennius' history, but there is no obvious alternative to the fourth century. The claim of citizenship gains possible support from a tribune's complaint in 270 that the garrison from Rhegium, which Polybius believed was led by a Campanian, was executed in violation of citizen rights. However, the precise origin of each soldier is not certain, and complaints might have applied only to a handful. See Val. Max. 2.7.15f; Polyb. 1.7.11; Dion. Hal. *Ant. Rom.* 20.4. Polybius insisted that the leader of the garrison was a Campanian named Decius, but Valerius contradicts this. Neither source claims that the entire garrison was Campanian.

rejects Ennius' near-contemporary authority on the grounds that no author could describe as citizens those who lacked tribal affiliation, but it is not self-evident that this principle should be allowed to overrule Ennius' authority.[68] Skepticism toward Ennius would be on stronger ground if the poet had used the term *quirites* – the fully integrated participants of the body politic – instead of *cives*, but he did not.[69] It is true that Livy (45.15.4) has a censor of 169 deny that a censor could remove a *homo* from the tribes without a vote of the people, since this would be to deprive him of *civitatem libertatemque*. A lacuna unfortunately prohibits clear understanding of whether Ti. Sempronius Gracchus accepted the claims of C. Claudius Pulcher here. Nor is it clear whether Claudius' position was an accepted legal fact or more a rhetorical assertion. What is clear, however, is that the status of citizenship had changed a great deal in the 170 years prior to this censorship. Whether Claudius' later views must reflect the reality in 338 should not be taken for granted.

Moreover, three sources come very close to refuting the idea that *civitas* depended on voting in the tribes. In the first place, line six of the *lex Latina Tabulae Bantinae* (*CIL* 1^2.582), orders all magistrates to forbid from the vote anyone convicted under its terms. This seems to imply that citizens could indeed lose the vote, and would thus contradict Claudius' assertion, unless one claims that the jurors' verdict represented a vote of the people. In the second passage, Pseudo-Asconius states that a plebeian enrolled in the *tabulae Caeritum* was expelled from the *album* of his century by the censors so that "he was a citizen only in so far as he paid ... *tributum*."[70] Grieve argued that this poor soul was only expelled from *his own* century, but could still vote in the enigmatic *niquis scivit* century.[71] Thus, he retained his suffrage. This, however, is not what the passage says. Pseudo-Asconius explicitly states that this man was a *civis* "only in so far as" (*tantummodo ut*) he paid *tributum*. He did not vote or serve in the legions, or perform any other civic function related to his century (tribes are left unmentioned). The fact that he remained a *civis* "only in so far as" he paid *tributum* is explicit evidence that contributing revenue was indeed sufficient grounds on its own to be a *civis*. In fact, the existence of such censorial procedures as *tribu movere* and *aerarium relinquere*, which Crawford takes as synonymous, would suggest that there were well-established ways for *cives* to exist as taxpayers without being enrolled in tribes.[72] Finally, Strabo (5.2.3) states that the Caeritani received citizenship without being enrolled "among the people" (εἰς τοὺς πολίτας), and instead were entered into the *tabulae Caeritum* along with others who lacked *isopoliteia*. The case of the Caeratani is notoriously convoluted and

68 Mouritsen (2007, 156–57), following Grieve (1983, 27).
69 Palmer (1970, 156–72, 189); Perret (1980); Stone (2005, 72); Smith (2006a, 200–2).
70 Ps-Asc. 189 Stangl: *esset civis tantummodo ut pro capite suo tributi nomine aera praeberet*.
71 Grieve (1983, 27), citing Festus, *Gloss. Lat.* 184L.
72 Crawford (2006).

Strabo is unspecific about the others who lacked *isopoliteia*, but he accepted that there was indeed a way for people to be *cives* without being enrolled "among the people" and thus presumably the tribes. Combined with references to the *civitas sine suffragio* as a real phenomenon, these sources suggest that absence of tribal registration is a problem, but not an insurmountable one. To insist otherwise – to insist that Claudius' reported speech in 169 is a sounder foundation for an argument than the plentiful evidence for citizenship without the vote – is to allow the tail to wag the dog. Mouritsen's alternative to accepting the *civitas sine suffragio* is the intriguing suggestion that the term was invented in the second century as part of the debate over allies' participation in Roman politics, and was subsequently adopted by Livy and other writers.[73] Yet this, too, is difficult to sustain. It is perhaps conceivable that an awkward and inaccurate moniker was produced by second century allies, but it strains belief that it survived contemporary Roman political debate, only to be adopted by Rome's own historians despite a war having been fought over the very real status of the *civitas*. Pity the poor man who had to convince a M. Porcius Cato or a M. Aemilius Scaurus that in fact some (but not all) allies enjoyed a 200-year-old category of citizenship that nobody had ever heard of before. Much like a *lectio difficilior*, the quirkiness of the institution is actually one reason to assume that nobody made it up – and thus nobody had to be convinced of it – as a fiction *ex nihilo*. The soundest course is thus to try to explain it within a fourth-century context.

Such an explanation has, however, been elusive due to the range of rights and constraints involved in *civitas sine suffragio*. How could one category contain such variety? Yet, the eclectic conditions of each community's status is only a problem if they are assumed to be of definitional importance. Mouritsen saw simply too much heterogeneity for one category to subsume. Sherwin-White wondered why there would be *cives sine suffragio* at all given their resemblance to other allies. The answer, I argue, lies in the obligations instead of the privileges. If *civitas sine suffragio* was defined purely by what Rome gained from these half-citizens – manpower and revenue – then the range of other conditions would be ancillary and inconsequential. In other words, the idiosyncrasies of Anagnia's rights would pose no categorical problem, so long as the burden to contribute revenue and manpower was consistent with that of other *cives sine suffragio*.[74] *That* was how the category was defined. Similarly, a difference in financial contributions would explain why otherwise similar categories coexisted. Table 4.1 (above) reveals that, although the provision of manpower was similar for *cives sine suffragio* as for Latins and other allies, the need to pay *tributum* would justify the creation of a separate status. In other words, if *cives sine suffragio* paid *tributum* (or some equivalent) while other allies did not, then there is a way to explain the status' existence separate from other forms of alliance: the different

---

73 Mouritsen (2007, 155).
74 Anagnia's exceptionalism: Mouritsen (2007, 154).

categories defined different obligations. This places the analytical burden on what the Romans were extracting from these other categories – much as the term *municeps* (one who bears an obligation) would suggest – instead of what they were granting.[75]

Why citizen, allied, or Latin?

Two questions remain. The first is why, if full citizenship so successfully furnished money and manpower, the Romans bothered to employ three other categories for conquered Italians. The second is to explain why some communities received one of these statuses while others received another.

The argument here is that, by the middle of the fourth century, enfranchisement had become the preferred method of extracting military and financial resources from conquered foes. While Rome was waging war in Latium and southern Etruria, that argument could work, but not all of the communities Rome dealt with after 343 were well-suited to the *dilectus-tributum* system so central to full citizenship. This cannot have been a simple matter of proximity; many *socii* and *cives sine suffragio*, after all, were nearer to Rome than the full citizens of the *tribus Falerna*.[76] One certain reason for the failure of the *dilectus-tributum* system to incorporate all communities was the Romans' desire in the 330s to limit the expansion of the tribes, even if the rationale for such restraint is difficult to pin down.[77] Once the decision was made to exclude certain peoples from the tribes, there simply had to be an alternative to the *dilectus-tributum* system. The designation of peoples as Latin, as peregrine *socii*, and as *cives sine suffagio* was the response. Each was employed to optimize the extraction of money, of manpower, or of both, depending on what a conquered community offered.

To begin this discussion, it is worth examining three categories – Latin colonist, *civis sine suffragio*, and independent *socius* – as ideal types. The benefit of this approach is that, in the absence of positive evidence explaining why each community received the status it did, the ideal type creates the hypothetical product of a posited rationale. This provides a kind of gauge. If the historical outcome aligns with the ideal type, then there is good reason to

---

75 Pinsent (1954); Pinsent (1957); Galsterer (1976, 67); Humbert (1978, 3–43); Hantos (1980, 113); Bispham (2007, 16–27) arguing that later sources were themselves unclear about the original, pre-Social War definition of the term.

76 The geography of the different statuses involved in the post-338 settlement is usefully demonstrated at Cornell (1995, 350); see also Staveley (1989, 420–21).

77 For prior attempts to explain this: Sherwin-White (1972); Humbert (1978); Salmon (1982). On language as a category in the Italian settlement: Oakley (1998, 552); Stewart (2017), each with further references. Stewart (2017, 193) argues that non-Latins were not fluent enough to exercise the franchise or use Roman law, though this suggestion fails to explain why the Romans would deprive them of the vote but not the law. Moreover, there would be no need to deprive them of the vote if they were not linguistically adept enough to use it in the first place.

believe that the rationale was indeed the one employed. So, imagine a rational choice model that sought to maximize the yield of manpower and revenue within political constraints. How closely do the results of that model align with the record from the 330s? The test will be whether the ideal allies look like the Samnites, whether the ideal *cives sine suffragio* look like Capuans, and whether the ideal Latins looks like colonists. Each can be treated in turn.

1 Non-citizen allies

Regular *socii* contributed only men. When searching for the ideal *socii*, the perfect candidate would be a community with excellent infantrymen – their currency of tribute – but, since these allies contributed no money, negligible financial reserves. Highland areas with transhumance pasturing and low rates of urbanization might well have been the ones ideally suited to allied status. Even if parts of Samnium, for example, were more economically advanced than sometimes thought, the nature of their overall wealth might have been difficult to measure, tax, and seize through a *tributum* system that focused on landed property.[78] Thus, the Romans did not bother collecting money from the Samnites, but, in the absence of *tributum*, created a flow of infantry who did not have to be paid *stipendium*. The unpaid condition of these allied troops also meant that far higher numbers could be enrolled: in the absence of military pay, the treasury did not need to leave sufficient tax-payers on the land to pay the *tributum* needed for *stipendium*, and so allies could probably be made to serve in higher numbers.[79]

The ideal allies, therefore, were in communities where revenue-raising would either be too difficult or too meagre, but where the supply of soldiers would be valuable. Samnium fits that type well.

2 Cives sine suffragio

From the perspective of the ideal type model above, the ideal candidate for any form of citizenship would be one with a high capacity to contribute revenue. Such a community could have its infantry accepted or rejected depending on demand, but would always contribute through its wealth. The resulting city would look something like Capua. To illustrate why Capuans would be such strong candidates for citizenship – which would have to be without the vote, since there was evidently some reluctance to enroll them in the tribes – imagine instead that they were made *socii*. In that hypothetical, the Capuans would deliver each year their unspectacular infantrymen to the Romans while offering no money.[80] This would mean that Rome would

---

78 Hoyer (2012, esp. 181–85).
79 Momigliano (1975, 45).
80 Frederiksen (1984, 192) even argues that the Romans rarely if ever deployed Campanian infantry. This seems to go too far: see Brunt (1971a, 17–18); Oakley (1998, 556). It is worth

be drawing from the Capuans a resource for which they were not renowned while leaving untapped the wealth for which they were synonymous.[81] That would be absurd. As *cives sine suffragio*, on the other hand, the Capuans were paying monetary contributions, and the Romans could take or leave their infantry according to circumstances. Of course, the Capuans' cavalry was first rate, but these knights were enfranchised *optimo iure* and thus incorporated into Roman forces. It is noteworthy, moreover, that the Romans mandated that at least some of these knights were to be financed through a *vectigal* at home in Capua. This subsidy allowed the Capuan elite to acquire the public horses that were necessary for equestrian status in early Rome.[82] Importantly, given the popular hostility to cavalry pay at the time, this was at no cost to the Romans themselves.[83] This levy meant that Capuans were paying a specific tax to fund their own fully enfranchised cavalrymen, in addition to the *tributum* they were paying to fund Rome's legions.

The *civitas sine suffragio* thus emerges as a way of creating allies whose contributions were not only in military service, but in money as well. Ideally, it would be applied to those who would be contributing more in revenue than they would be deriving in pay.

3 Latin colonists

The exercise of drafting ideal types is most difficult here, and although fiscality is the focus of this discussion, there were of course other factors at play.

Many colonies were no doubt originally established as Latins because they were well beyond the limits of the tribes and thus looked more like the members of the Latin league than they did citizens in the *ager Romanus*. Even after the extension of tribes through Latium and southern Etruria, however, many colonies retained their old Latin status. The hillside towns of Setia, Signia, and Norba were not incorporated as citizens in the adjacent *tribus Scaptia* or *Pomptina*, nor were the *tribus Stellatina* and *Sabatina* extended to incorporate the nearby colonies of Nepet and Sutrium. There must have been a reason. Colonies on frontiers might have needed their own enlistment mechanisms to meet sudden threats from the Samnites, Etruscans, or Sabines, but this would not explain the retention of Latin status

---

remembering that Campanian mercenaries, whether or not from Capua itself, were held in the highest regard by Sicilians (Diod. Sic. 13.80). It is nonetheless clear that the tradition associated Capua's military capacity with its cavalry rather than its infantry.

81 On the quality of Capuan and Campanian infantry, see the preceding note.

82 Plin. *HN* 33.29; Wiseman (1970, 71).

83 On the *vectigal*, see Galsterer (1976, 74); Frederiksen (1984, 192). On hostility to cavalry pay in the year 342, see Livy 7.41.8; Oakley (1998, 385–87). Humbert (1978, 173–76) considered the citizenship of the knights' "potential" citizenship and more like *isopoliteia*, but this neglects the immediately practical consideration of where these cavalrymen would serve and how they would be levied.

by some more secure colonies like Ardea, which were nestled among the tribes. There may well have been strategic reasons for retaining Latin status among certain colonies commanding certain positions, but such armchair strategizing is beyond the scope of the present paper.

The problem needs to be framed properly. Colonies close to Rome could have been enfranchised, but they were not. Thus, it needs to be asked whether they were established in such a way that the fiscality of citizenship did not work there; that is, it needs to be asked whether they would have been receiving more in *stipendium* than they could pay in *tributum*. This would have been the case if the colonies' lands were not worth all that much, or if the mobilization rate was high enough to leave too few taxpayers at home.

Just as in the allied communities, the usual balance between taxpayers and conscripts did not need to be observed in colonies because nobody had to fund *stipendium*. This meant that, since nobody had to be left at home to provide *tributum*, higher rates of enlistment would have been possible in these communities than in Rome. An ideal candidate for Latin status would, therefore, be a community which was required to enlist a disproportionate share of the eligible population into the legions, since the number remaining to pay tax would be irrelevant. If Latin colonies were indeed required to deliver more soldiers per capita than others, this would explain why, in 209, it was the Latin colonies that were first to be exhausted by the manpower demands of the Second Punic War.[84] With a more intensive ratio of active to inactive soldiers, colonies may well have offered a means of raising more manpower at a very low cost.

If, moreover, Rome was indeed enduring fiscal difficulties in the fourth century, then the Latin status must have offered real appeals. Enfranchisement only had the advantage of raising revenues if enough landowners with enough property remained home to pay more in *tributum* than the soldiers were receiving in *stipendium*. This is no doubt speculative, but colonies could be created in such a way that relatively high mobilization rates or relatively low landholdings left the balance between *stipendium* and *tributum* unviable. In this case, by removing the colonists from the fiscal system altogether, Latin status would ameliorate the treasury's problems.

There is no doubt that Latin colonies increased manpower without increasing costs. Something in the order of 70,000 colonists were sent out between 334 and 263, and while Cornell is no doubt right that they could not all have come from the Roman citizenry, a great many probably did.[85] Colonization allowed the state to shift some of these Romans from the column of costly *stipendium*-earners to the column of unpaid Latins. Importantly, this accounting maneuver presented no loss in revenue. Because *tributum* was a tax on property, the land that the colonist left behind him in Roman territory would remain in the tax base, only now assessed as the

---

84 Livy 27.9.7.
85 Cornell (1995, 367). See also Roselaar in this volume, with a slightly different position.

property of its new owner. Thus, every transfer of a *stipendium*-eligible Roman to the status of a Latin colonist was the removal of a soldier from the debit column of the fisc's ledger. The opportunity cost was that Latin colonization did not produce any new payers of *tributum*; had Setia or Norba been enfranchised, they would have increased costs through *stipendium*, but they would also have increased revenue through *tributum*. Again, however, the net effects of this would depend on the ratio of enlisted soldiers to taxed land. Citizenship would still have been costly for Rome if colonies were instituted in such a way that, whether because of limited landholdings or high enlistment rates, they were drawing more in *stipendium* than they were paying in *tributum*. Whether this was the case is ultimately unknowable, but the exhaustion of the colonies in 209 suggests that this may indeed have been the case.

These examples suggest a rough fiscal rubric for organizing the defeated. If a community was a Latin one that could be integrated into the tribes, then they should be full citizens and subjected to the *dilectus* and *tributum*. If, however, they were outside that group, then they would need some other status. A decision (probably impressionistic) then had to be made as to whether they would contribute men as *cives sine suffragio* or as allies. That decision was based on what resources a population could offer. Provided that it was diplomatically feasible (see below), those with landed wealth would ideally receive the *civitas sine suffragio*, since this would provide Rome with access to new revenue as well as to troops. Those with more valuable men than capital, on the other hand, would provide troops at whatever rate Rome determined, but in a form insulated from the *aerarium* because they were unpaid. Romans were thus happy to provide citizenship to those who could pay more in *tributum* than their men would cost in *stipendium*, but preferred to requisition unpaid troops from those whose ability to pay was not assured. Integrating poorer communities into the *stipendium-tributum* system was a recipe for fiscal disaster, so they were made allies.[86] Citizenship with or without the vote was best kept for those who could pay more than they received in *stipendium*.

This, I argue, was the basic template. Yet, despite how convenient it would be to argue that Roman bean counters employed this model to organize Italy along exclusively fiscal lines that clearly was not the case. Exceptions abound. The *cives sine suffragio* in Campania were well known for their wealth, but could the same be said for Privernum, Fundi, and Formiae, or for the Hernican cities granted the half citizenship late in the fourth century?[87]

---

86 Galsterer (1976, 82) notes that, once Romans were used to a certain ratio of citizen to allied troops in the army, they were required to create citizens and allies in similar proportions. He provides no reason why such a ratio would have to be preserved besides custom, but the need to keep poorer communities out of the *dilectus-tributum* system might be one explanation.

87 Privernum, Fundi, and Formiae: Livy 8.14.10, 8.21.9–10; Vell. 1.14.3. Hernican cities: Livy 9.43.24. The presentation of the data in map form at Cornell (1995, 350) is revelatory.

The question is really whether they were wealthy *enough*, but the evidence is unavailable. On the other hand, Etruscan cities were definitely wealthy, but were rarely enfranchised.[88] There were also exceptions in Latium. Praeneste and Tibur, for example, were the two largest Latin cities aside from Rome itself, and thus held vast reserves of taxable property.[89] Yet, the Romans repeatedly baulked at enfranchising them. Praeneste was defeated in 380, but unlike Tusculum a year earlier, was not enfranchised.[90] Furthermore, whatever led to that divergent policy must have still been in effect after 338, when, despite mass enfranchisement throughout much of Latium, Praeneste and Tibur were kept as independent Latin allies rather than citizens. Salmon and Humbert were perhaps right that these two cities were simply too large and respected to insult with civic eradication, and the absence of enfranchisement among the prestigious old cities of Etruria might support that thesis.[91] On the other hand, even in this case, the Romans made certain to expand the tax base. Praeneste, for example, might not have been "made Roman," but a large swath of territory was converted to *ager Romanus* and thus, to the extent that it was occupied by Roman *assidui*, became subject to *tributum*.

Such land seizures do not, however, negate the problem that Praeneste is an exception to the argument above. As Toynbee showed, land seizure was a relatively minor part of the overall settlement of Italy; it represented just one-ninth of the territory annexed through enfranchisement, and it is self-evident that the Romans could have raised more revenue by absorbing all of Praeneste than by absorbing just part of it.[92] The point thus remains that a myopic focus on public revenue cannot provide a complete explanation for the organization of Italy. Only a fool would claim as much. Instead, the rubric above provided a guide *so long as diplomatic and strategic concerns permitted*. The Romans apparently dared not impose *civitas sine suffragio* on the most powerful neighbors, even if well-suited to it. Given the hostility to the institution, this is understandable. This may also explain why *civitas sine suffragio* disappeared as a fiscal and diplomatic weapon. Perhaps the diplomatic costs were too high. Or perhaps the need for revenue was met by the existing base, and it was decided to employ the *formula togatorum* in future. Imposing half citizenship fell out of fashion.

---

88 My thanks to Guy Bradley for pointing out the absence of enfranchisement in Etruria.
89 Beloch (1926, 178) helpfully displayed as a map and a graph at Cornell (1995, 206–7).
90 Livy 6.29.7–10; Hantos (1983, 52).
91 Humbert (1978, 190); Salmon (1982, 54); Cornell (1995, 322–23, 349). Oakley (1998, 202) acknowledges that imposing *civitas sine suffragio* on a powerful Etruscan city would have been no simple matter.
92 Toynbee (1965, 133), following the estimates of Afzelius (1975, 153). For discussion, see Roselaar (2010, 33–36); Bourdin (2012, 499–513) (for an earlier period). Despite criticisms, Afzelius' order of magnitude remains unimpeachable. Confiscation of territory was not a new development in the fourth century, as Livy records similar confiscations in the early fifth century: Armstrong (2016c, 153) with examples.

Despite its disappearance, the *civitas sine suffragio* remains the great novelty of the Latin settlement.[93] The "half-citizens" contributed men in the same way as the allies but also contributed revenues like full citizens. In this sense, the term *civis sine suffragio* is misleading in the way that all great euphemisms are. An equally accurate, but less flattering label would be something like *socius cum tributo* or *socius aerarius*: an ally who also pays tax. These terms would be no less accurate than *civis sine suffragio*, but, advertising the truth of their burdens, would presumably also be more galling. In any case, the civic rights implied by the term *civitas* probably came as the most meagre consolation. Though many had to shoulder the burdens, few if any could access the benefits. Again, it is not hard to see why such a demeaning status was not imposed on certain powerful or prestigious communities and why it was eventually abandoned.

The exception that proves the rule in all this is the citizen colony. Coastal colonies (*coloniae maritimae*) were frequently in non-Latin areas and the citizen colonists, thanks to paltry plots of land, were poor candidates for citizenship in the model above.[94] Without much assessed wealth, they would not be eligible to pay much *tributum*, and frequently outside the tribes, they could not take part in the *dilectus*. And yet they held full citizenship. This would undercut much of the model above, if it were not for the fact that these citizen colonies were also granted *vacatio militiae*, or exemption from conscription.[95] These colonists lived in what were essentially forts and were required to remain in place to provide local defense.[96] Thus, colonists from a place like Antium never had to be distributed through the *dilectus*. Similarly, the 500 colonists reportedly sent to Sardinia in 378 could not have attended the *dilectus* each year, and so Diodorus' explicit reference to their untaxed status (ἐπὶ ἀτελείᾳ) surely reflects a broader exemption from both *tributum* and *dilectus*.[97] Moreover, since maritime colonists did not receive *stipendium*, they did not cost much and so their low contribution through *tributum* did not matter. They could thus be part of the enfranchisement process running from Veii onwards because the Romans never wanted them to contribute infantry (and earn *stipendium*) like everybody else. Their dislocation from the *dilectus-tributum* system made their incompatibility with it moot. So instead of creating yet another category of Italian community, these colonists were simply absorbed as citizens, but with the *vacatio militiae*. Had they contributed infantry, they presumably would have been Latins or allies.

---

93 There is evidence for Caere receiving the *civitas sine suffragio* in 353, but I adhere to the position that, whatever Caere's status in the mid-fourth century, it did not receive the "half citizenship" until the 270s, and hence that the institution did not predate 338. For discussion, see Oakley (1998, 199–202).

94 The colonists received just two *iugera*, which implies that they must have had access to other land. For the demographic details, see de Ligt (2014).

95 Livy 27.38.3–5; Capogrossi Colognesi (2014, 94).

96 Sewell (2014, 130); Jaia (2013); de Ligt (2014, 61–62).

97 Diod. Sic. 15.27.4.

Conclusions

By the middle of the fourth century, enfranchisement had emerged as the Romans' preferred way to convert conquest into a flow of revenue and military manpower. Each new *assiduus* was not only a new potential conscript, but a new taxpayer. He could either increase the total revenue of the state or ease the burden on other payers by raising the denominator over which the existing revenue was divided. The problem by the 330s, however, was that not every potential tributary was well-suited to the *dilectus-tributum* system through which citizen contributions were processed. Some were unable to contribute manpower and money at an efficient ratio, some were strategically better-suited to mustering autonomously, and many lacked the language skills to be dispersed throughout the legions by the *dilectus'* atomizing processes. The result was a set of categories defined more by what the Romans wanted to take from each community – money and/or military manpower – than by juridical, military, or cultural statuses.

There are four major advantages to the view that extraction was the primary determinant of Rome's categorization of erstwhile enemies. The first is that, by avoiding a focus on the benefits of each status, one can avoid the impossible task of assessing what the rights of citizenship or the *ius Latini* might have meant for those who lived far from Rome and for whom they were rarely if ever actionable. This, in turn, minimizes the problem of how and why the conditions of allies and citizens could vary so widely within the one category: if the *civitas sine suffragio*, for example, was defined by the provision of money and manpower, then so long as those extractive conditions were being met, the perplexing variety of ancillary benefits or restrictions are no longer central to the taxonomy of communities.[98] The second advantage is that this view is surely more plausible. The approach of this chapter assumes that an expansionist and victorious hegemon, acting unilaterally and against wholehearted resistance, was more interested in what resources it was squeezing from the vanquished than in what rights it was conceding.[99] The third advantage is that the subordination of cultural and political integration accords much better with subsequent history.[100] The extractive explanation, in other words, negates the problem of why the Romans "incorporated" so many communities with so little effort to

---

98 On variety among *cives sine suffragio*, see: Stavelely (1990, 427–29); Mouritsen (2007, 154–55); Scopacasa (2016, 43–44). Colonies also exhibited a range of conditions on the ground: Scopacasa (2016, 47–49). Yet, it would be absurd to dismiss the existence of the category of "colony" because it is understood that the (legal and extractive) commonalities overrode the differences.

99 Humbert (1978, 191).

100 See, most recently, Armstrong (2016c, 293–94) for "emerging identity."

"Romanize" them or to affect their domestic institutions.[101] It minimizes the problem that, even after two centuries of operation and the successes of military expansion, formation of a pan-Italian Roman community is notoriously difficult to find in the evidence. Romans, instead, were primarily interested in the immediate problem of acquiring resources. Finally, this argument acknowledges that an army could not be raised without money, and hence avoids the trap of examining the soldiery without considering the treasury that committed to pay it.

The resulting portrait is of something much more akin to a tributary empire than is usually countenanced. Rome organized a network of extractive relationships defined in large part (even if not exclusively) by who was delivering what resource and how. Parts of that network comprised a set of tributary alliances, through which contingents of young men were supplied to the Roman army. Other parts involved the imposition of a citizenship that, given that its benefits were probably enjoyed by few but its burdens endured by many, can similarly be described as a form of tribute. The *civitas sine suffragio* in particular reveals Roman thinking about citizenship and imperial extraction. This implied a conception of citizenship which was as much about burdens as about privileges, and which viewed the institution not as a unified whole, but as a sort of basket containing a collection of rights and obligations, some of which could be removed or added without changing the nature of the basket itself: the *civitas*. This made the use of citizenship as a burden much more obvious. Stitched together with the Latins and the non-citizen allies, these citizens created precisely what the Romans had in mind. The goal was not to build nice buildings, to attract the finest scholars, or to shape some Italian order built on Roman law. They sought simply to create a turbo-charged war machine fuelled by revenue and young men, and in this, they were entirely successful.

---

101 The bibliography is vast, but Mouritsen (1998) offers the standard account of an Italy with little to no integration, with *Mouritsen* (2007, 149–50). On fragmentation in the fourth century, see Kent (2012a) and most recently Helm (2017, esp. 207) on "pacification without integration." For the larger debate, see also Jehne and Pfeilschifter (2006) – with the telling title, *Herrschaft ohne Integration* – and the collection within Roselaar (2012).

5 Organized chaos

Manipuli, socii, and the Roman army c. 300*

Jeremy Armstrong

Introduction

The period from 338 to 264 is traditionally seen as one of immense change for the Roman army, and for good reason. The year 338 supposedly marked the end of the "Great War" against the Latins and the beginning of the "Latin Settlement," which brought the entirety of Latium under direct Roman control. The terms of the "Settlement" included a massive expansion of both Roman citizenship, through the inclusion of various settlements as *civitas* and *civitas sine suffragio*, and the number of Roman allies. As a result, Rome's available military manpower increased exponentially in this period. And the community was not done yet: once whetted, Rome's appetite for manpower seems to have grown as the city continued to expand, via both conquest and diplomacy, both northward and southward in the following decades.[1]

Alongside the growth in manpower came the growth of Rome's territorial empire, including both *ager publicus* and many new colonies. Between 338 and 264, Roman territory expanded from roughly 5,500 km² to 26,800 km².[2] Although the Romans had been interested in land previously, with something akin to a "mini-agricultural revolution" occurring back in the sixth through fifth centuries,[3] this was taking it to a completely new level.[4] The locations of Rome's new colonies in the second half of the fourth century also reveal Rome's expanding interests. Along with new settlements pushing Roman interests into southern Italy, and particularly Samnium, Rome founded a series of *coloniae maritimae* along the Tyrrhenian coast – hinting

* Many thanks to the peer reviewers for their incredibly helpful comments, as well as the assistance of my co-editor Mike Fronda. Special thanks as well to Ronald Rocco for the many conversations which helped inspire parts of this piece. All errors and omissions naturally remain my own. All dates are BC unless otherwise noted.
1 Classically, see Harris (1979) for discussion. See also, and more recently, Rosenstein (2012b).
2 Morel (2007, 499).
3 Forni (1989); Fulminante (2014, esp. 228–29).
4 Roselaar (2010); see also VanDerPuy in this volume.

DOI: 10.4324/9781351063500-5

This chapter has been made available under a CC-BY-NC-ND license.

at new, naval aspirations.[5] Indeed, this period saw the advent of a more for-mal Roman navy with the creation of the *duoviri navales* in 311.[6]

The period also seems to have seen changes in military infrastructure. The new *duoviri navales* would have needed ships, which may have been either purchased or organized, and possibly maintained, by the state.[7] The period is also famous for its military roads, most notably the *Via Appia*, which would have facilitated the movement of troops south toward Magna Graecia and Samnium, and the creation of forts, many of which were associated with the new *coloniae maritimae*.[8] And all of this may have been paid for, or at least connected with, the new Roman coinage which began to circulate around this time. Indeed, the economic aspects of Roman expansion, discussed else-where in this volume (see Tan and VanDerPuy especially), were both vital and profound. Additionally, the period is associated, both in the literature and the archaeology, with changes in military equipment – including the use of the *pilum*, oblong *scutum*, and Montefortino-style helmet.[9]

It should also not be forgotten that the system which would become Rome's standard military command structure for the Republic – with two consuls and assorted praetors wielding *imperium* – had only recently been put in place as well. The Licinio-Sextian rogations which had (re)introduced the consulship and praetorship are traditionally dated to the 360s, and the *lex Genucia* of 342 suggests the system was still being "tinkered with" in the second half of the century. Although *imperium* itself may have had regal origins, the offices it imbued were clearly evolving (see Drogula in this vol-ume),[10] and the men occupying them were increasingly part of a new group of wealthy, but not necessarily "ennobled," elites who were emerging onto the military and political stage.[11]

In sum, by 264, Rome's army may have included an entirely new set of men, mobilized under a new set of arrangements, wearing a new type of equipment, living in territory which may have only recently been connected to Rome, commanded by a new set of magistrates, and fighting for a new set of goals and objectives. It is perhaps unsurprising then that this period is also often associated with a change in organization and the introduction of the

---

5 Salmon (1963); Mason (1992).
6 Livy 40.18.7. See Harris (2017) for recent discussion.
7 Pitassi (2011, 84).
8 See Jaia and Molinari (2011, 91–94) for further discussion.
9 Armstrong (2017b).
10 The nature of both the consulship and praetorship in the mid-fourth century is obviously heavily debated, but even the opening of the offices to plebeians suggests at least some development during this period. See Armstrong (2017a) for discussion.
11 While the Licinio-Sextian rogations seem to have opened the consulship to plebeians, the *lex Genucia* is thought to have reserved one of the positions for them. As a result, in the period after 342, roughly half of the consuls would have been *novi homines* by definition. This is in addition to the increased incorporation of Latium's elite in the second half of the fourth century as part of the "Latin Settlement."

so-called "manipular legion" – a supposedly new, heterogeneous military system made up of small groups (*manipuli*) of soldiers, variously equipped, which supposedly replaced an archaic Roman phalanx.[12] Whether it was brought about by contact with new enemies (the Gauls or the Samnites), major defeats (as at the Caudine Forks), or was simply the end result of the myriad other changes which were occurring in the army at this time, the ancient sources seem to suggest (and modern scholars often argue) that during the late fourth century Rome began to utilize a new manipular structure, which would become synonymous with Roman warfare during Rome's subsequent expansion across the Mediterranean. This is the narrative which our sources, followed by modern scholars, have suggested for the period from 338 to 264: a radical, and successful, reinvention of the Roman war machine.

But is this narrative correct? Although the period in question is usually seen as being on the cusp of being properly "historical" – with Hieronymous of Cardia and Timaeus of Tauromenium both writing histories including Rome in this period, and indeed the Romans themselves seeming to get a little more interested in record keeping[13] – we are still an eventful 100 years away from the first native Roman historians and c. 150 years from Polybius and the first extant "eyewitness" account of the Roman army. While we can talk about the army which emerged from this period with some confidence, the events, and the nature of the Roman army *during* this period, are still (arguably quite) debatable.[14] Additionally, although the narrative for change in the late fourth century is a compelling one, it also raises some questions – most notably, how was Rome able to deploy this supposedly "new" army so effectively? The army which emerged from the fourth century allowed Rome to move from being a regional power to being able to compete with, and ultimately defeat, a major Hellenistic army like that of Pyrrhus, and then the Carthaginians only a few years later. Rome famously went from struggling against local enemies in the mid-fourth century, to competing on the Mediterranean stage by the early third. However, could an army with so many new features and variables actually prove this successful immediately? And indeed, is this much change even possible in a military context? Although innovation is often touted as the key to success in the modern military environment, comparative evidence suggests that most militaries are fundamentally conservative.[15] Such large-scale and wholesale changes

---

12 Traditionally, Mommsen (1854–56, 2.402–29.); Delbrück (1975, 272–74); more recently: Roth (2009, 28).

13 Famously, Appius Claudius Caecus is often considered the "Father of Latin prose," as Cicero (*Sen.* 16; *Brut.* 61) records that he wrote down his speech opposing peace with Pyrrhus in 279 while Pomponius attributed a legal text to the great censor (Dig. 1.2.2.36). See Cornell (1989, 398) and Rosenstein (2010) for discussion.

14 Rawson (1971a, 13) suggested the evolution of the Roman army in the period prior to Polybius was a subject of "almost inextricable confusion," a point with which Harris (1990, 507) concurred almost 20 years later, and a position which most scholars hold today.

15 See, famously, Keegan (1993).

to Rome's military in such a short period of time would be truly remarkable. So, is there another explanation?

This chapter will argue that while Roman society, and the Roman army, did indeed undergo some significant changes during the second half of the fourth century, in the military realm at least the most significant of these likely related to scale. To put it simply, instead of the Roman army undergoing a radical *reinvention* during the period from 338 to 264, we should think of it as simply experiencing a dramatic *expansion*. While the Roman army c. 300 undoubtedly contained new soldiers and features, its core structures and relationships actually may have had far more in common with Rome's archaic forces than usually thought. Indeed, far from being a late fourth-century innovation, the Roman army had always likely been "manipular" to a certain extent, and this core characteristic may have been the key to its success and the ability to expand and integrate new soldiers in later periods.[16] The early Roman army had always been a patchwork of small, variously equipped, heterogeneous units. The biggest change during the period c. 300 was how Rome was able to redeploy and expand her existing systems and relationships in a new imperial context.

Bridging the gap

At the basis of this revised interpretation is a change in how we understand the Roman army in the fifth and early fourth centuries. The traditional narrative for Rome's early military development can be found in texts like the first-century *Ineditum Vaticanum*,[17] which supposedly records the conversation between a Kaeso (possibly Caeso Fabius) and the Carthaginians before the start of the First Punic War:

> This is what we Romans are like . . . with those who make war on us we agree to fight on their terms, and when it comes to foreign practices we surpass those who have long used them. For the Tyrrhenians used to make war on us with bronze shields and fighting in phalanx formation, not in maniples; and we, changing our armament and replacing it with theirs, organized our forces against them, and contending thus

16 As Smith has noted, however, this should probably be taken as "manipular" in the broadest sense. Smith (2006a, 289) suggests this is because "the fact that the maniple had two centuries would seem to imply that it was not the original element, but one produced by aggregation." As argued later in this chapter, the existence of centuries as discrete units within the maniples may not be as secure as usually thought. Nevertheless, such skepticism is warranted.

17 The date of the *Ineditum Vaticanum* (*FGrHist* 839 F1) is contested. The source (Synesios, Codex vaticanus, 435, fol. 220) has been dated, variously, anywhere from the first century BC (arguably the current consensus) to the first and second centuries AD. The historian referenced in the particular passage though, has been dated as early as the late third or second century. See Beck (2011c) for discussion.

against men who had long been accustomed to phalanx battles we were victorious. Similarly, the Samnite shield was not part of our national equipment, nor did we have javelins, but fought with round shields and spears; nor were we strong in cavalry, but all or nearly all of Rome's strength lay in infantry. But when we found ourselves at war with the Samnites we armed ourselves with their oblong shields and javelins, and fought against them on horseback, and by copying foreign arms we became masters of those who thought so highly of themselves.[18]

This is usually supplemented by passages like that in Diodorus Siculus, who noted that:

> For example, in ancient times, when [the Romans] were using oval shields, the Etruscans, who fought with round shields of bronze and in a phalanx formation, impelled them to adopt similar arms and were in consequence defeated. Then again, when other peoples were using shields such as the Romans now use, and were fighting in maniples, they had imitated both and had overcome those who introduced the excellent models. From the Greeks they learned siegecraft and the use of engines for war for demolishing walls, and had then forced the cities to do their bidding.[19]

Thus, by the late Republic at least, the Romans seem to have thought that they had once fought in a phalanx, following an Etruscan model, but then shifted to the manipular formation, likely sometime during the Samnite wars.

In the nineteenth century, this evidence was analyzed and interpreted by scholars like Mommsen and Delbrück, with both suggesting that, in addition to perhaps mirroring the Samnites' formations,[20] Rome's fourth-century

18 Trans. Cornell (1995, 170); text von Armin (1892, 121) (= *FGrHist* 839 F1): Ἡμεῖς εἶπεν οὕτως πεφύκαμεν· – ἐρῶ δέ σοι ἔργα ἀναμφισβήτητα, ἵνα ἔχῃς ἀπαγγέλλειν τῇ πόλει – τοῖς πολεμοῦσιν εἰς τὰ ἐκείνων ἔργα συγκαταβαίνομεν, κἂν τοῖς ἀλλοτρίοις ἐπιτηδεύμασι περίεσμεν τῶν ἐκ πολλοῦ αὐτὰ ἠσκηκότων. Τυρρηνοὶ γὰρ ἡμῖν ἐπολέμουν χαλκάσπιδες καὶ φαλαγγηδόν, οὐ κατὰ σπείρας μαχόμενοι·ὁμεῖ ἡμεῖς μεθοπλισθέντες καὶ τὸν ἐκείνων ὁπλισμὸν μεταλαβόντες παρεταττόμεθα αὐτοῖς· καὶ τοὺς ἐκ πλείστου ἐθάδας τῶν ἐν φάλαγγι ἀγώνων οὕτως ἀγωνιζόμενοι ἐνικῶμεν. οὐκ ἦν ὁ Σαυνιτικὸς ἡμῖν θυρεὸς πάτριος, οὐδ᾿ ὑσσοὺς εἴχομεν, ἀλλ᾿ ἀσπίσιν ἐμαχόμεθα καὶ δόρασιν· ἀλλ᾿ οὐδ᾿ ἱππεύειν ἰσχύομεν, τὸ δὲ πᾶν ἢ τὸ πλεῖστον τῆς Ῥωμαικῆς δυνάμεως πεζὸν ἦν. ἀλλὰ Σαυνίταις καταστάντες εἰς πόλεμον, καὶ τοῖς ἐκείνων θυρεοῖς καὶ ὑσσοῖς ὁπλισθέντες ἱππεύειν τε αὐτοὺς ἀναγκάσαντες, ἀλλοτρίοις ὅπλοις καὶ ζηλώμασιν ἐδουλωσάμεθα τοὺς μέγα ἐφ᾿ ἑαυτοῖς πεφρονηκότας...

19 Diod. Sic. 23.2: τὸ μὲν γὰρ παλαιὸν αὐτῶν θυρεοῖς τετραγώνοις χρωμένων, Τυρρηνοὶ χαλκαῖς ἀσπίσι φαλαγγομαχοῦντες καὶ προτρεψάμενοι τὸν ὅμοιον ἀναλαβεῖν ὁπλισμὸν ἡττήθησαν. ἔπειτα πάλιν ἄλλων ἐθνῶν θυρεοῖς χρωμένων οἷς νῦν ἔχουσι καὶ κατὰ σπείρας μαχομένων, ἀμφότερα μιμησάμενοι περιεγένοντο τῶν εἰσηγησαμένων τὰ καλὰ τῶν παραδειγμάτων. παρὰ δὲ τῶν Ἑλλήνων μαθόντες πολιορκεῖν καὶ ταῖς μηχαναῖς καταβάλλειν τὰ τείχη, τὰς πόλεις τῶν διδαξάντων ἠνάγκασαν ποιεῖν τὸ προστατόμενον.

20 As Polybius (6.25.11) suggested, ἀγαθοὶ γάρ, εἰ καί τινες ἕτεροι, μεταλαβεῖν ἔθη καὶ ζηλῶσαι τὸ βέλτιον καὶ Ῥωμαῖοι ("for they [the Romans] are as good as any others in adopting new

wars in mountainous terrain, against both Gauls and Samnites, may have helped lead to a "breaking up" or articulation of a traditional hoplite phalanx.[21] As a phalanx needed to maintain tactical cohesion, which would have been difficult in broken terrain, the Romans broke their single large phalanx into smaller phalanges. Over time, the various facets of the articulated phalanx, which may have already featured different panoplies based on the centuriate system, gradually took on different functions as well – becoming more specialized. The culmination of this process was the highly structured manipular legion described by Polybius in the second century, with its lines of *velites, hastati, principes,* and *triarii,* which would normally line up in the triple *acies, quincunx* formation.[22]

This basic model, albeit with various adjustments, has gone largely unchallenged until recently. However, with the gradual dismantling of Rome's archaic phalanx (which is increasingly thought to be incompatible with Roman society during that period) in favor of a more disparate, clan-based approach to warfare in Rome's early period, questions are being raised.[23] The model proposed by Mommsen and Delbrück, and indeed supported by the Romans themselves, seems to have been predicated on a desire to bridge the perceived gap between a "known" second-century manipular legion and a hypothesized hoplite phalanx of the archaic period. Given the two end points – with a highly structured but heterogeneous and fragmented manipular army in the second century, and a highly structured, but more homogenous and densely packed hoplite phalanx in the archaic period – the Romans (and modern scholars) quickly identified the key differences and argued for a process of articulation and differentiation, effectively connecting the dots. However, there is now strong reason to doubt that the Roman army was organized and fought as a hoplite phalanx.[24] Rather, the starting point of the Roman army, from which the later manipular form developed, was likely to be far more fragmented than traditionally thought.

Additionally, scholars have increasingly recognized that even the more "reliable" endpoint in this process of evolution, the legion of the second century, may not be as stable as previously thought. Our primary evidence for this legion and its structure is Polybius, who presented himself as a sober and reliable eyewitness. But as scholars have explored Polybius' work in ever more detail, the complexity of his account suggests that a reader would be wise to take his military descriptions with some care and caution – and always in the

---

fashions and instituting what is better"). See also Diod. Sic. 23.2.1. This principle remains the staple tenant of most models of Roman military development, although its accuracy is uncertain.

21 See, most notably, Delbrück (1975, 272–74).

22 See Sage (2008, 72–74) for a summary.

23 While many have wondered how a Roman phalanx could have functioned in a society dominated by clans, Harris (1990, 508) was one of the first to suggest it was actually "fictitious." For recent revisions, see Armstrong (2016c) as well as Rocco (2017).

24 Rich (2009); Rosenstein (2010); Armstrong (2016c, esp. 111–28).

context of his wider vision for Roman history and imperialism.[25] Polybius was not a mere reporter of facts. One key aspect of this "Polybian lens" is its highly structured and consciously schematic nature, as Polybius tended to focus on ideals instead of the messy reality. This is arguably taken to another level in his descriptions of Rome's political and military systems, which Polybius consciously praises and offers as *exempla*.[26] It was not in Polybius' interests to discuss where Rome's systems were irrational or disordered. Additionally, one must also remember that the "manipular structure" which Polybius described is now thought to have been on its way out by the time he was writing, being gradually replaced by a cohort-based organization.[27] So, taken as a whole, the Polybian legion may very well have represented a highly stylized and idealized version of what was, in reality, a more untidy and fluid system.

This is most certainly not to say, however, that the Polybian legion of c. 150 was not vastly more organized and regimented than the legion of c. 300 – or, indeed, that he was fundamentally incorrect in his description. The remains of Roman camps from Numantia, and elsewhere in Spain, show that, by the second century, the Romans certainly seem to have exhibited – at least in some instances – an order which roughly aligns with Polybius' system.[28] Indeed, as McCall illustrates later in this volume, in the aftermath of the Second Punic War, the Roman legion seems to have developed its own modus operandi, which allowed it to function with limited input from the top. This modus operandi would have likely relied on, and indeed encouraged the development of, systems like that described by Polybius. Nevertheless, I would suggest that it is a distinct possibility that the Roman army was not *quite* as structured, systematized, and organized as Polybius suggested – and that we should perhaps expect, or at least account for, the possibility of inconsistency and variability. We are, after all, talking about thousands of individuals, recruited from an increasingly wide range of contexts and communities in Italy, who were, at best, semi-professional soldiers. Integration is unlikely to have been swift or easy for new legions. Indeed, although the specific origin of soldiers in Roman units is rarely (if ever) given by our sources, allied troops – and even soldiers from *civitas sine suffragio* – do seem to have retained their civic and ethnic associations within Roman armies.[29] In the third and early second centuries, soldiers were still providing

25 Walbank (2002, 14–15).
26 Eckstein (1995, 173). Indeed, as an addition, it is noteworthy that Polybius never properly discussed the allies in his description of the army – even though the *socii* likely comprised more than half of Rome's armed forces by this period.
27 See Goldsworthy (1996, 13–28) for discussion. This process of replacement arguably began back in the Second Punic War, if not earlier, as Scipio Africanus supposedly deployed his army in cohorts in the battle of Carthago Nova (see Bell 1965) – although, as will be suggested later with *manipulus*, it is uncertain whether the term *cohors* should always be read in a technical fashion.
28 Dobson (2008).
29 See Tan, in this volume, for discussion.

their own equipment, and while military leaders may have requested and expected some uniformity, this would have been difficult to enforce. And there are concerns with the practicality of the Polybian *dilectus* (a point we will return to later) – which supposedly separated individuals from their preexisting social frameworks, and forced them to operate and fight as part of an entirely new group (in some periods with little to no training), where the only initial connection was a shared citizenship.[30] One wonders how this system, although a wonderful model of civic interrelation, would have worked in real terms.

But even accepting Polybius at something resembling "face value" for the second-century Roman army, we are still left with a very different – and arguably much smaller – gap to bridge. In the archaic period, we seem to have a heterogeneous and fragmented collection of clans and groups, all fighting (at least sometimes) under the banner of Rome, and in the second century, we have a heterogeneous and fragmented collection of soldiers and allies, at least some seemingly organized into "maniples," all fighting under the banner of Rome. The bridge linking these two points may be more about identity than formation. The question therefore emerges, what is a "maniple," and what makes it different from a clan or other group?

De Manipulorum Natura

The word *manipulus* (literally "a handful") had a number of different etymologies associated with it. The military connotation is present from its earliest extant usage in Plautus, but beyond that things are vague.[31] Varro suggested that the word was derived simply from a group or handful of men, or literally "hands" (*coniungit plures manus*)[32] – a rather generic meaning which seems to have had long use, as the term was deployed in a similar way by the fifth century AD military writer Vegetius to describe a group of ten men who fought together.[33] However, Varro also offered an association with a standard (*manipulus exercitus minima manus quae unum sequitur signum*)[34] – a version followed, albeit with some minor variations, by Plutarch and Ovid.[35] So the basic meaning seems to have been a small group of soldiers who fought together and were in some way associated with a standard.

---

30 See also Helm and Tan on this topic in this volume. Jehne (2006, 255–56) argued that the Polybian *dilectus* consciously worked to replace local allegiance with Roman. His argument is likely correct, although it is uncertain both how early this began and how practical and successful it was.
31 Plaut. *Curc.* 585. See Rocco (2017, 81–84) for additional discussion.
32 Varr. *Ling.* 6.85.
33 Veg. 2.13.
34 Varr. *Ling.* 5.88.
35 Ov. *Fasti* 3.115–17; Plut. *Rom.* 8.

This definition is confused, however, by the number of other terms deployed in our two main sources for the legion (Livy and Polybius) to refer, apparently, to similar things.[36] For instance, after Livy speaks of *manupuli* in his famous discussion of the army in Book 8, he goes on to discuss the *ordines* and *vexilla*; his change in terminology implies a difference between the units/terms:

> The Romans had formerly used small round shields; then, after they be-
> gan to serve for pay, they made oblong shields instead of round ones; and
> what had before been a phalanx, like the Macedonian phalanxes, came
> afterwards to be a line of battle formed by *manipuli*, with the rearmost
> troops drawn up in a number of *ordines*. The first line, or *hastati*, com-
> prised fifteen *manipuli*, stationed a short distance apart; the *manipulus*
> had twenty light-armed troops (*leves*), the rest of their number carried
> oblong shields; moreover those were called *leves* who carried only a spear
> (*hasta*) and javelins (*gaesa*). This front line in the battle contained the
> flower of the young men who were growing ripe for service. Behind these
> came a line of the same number of *manipuli*, made up of men of a more
> stalwart age; these were called the *principes*; they carried oblong shields
> and were the most showily armed of all. This body of thirty *manipuli*
> they called *antepilani*, because behind the standards there were again
> stationed another fifteen *ordines*, each of which had three sections, the
> first section in every company being known as *pilus*. The *ordo* consisted
> of three *vexilla* or "banners"; a single *vexillum* had sixty soldiers, two
> centurions, one *vexillarius*, or colourbearer; it numbered a hundred and
> eighty-six men. The first banner led the *triarii*, veteran soldiers of proven
> valour; the second banner the *rorarii*, younger and less distinguished
> men; the third banner the *accensi*, who were the least dependable, and
> were, for that reason, assigned to the rear most line.[37]

36 Whenever Livy and Polybius are mentioned in the same breath, some slight comment on their relationship is probably warranted. While Livy almost assuredly knew Polybius' ver-sion of the manipular legion, it is unlikely he was working directly from his text in Book 8. First, there are clear and well-established differences in their descriptions. Most notably, there is a discrepancy in the number of units in the legion. Livy gives 45 while Polybius gives 30, although they do contain the same number of "maniples," 30, a number which has symbolic resonances in Roman/Latin society, for instance with the *Prisci Latini*. Sec-ond, Livy's digression on the army in book 8 is located in his narrative of events in the 330s, and so it is unlikely that he had Polybius' text discussing the second century open in front of him. See, traditionally, Tränkle (1977) and more recently Levene (2010, 126–63) for in-depth discussion of Livy's use of Polybius.

37 Livy 8.8.3–9 (adapted from Foster's Loeb translation): *Clipeis antea Romani usi sunt; dein, postquam stipendiarii facti sunt, scuta pro clipeis fecere; et quod antea phalanx similis Mac-edonicis, hoc postea manipulatim structa acies coepit esse: postremi in plures ordines instrue-bantur. Prima acies hastati erant, manipuli quindecim, distantes inter se modicum spatium; manipulus leves vicenos milites, aliam turbam scutatorum habebat; leves autem qui hastam tantum gaesaque gererent vacabantur. Haec prima frons in acie florem iuvenum pubescen-tium ad militiam habebat. Robustior inde aetas totidem manipulorum, quibus principibus est nomen, hos sequebantur, scutati omnes, insignibus maxime armis. Hoc triginta manipulorum*

Livy therefore suggests that the Roman army, at this time, had 30 maniples (15 maniples of *hastati* and 15 maniples of *principes*) and 15 *ordines*, with each *ordo* composed of three *vexilla* (one of *triarii*, one of *roarii*, and one of *accensi*). Each *manipulus* seems to have had 20 light infantry attached, although the total number of soldiers in each *manipulus* is not given. The only total unit numbers provided are for the entire army (5,000 infantry and 300 cavalry – *scribebantur autem quattuor fere legiones quinis milibus peditum, equitibus in singulas legiones trecenis*, Livy 8.8.14) and for the *ordines* and *vexilla*, with each *vexilla* containing 60 soldiers and two centurions, plus a *vexillarius*.[38] Livy suggests this should add up to 186 men in an *ordo* – his calculation evidently not including the *vexillarii* for some reason.

Famously though, Livy's figures have created some issues with getting the numbers to "add up."[39] Subtracting the total he gives for the *ordines* (186 × 15 = 2790) from the total for the army (5,000), leaves only 2,210 troops to be spread among the 30 maniples of the *hastati* and *principes*, or roughly 74 per unit. This has usually been taken to be far too few, and one reason why most modern scholars instead rely on Polybius' numbers for the unit types. Polybius suggested, referring to the army of the second century:

> They divide them so that the senior men known as *triarii* number six hundred, and *principes* twelve hundred, the *hastati* twelve hundred, the rest, consisting of the youngest, being *velites*. If the legion consists of more than four thousand men, they divide accordingly, except as regards the *triarii*, the number of whom is always the same.[40]

---

agmen antepilanos appellabant, quia sub signis iam alii quindecim ordines locabantur, ex quibus ordo unusquisque tres partes habebat – earum unam quamque primam pilum vocabant; tribus ex vexillis constabat ordo; sexagenos milites, duos centuriones, vexillarium unum habebat vexillum; centum octoginta sex homines erant; primum vexillum triarios ducebat, veteranum militem spectatae virtutis, secundum rorarios, minus roboris aetate factisque, tertium accensos, minimae fiduciae manum; eo et in postremam aciem reiciebantur.

38 Many modern scholars, for instance Sage (2008), blur these distinctions and simply translate both *manipulus* and *vexillum* as "maniple."

39 There is debate on whether Livy meant to include the *ordines* in his count of the maniples. For instance, Walters and Conway (1918, 14) suggested that Livy's description should be read as having 10 maniples of *hastati*, 10 maniples of *principes*, and 10 *ordines* – which included *triarii*, *roarii*, and *accensi*. In this model, each of the ordinmenes has 600, with each maniple of *hastati* and *principes* containing 160, adding up to Livy's total of 5,000. This argument, however, is largely driven by an attempt to "make the numbers add up" – which I would suggest is unwise. There was likely variability, even in later periods. As Polybius (6.20–21) explicitly noted, although the legion in his day usually numbered roughly 4200, this was not always a given.

40 Polyb. 6.21.9–10: διαιροῦσι δ' αὐτοὺς τὸν τρόπον τοῦτον ὥστ' εἶναι τοὺς μὲν πρεσβυτάτους καὶ τριαρίους προσαγορευομένους ἑξακοσίους, τοὺς δὲ πρίγκιπας χιλίους καὶ διακοσίους, ἴσους δὲ τούτοις τοὺς ἀστάτους, τοὺς δὲ λοιποὺς καὶ νεωτάτους γροσφοφόρους. 10ἐὰν δὲ πλείους τῶν τετρακισχιλίων ὦσι, κατὰ λόγον ποιοῦνται τὴν διαίρεσιν πλὴν τῶν τριαρίων. τούτους αἰεὶ τοὺς ἴσους.

In a later section (6.24.3), Polybius then goes on to indicate that each class was further subdivided into ten subgroups (ἑκάστην εἰς δέκα μέρη), resulting in most modern scholars assuming an army with maniples of 120 men each for the *principes* and *hastati,* and 60 men each for maniples of the *triarii.*

There remains some ambiguity though. Polybius never explicitly stated that the ten subgroups of each class were of equal size or indeed all the same type. In fact, one can easily read Livy's varied structure of *manipuli, ordines,* and *vexilla* in Polybius' description. Rather than relying on one term, Polybius uses a range of Greek terms for the subdivisions of Roman infantry when discussing the distribution of light infantry: "[the light infantry] are divided equally among all the units; these units are called τάγμα or σπεῖρα or σημαία, and their officers are called centurions (κεντυρίωνας) or *ordinum ductores* (ταξιάρχους)."[41] Some of these terms are quite specific, with the τάγμα and σπεῖρα both representing units of the Hellenistic phalanx,[42] and while the construction is slightly ambiguous (...καὶ τάγμα καὶ σπεῖραν καὶ σημαίαν...),[43] he does seem to be suggesting that the Roman army was divided up into three different types of units equivalent to Livy's *ordines,* *manipuli,* and *vexilla,* and not offering a selection of names which might apply to the *manipulus* alone. Thus, in the so-called "manipular legions" of both Polybius and Livy, the maniple seems to have represented but one of three different unit types to be found within the army, and one which was not necessarily regular in size.

Additionally, it is worth noting that the traditional division of *manipuli* into two centuries (*centuriae*) is not directly supported by the evidence. In passage 8.8, Livy only associates centurions with the *vexilla* of the *ordines,* not the *manipuli.* And while Polybius says that the Roman army had centurions, the unit or division they are associated with is not entirely clear:

> Finally these officers (κεντυρίωνας) appoint from the ranks two of the finest and bravest men to be standard-bearers (σημαιαφόρους) in each maniple (σπεῖραν). It is natural that they should appoint two commanders (ἡγεμόνα) for each military unit (τάγμα); for it being uncertain what

41 Polyb. 6.24.4–5: τῶν δὲ γροσφομάχων τοὺς ἐπιβάλλοντας κατὰ τὸ πλῆθος ἴσους ἐπὶ πάντα τὰ μέρη διένειμαν. καὶ τὸ μὲν μέρος ἕκαστον ἐκάλεσαν καὶ τάγμα καὶ σπεῖραν καὶ σημαίαν, τοὺς δ' ἡγεμόνας κεντυρίωνας καὶ ταξιάρχους.

42 The τάγμα being traditionally smaller than the σπεῖρα, thought to have 256 men, although the term τάγμα in particular was also quite fluid and was sometimes used to refer to an entire army or force.

43 It is possible, although unlikely, that Polybius was offering three words which each might apply to the Roman *manipulus* – and indeed, this is how many modern scholars have translated this. For instance, the modern 2011 Loeb version (Paton, revised by Wallbank and Habicht) offers "the *velites* are divided equally among all the companies; these companies are called *ordines* or *manipuli* or *signa,* and their officers are called centurions or *ordinum ductores,*" while Paton's original 1922 Loeb translation has "the *velites* are divided equally among all the companies; these companies are called *ordines* or *manipuli* or *vexilla,* and their officers are called centurions or *ordinum ductores.*"

may be the conduct of an officer or what may happen to him, and affairs of war not admitting of pretexts and excuses, they wish the maniple (σπεῖραν) never to be without a leader and chief. When both leaders are on the spot, the first commands the right half of the maniple and the second the left, but if both are not present the one who is commands the whole.[44]

Thus, the σπεῖραι (the Greek word most commonly associated with *manipuli*) were argued to have two σημαιαφόρους, or standard-bearers, who seem to have acted as commanders, but it is unclear how the centurions (κεντυρίωνας) fit in – seemingly above or outside of this level. In fact, centurions and centuries are only directly associated with *manipuli* in a much later section of Livy (42.34) and in a somewhat dubious context: the reported speech of the tribune Spurius Ligustinus detailing his military achievements.[45] Indeed, taking things a step further, the evidence for the "century" as an actual military unit is – perhaps surprisingly – virtually non-existent in the Republican period.[46] The mythical, tribal army of Romulus was supposedly divided into centuries,[47] and the later army of Servius Tullius and the early Republic was based on the centuries of the *comitia centuriata* – although these divisions seem to have primarily offered a mechanism for recruitment and political representation, and did not seem to function as tactical units on the battlefield.[48] The position of the centurion is obviously very well attested in a range of sources, but the units they commanded did not seem to be uniform and took much of their character and identity from the commanding officer.[49] It is therefore entirely possible that the office of centurion developed out of a political context (i.e., representatives or officers of the centuriate assembly) and were not, in fact, named after the units

44 Polyb. 6.24.6–8. (adapted from Paton's Loeb translation): οὗτοι δὲ καθ᾽ ἑκάστην σπεῖραν ἐκ τῶν καταλειπομένων ἐξέλεξαν αὐτοὶ δύο τοὺς ἀκμαιοτάτους καὶ γενναιοτάτους ἄνδρας σημαιαφόρους. δύο δὲ καθ᾽ ἕκαστον τάγμα ποιοῦσιν ἡγεμόνας εἰκότως· ἀδήλου γὰρ ὄντος καὶ τοῦ ποιῆσαι καὶ τοῦ παθεῖν τι τὸν ἡγεμόνα, τῆς πολεμικῆς χρείας οὐκ ἐπιδεχομένης πρόφασιν, οὐδέποτε βούλονται τὴν σπεῖραν χωρὶς ἡγεμόνος εἶναι καὶ προστάτου. παρόντων μὲν οὖν ἀμφοτέρων ὁ μὲν πρῶτος αἱρεθεὶς ἡγεῖται τοῦ δεξιοῦ μέρους τῆς σπείρας, ὁ δὲ δεύτερος τῶν εὐωνύμων ἀνδρῶν τῆς σημαίας ἔχει τὴν ἡγεμονίαν· μὴ παρόντων δ᾽ ὁ καταλειπόμενος ἡγεῖται πάντων.

45 This is quite a problematic speech on a number of levels, but particularly with regards to the terminology associated with the units, as Livy describes *ordines* of *hastati* as well as centuries of *principes*.

46 Things are little better in the Empire, as the names of most legionary *centuriae* were derived from their commander, with the few remaining ones named after their cohort and rank. See Speidel (1990) for discussion.

47 Livy 1.36.

48 Dion. Hal. *Ant. Rom.* 4.19. See Armstrong (2008) for discussion.

49 For instance, the sources agree that, at least from the Flavian period, the units led by centurions would have varied in size depending on which cohort they belonged to – with the units commanded by the centurions of the first cohort being double in size: Goldsworthy (1996, 13–15).

they supposedly commanded[50] – in this way being perhaps similar to the *tribuni*, who are usually never argued to have commanded a *tribus*.[51] The tactical division of "century" may have been a theoretical construct which simply referred to the troops under the command of a "centurion."[52]

Trying to make sense of this muddle has caused scholars headaches for centuries. Part of the issue may be that the terms used in our sources were never intended to be technical. As we see in the words used for military equipment, ancient writers often demonstrated a remarkable flexibility with what modern scholars would consider "technical language" (particularly with regards to the military) and a frustrating lack of concern with using the exact "right word."[53] Additionally, the most detailed and contemporary account, that of Polybius, struggles with the translation of the terminology and military model into Greek/Hellenistic forms. Some clarity may still be achieved though. Highlighting the active etymologies offered by Varro for the various units – the "bringing together" of things – it is possible that the unit terms reflected more abstract or relational identifications. For instance, while both *manipulus* and *vexillium* denote units associated with a standard of some sort, *manipulus* seems to be used to indicate a group of "individuals" following a standard, while *vexillium* (and *cohors*) seem to represent units, also following a standard, but made up of other smaller units.[54] In this interpretation, *ordo* may have a similar meaning to τάγμα in the Greek – generally referring to a group of soldiers in a formation or rank. Although hypothetical, if this is the case, it would make some sense of both Polybius' association between of ταξίαρχος and "centurion," as well as their connection with the *triarii*.

50 In his etymology, Varro (*Ling.* 5.88) seems to be trying to work back from the office of the centurion, rather than referring to a distinct or well-defined unit: *centuria qui sub uno centurione sunt, quorum centenarius iustus numerus.* Alternatively, Festus (*Gloss. Lat.* 53L) offers the traditional, idealized grouping of 100 men (*in re militari centum homines*), which seems to have never existed outside of the mythical army of Romulus.
51 The label *tribuni militum* – referring to both the regular, low-level officers and the enigmatic officials with consular *potestas* from the late fifth and early fourth centuries – has always been a problematic, particularly given the largely civic character of the *tribus* and the *comitia tributa.* Although the origin is likely beyond us, the label suggests that Rome's tribes had some sort of military association in the early Republic. And indeed, one could speculate that they represented a key aspect of military organization – perhaps preserved in recruiting practices even in the late Republic. There is no solid evidence, however, that a "tribune" was ever in command of a single *tribus.*
52 This removes, perhaps, the point raised by Smith (2006a, 289), who argued that "the fact that the maniple had two centuries would seem to imply that it was not the original element, but one produced by aggregation."
53 For instance, in the terms used for javelin (and specifically the Roman *pilum*), a wide range exists in our sources, even within the exact same context or passage; see Armstrong (2017b) for discussion.
54 *Vexilla* could be quite small, for example Pliny the Younger gave a *vexillum* of 16 men to the procurator Maximus in Paphlagonia (*Ep.* 10. 27.28). Still, they generally seem to have been composed of men drawn from different cohorts/units; see Goldsworthy (1996, 27) and Speidel (1982, 850–60).

But, returning to the question which preceded this section, what makes a maniple different from a clan or similar military grouping? The answer seems to be, "not very much," at least structurally speaking. Roman *manip-uli* seem to have been small, self-contained armies, likely with two "commanders," and containing both light and medium/heavy infantry, following a standard. As Harris suggested, the early maniples may indeed have had "the character of semi-private warbands."[55] Indeed, even the behavior of Roman armies and maniples seems to have been surprisingly clan-like during the third and early second centuries, featuring significant amounts of raiding conducted by small units.[56]

Filling the legions – *cives* and *socii*

The largest remaining difference between the Roman *manipuli* and archaic clans seems therefore to be their composition and how they were recruited. While Rome's hypothesized, archaic clan-based forces were presumably recruited and organized along family lines and via patron-client bonds, the Roman manipular army is usually seen as a model of republicanism. In Polybius' account of the *dilectus*, soldiers were levied individually and assigned to a particular legion, troop type, and presumably maniple, as part of this process. Indeed, Polybius suggests that by the middle of the second century, the Roman *dilectus* was an incredibly egalitarian affair – at least for the citizen infantry:

> The division and appointment of the tribunes having thus been so made that each legion has the same number of officers, those of each legion take their seats apart, and they draw lots for the tribes, and summon them singly in the order of the lottery. From each tribe they first of all select four lads of more or less the same age and physique. When these are brought forward the officers of the first legion have first choice, those of the second choice, those of the third, and those of the fourth last. Another batch of four is now brought forward, and this time the officers of the second legion have first choice and so on, those of the first choosing last. A third batch having been brought forward the tribunes of the third legion choose first, and those of the second last. By thus continuing to give each legion first choice in turn, each gets men of the same standard. When they have chosen the number determined on — that is when the strength of each legion is brought up to four thousand two hundred, or in times of exceptional danger to five thousand — the old system was to choose the cavalry after the four thousand two hundred infantry, but they now choose them first, the censor selecting them according

---

55 Harris (1990, 508). See Rocco (2017) for a more detailed discussion of the possible origins from Rome's gentilicial groupings.
56 See Rawlings (2016) for detailed discussion.

to their wealth; and three hundred are assigned to each legion...The youngest soldiers or *velites* are ordered to carry a sword, javelins, and a target (*parma*)... The next in seniority called *hastati* are ordered to wear a complete panoply... The *principes* and *triarii* are armed in the same manner except that instead of the *pila* the *triarii* carry long spears (*hastae*).[57]

This passage would seem to suggest that most soldiers "started at the bottom" and gradually worked their way through the various troop types and maniples during their tenure in the army, initially fighting as a *veles*, then moving to the *hastati* and *principes*, and finally the *triarii*.[58] The various divisions of the army were therefore permeable and related to experience. This is also, arguably, supported by Livy's account of the Roman army in the late fourth century, given in full in the previous section, where he notes,

The first line, or *hastati*, comprised fifteen maniples, stationed a short distance apart; the maniple had twenty light-armed soldiers, the rest of their number carried oblong shields; moreover those were called "light-armed" who carried only a spear and javelins. This front line in the battle contained the flower of the young men who were growing ripe for service. Behind these came a line of the same number of maniples, made up of men of a more stalwart age; these were called the *principes*; they carried oblong shields and were the most showily armed of all.[59]

---

57 Polyb. 6.20–23: γενομένης δὲ τῆς διαιρέσεως καὶ καταστάσεως τῶν χιλιάρχων τοιαύτης ὥστε πάντα τὰ στρατόπεδα τοὺς ἴσους ἔχειν ἄρχοντας, μετὰ ταῦτα καθίσαντες χωρὶς ἀλλήλων κατὰ στρατόπεδον κληροῦσι τὰς φυλὰς κατὰ μίαν καὶ προσκαλοῦνται τὴν ἀεὶ λαχοῦσαν. ἐκ δὲ ταύτης ἐκλέγουσι τῶν νεανίσκων τέτταρας ἐπιεικῶς τοὺς παραπλησίους ταῖς ἡλικίαις καὶ ταῖς ἕξεσι. προσαχθέντων δὲ τούτων λαμβάνουσι πρῶτοι τὴν ἐκλογὴν οἱ τοῦ πρώτου στρατοπέδου, δεύτεροι δ' οἱ τοῦ δευτέρου, τρίτοι δ' οἱ τοῦ τρίτου, τελευταῖοι δ' οἱ τοῦ τετάρτου. πάλιν δ' ἄλλων τεττάρων προσαχθέντων λαμβάνουσι πρῶτοι τὴν αἵρεσιν οἱ τοῦ δευτέρου στρατοπέδου καὶ ἑξῆς οὕτως, τελευταῖοι δ' οἱ τοῦ πρώτου. μετὰ δὲ ταῦτα πάλιν ἄλλων τεττάρων προσαχθέντων πρῶτοι λαμβάνουσιν οἱ τοῦ τρίτου στρατοπέδου, τελευταῖοι δ' οἱ τοῦ δευτέρου. [καὶ] αἰεὶ κατὰ λόγον οὕτως ἐκ περιόδου τῆς ἐκλογῆς γινομένης παραπλησίους συμβαίνει λαμβάνεσθαι τοὺς ἄνδρας εἰς ἕκαστον τῶν στρατοπέδων. ὅταν δ' ἐκλέξωσι τὸ προκείμενον πλῆθος — τοῦτο δ' ἐστὶν ὀτὲ μὲν εἰς ἕκαστον στρατόπεδον πεζοὶ τετρακισχίλιοι καὶ διακόσιοι, ποτὲ δὲ πεντακισχίλιοι, ἐπειδὰν μείζων τις αὐτοῖς προφαίνηται κίνδυνος — μετὰ ταῦτα τοὺς ἱππεῖς τὸ μὲν παλαιὸν ὑστέρους εἰώθεσαν δοκιμάζειν ἐπὶ τοῖς τετρακισχιλίοις διακοσίοις, νῦν δὲ προτέρους, πλουτίνδην αὐτῶν γεγενημένης ὑπὸ τοῦ τιμητοῦ τῆς ἐκλογῆς· καὶ ποιοῦσι τριακοσίους εἰς ἕκαστον στρατόπεδον...καὶ τοῖς μὲν νεωτάτοις παρήγγειλαν μάχαιραν φορεῖν καὶ γρόσφους καὶ πάρμην...τοῖς γε μὴν δευτέροις μὲν κατὰ τὴν ἡλικίαν, ἀστάτοις δὲ προσαγορευομένοις, παρήγγειλαν φέρειν πανοπλίαν...ὁ δ' αὐτὸς τρόπος τῆς καθοπλίσεώς ἐστι καὶ περὶ τοὺς πρίγκιπας καὶ τριαρίους, πλὴν ἀντὶ τῶν ὑσσῶν οἱ τριάριοι δόρατα φοροῦσιν.

58 This progression is explicitly presented in Livy 42.34 in the life of Spurius Ligustinus.

59 Livy 8.8.5–6.

However, there are some problems with a strict egalitarian reading. Livy suggests that the *principes* had the "showiest equipment" (*insignibus maxime armis*) and, returning to Polybius, the light infantry may have also been selected (at least in part) based on wealth – as he noted "they choose the youngest *and poorest* to form the *velites*" (my own emphasis). And, of course, the cavalry were also selected by wealth.

So, while both Livy and Polybius emphasize age, it seems clear that they thought wealth also played a key role in how the army was levied and organized. In many ways, this makes practical sense, as soldiers would still have been required to supply their own equipment as late as the early second century, let alone in the fourth. As suggested previously, one could request soldiers show up with a particular panoply, but what they actually brought would obviously be dependent on what they could afford and/or what their family owned. Soldiers needed to be placed in *acies* and maniples which were economically appropriate. Thus, it may be possible to read the comments about age in Livy and Polybius' accounts of the legion not as prescriptive rules, but as descriptions of the norm – the older soldiers may have often been wealthier ones, having had more time and campaigns to accumulate wealth and equipment. Alternatively, or indeed additionally, age may have been a subtle literary device used to suggest the increased status of the units – something which is by no means uncommon in our sources in other contexts.[60]

Of most relevance for the current discussion, though, is the fact that the system described by Polybius for recruitment likely represents an idealized model which did not function exactly as he described, even in the second century.[61] If wealth and experience were considerations in how troops were selected and distributed between the legions, and within each legion, the *dilectus* must have been a bit more complex than he suggested. Selecting based on "age and physique" alone would not have been practical.[62] At the very least soldiers would have had to have been initially grouped by wealth or census class, as well as (or instead of) "age and physique."[63] This would

---

60 Associations between age and status were common in antiquity, particularly with regards to warfare. Thus, Athenian writers were generally negative toward light-armed troops and, for instance, the *psilos* meant both shaven and smooth as well as a light infantryman. See Trundle (2010, 143) for discussion.

61 The practicalities of the Polybian *dilectus* have long been debated. For instance, Brunt (1971a, 625–34) argued convincingly that Polybius' *dilectus* could not have occurred as described given the total numbers involved in the third and second centuries. For a more recent summary of the issues, see de Ligt (2007, 114–17).

62 As Gabba (1976, 127) noted, "in fact, when citizens took to presenting themselves at the levy without distinction of class or century, a procedure to which in fact Polybius alludes in his description, how would it have been possible to operate a preliminary choice if there had not been a list compiled by the censors in which citizens were included on the basis of fixed qualifications?"

63 One possible solution to this particular issue would have been to run the *dilectus* by century, as indeed Hill (1952, 360) suggested over 60 years ago. However, this is not explicitly supported by Polybius and would likely introduce a range of other complexities, given the

also presumably bring family considerations into play, given the nature of property and wealth in Rome (typically being held by the *paterfamilias*), hinting that, even in the second century, recruitment was based, in part, on family status and affiliation. The process of the *dilectus* also seems to have afforded commanders and their officers significant autonomy and latitude, as suggested by the many controversies which often arose around it and the way in which ranks and assignments were distributed in a decidedly ad hoc and personal manner.[64]

There have also long been questions about the practicalities of mobilizing Rome's citizens – which numbered, according to the census for 294, over 262,000[65] – for the *dilectus* each year during this period.[66] While, in theory, any Roman citizen between the ages of 17 and 60 could be called up, regularly summoning all of these men to Rome, from all over Italy, to select less than 10% of them for the legions has long seemed impractical.[67] As a result, Gabba famously argued that Polybius' *dilectus* was likely tribal in nature and thus each iteration pulled individuals from a particular region, at least before 241.[68] And even after 241, and the reorganization of the tribes, recruitment may have still been a localized phenomenon. As Fronda notes elsewhere in this volume, both P. Sulpicius (in 200) and T. Flamininus (in 198) specifically recruited Scipionic veterans in the region of Brundisium before setting sail across the Adriatic.[69] Here, commanders were clearly not drafting Romans indiscriminately through an egalitarian *dilectus*, but were targeting specific men, from a specific region, connected via specific relationships, to fill their legions. Thus, soldiers were unlikely to be levied and broken up by legion in exactly the way Polybius suggests – *en masse*, and in a largely undifferentiated fashion, from the entire citizen body. Other factors, and most importantly relationships, seem to have played significant roles in how they were recruited as well.

---

number of centuries (193, incl. that for the *capite censi*) in the system. That being said, it is arguably possible – although unlikely – that Polybius' reference to soldiers being selected with "the same age and physique" (ταῖς ἡλικίαις καὶ ταῖς ἕξεσι, Polyb. 6.20.3) may refer somehow to census rating and not physical characteristics, along the lines argued above.

64 Dobson (2008, 47–66). See Terrenato (2019, 163–65) for recent discussion.
65 Livy 10.47.2.
66 See Nicolet (1980, 97–98) for discussion.
67 Even allowing, quite conservatively, for half of Rome's citizen body being exempted form service (listed members of the *capite censi*, being above 60 years of age, or having served their quota of campaigns), summoning over 130,000 men annually to Rome, in order to recruit 16,800–20,000 infantry and 1,200 cavalry, poses serious logistical and organizational questions.
68 Gabba (1976, 127). Cf. Nicolet (1980, 96–101); Jehne (2006, 250–58). More recently, de Ligt (2007, 116) argued for a version of the *dilectus* which included two stages, "a first stage in which recruits were enlisted by the local authorities and a second one in which the men thus selected were distributed among the legions. Polybius' description refers solely to the second stage."
69 Livy 31.14.1, 32.9.1.

The importance of maintaining relationships during the *dilectus* process is arguably further supported by evidence for local affiliations being maintained in Rome's legions. In 282, the Romans deployed a detachment of Campanians and Sidicini under the command of the Campanian noble Decius Vibellius.[70] As Rosenstein has suggested, "although some ancient sources call them mercenaries in Roman employ, the Campanians were Roman citizens, and the Sidicini were allies. So in all likelihood these were legionaries accompanied by allies, an interpretation rendered more likely in view of the similar garrison installed at Thurii in the same year by the same consul and for the same reason."[71] One could further support this argument with examples like the cohort from Praeneste, which was trapped in Casilinum alongside Roman legionaries in the Second Punic War after showing up late to the muster point, and similar anecdotal evidence.[72] We also have evidence that the *dilectus* could favor some individuals over others, most notably veterans, and that these may have returned to their old units and posts.[73] And, of course, generals were known to recruit armies from their own clients, particularly in the late Republic, although it is attested earlier as well – for instance during the years of the Second Punic War and with Scipio Aemilianus in 134.[74] Thus, we seem to have legions with strong preexisting relationships and affiliations – it was, in many ways, an army composed of existing groups and entities – and not just an undifferentiated mix of "Romans." And while the increasing length of campaigns during the Second Punic War and early second century may have slowly allowed a new social order to develop – the "Roman army as society" – it is likely that Rome's armies of the fourth and early third centuries maintained the social and power dynamics of contemporary society.

One must also consider the importance of the allies in shaping the character of Rome's forces, as the allies formed an increasingly large and vital part of the Roman army during this period. There is strong evidence for allies maintaining their preexisting organizations, command structures, and affiliations – as seen in the instance of the Sidinician contingent above and with the 300 Campanian cavalry sent to garrison Sicily in the Second Punic War.[75] As Polybius notes, consuls were able to issue appropriate orders to the allies, but do not seem to have controlled them directly.[76] These would have been led by local commanders, as we see in Plutarch's account of the Frentanian Oplax, who commanded a unit of cavalry against Pyrrhus.[77]

---

70 Dion. Hal. *Ant. Rom.* 20.4.1–2; Val. Max. 2.7.15f.
71 Rosenstein (2012b, 55).
72 Livy 23.17.
73 App. *Pun.* 75.351. See Hoyos (2007, 64) for discussion.
74 Second Punic War: Livy 23.4, 23.14, 23.17, 25.5, 27.38. See Gruen (1995, 376) for discussion of Scipio Aemilianus.
75 Campanian cavalry: Livy 23.4. A certain amount of allied autonomy is also suggested by Polybius (6.26.5–9) in his account of the *dilectus* and mobilization.
76 Polyb. 6.12.
77 Plut. *Pyrr.* 16.10.

Allied contingents, presumably speaking their own language and equipped in native fashion, seem to have functioned quite independently – at least tactically – within the Roman military system. And, of course, this was all likely based on long-standing precedent for the organization of allied armies in Latium and central Italy, through the Latin League and similar structures, where communities and clans retained their identity and autonomy in a federal system.[78] If the history of the Latins and the Latin League is to be believed (admittedly a big "if"), then these sorts of armies – possibly regularly led by Rome or Roman leaders in the fifth and fourth centuries – were not a new development in the late fourth century. Indeed, as epigraphic evidence indicates (esp. *CIL* 1^2.5, see below), Rome was not the only Italian power to have "*socii*" at its disposal during this period. What *was* new was both the scale of operation, which was now much larger, and the nature of the power dynamic which underpinned the relationship, with the seat of power increasingly moving away from more neutral assemblies, like that of the Latin League at the *lucus Ferentina*, and toward the assembly of Rome.

In this context, Rome's ability to accommodate troops and units from a wide range of cultures in ever-increasing numbers is certainly impressive, but perhaps not surprising. Roman armies, particularly above a certain size, had likely always been composed of a heterogeneous mix of units which seem to have maintained their original identities, and were likely tactically differentiated. This did not change as Rome's empire grew. Although modern historians usually assume that the Romans "required" allies to come equipped in arms and armor similar to that of the Romans, this would have been both an impractical and unenforceable request. Allied soldiers fighting for Rome would have likely served using their native equipment, following their local commander, and fighting in a manner similar to that which they had used previously and independently.[79] While the fourth century saw an increasing "homogenization" of military equipment across the Italian peninsula, and likely the mode of combat as well, there were still regional and ethnic differences.[80] Rome could not expect her allies to field units which looked exactly like her own, and indeed did not seem to want that – as the allies, many of whom were from the plains of Campania, seem to have been asked to bring a larger contingent of cavalry. The Romans embraced the diversity which their allies brought, rather than try to stamp it out and standardize things.

Organized chaos

The "manipular legion" has always been, in many ways, defined by its complexity. With a range of different divisions and equipment types, even within the citizen contingent (not to mention the variety likely present among the

---

78   Alföldi (1965, 35–38).
79   See also Kent (2012a).
80   See particularly Burns (2003, 2006). More recently, also Taylor (2014).

allies), scholars have long struggled to explain how it was organized and functioned. I would suggest that this is because its fundamental nature, as described in Polybius and Livy, has been misunderstood. The manipular legion they described was not a rational, imposed system. The Romans did not sit down and create an elaborate military structure, with various parts and pieces designed to work together in perfect, mechanistic harmony, into which they then simply plugged soldiers. Nor was it designed to allow a particular tactical formation and way of battle – epitomized by the complex "dance" of units, flowing in and out of gaps, in an ordered fashion, often proposed by scholars.[81] Rather, the manipular system can be seen as the physical manifestation of Roman *imperium* in the middle Republic. The army was not yet (if indeed it ever was) a "melting pot" of "*Romanitas*." The army of the early and middle Republic was instead a patchwork quilt, with its constituent parts displayed quite obviously and visibly. However, as with most quilts, the individual patches which make it up are not the most important aspect. These can be changed, swapped, and indeed can vary significantly in size and quality. What matters is the overall pattern and relationship, and how they are stitched together.

Both Rome and the army of the middle Republic were founded on the principle of integration. Within the main citizen body, Roman society was famously composed of patricians and plebeians, and within these larger divisions, we also have family and clan divisions which had a long tradition of acting independently, particularly in the realm of war – something which was not a relic of the past, but which seems to have continued throughout the fourth century.[82] Much of the narrative of the early Republic is concerned with the gradual unification of Roman society and the slow amalgamation of these groups – a process which was still ongoing during the period in question with the gradual ending of the Struggle of the Orders. It is unlikely to be a coincidence, though, that as Rome began to display a certain level of internal cohesion and self-awareness within both the elite and the wider body politic during the fourth century, the community also began to explore ways to extend itself. In the 330s, Rome massively expanded her citizen body beyond the confines of the *ager Romanus antiquus* and integrated most of the Latin communities as either *cives* or *cives sine suffragio*, embarking on yet another phase of expansion, unification, and reinterpretation of what it meant to be "Roman." This expansion of citizenship was accompanied by

---

81 This is not to say this never occurred – as, for instance, we see in the campaigns of Scipio Africanus. Merely that it was likely rare and required an extraordinary level of cohesion and leadership.

82 This is arguably attested by inscriptions like that by Caso Cantouio (*CIL* 1².5), found near Alba Fucens but now lost, probably of the very late fourth century, which notes "Caso Cantouius of Aproficulum (La Regina) captured (this) by the finis Gallicus in the city of Caiontonia (?), and his socii brought it as a gift to Angitia on behalf of the Marsic legiones." – translation and reconstruction by M. Crawford (forthcoming). Photograph of the inscription can be found in *CIL* 1² p. 859, pl. 2, 1.

a massive increase in Rome's allies, where groups and communities from across the Italian peninsula – from the Etruscans in the north to the Greeks and Samnites in the south – were regularly brought together in a single army. At the same point that Rome was reinterpreting what it meant to be Roman, she was having to field ever more disparate and heterogeneous armies.

The period of change (c. 338–264), highlighted at the opening of this chapter, was very real. However, the mechanisms by which Rome negotiated this change – on the battlefield at least – were not new. As suggested, much of Rome's history had been defined by the integration of different clans and groups, both politically and militarily. Rome's manipular legion arguably demonstrates this through its various units and divisions. The *manipuli*, rather than being proscriptive divisions of *hastati* and *principes* which needed to be filled, were likely how Rome categorized and integrated groups of *cives* and *socii* who were organized around a banner, usually contained a group of *leves* in addition to medium/heavy infantry, and were probably roughly 100 men in strength. Given that these groups were likely based on clans, with a long history of fighting independently, and who may have also only recently been integrated into the Roman social, political, and military matrix, it is not unsurprising to think they would have preferred to fight a little apart from other such groupings.[83] But, of course, Rome's army was not only composed of clans – although these do seem to have represented some of the most important entities with the longest tradition of warfare – as communities were also allied, most notably those from Campania. Additionally, not all clans may have fought in the same way. While the *scutum* and *pilum* were increasingly common in Italian warfare during this period, most notably among the armies of southern Latium and parts of Campania, there was a long tradition of fighting with thrusting spears (*hastae*) as well.[84] Rome's *ordines* and *vexilla* may have been a way for the Romans to integrate and coordinate infantry which came from military structures where this was the norm. As a result, rather than representing a highly specialized system, like the Hellenistic armies of the time, which required particular pieces working in harmony (phalanx, peltasts, cavalry, etc.) to function, the Roman army seems to have been based on the opposite philosophy – having a place and role for every available piece. A key element within this system was the space which allowed each unit to function independently. Rather than offering tactical gaps through which they could advance or retreat, the classic quincunx formation may have simply been a result of the varied units of Rome's army not wanting to get too close to their neighbors. Over time, the army may have found ways to use these gaps in an organized fashion, as exhausted units may have sought respite and fresh units from the rear may have been keen to engage, but the fundamental reason for them is unlikely to have been so rational. What mattered, was that they were all on the battlefield fighting for "Rome."

---

83 On this see also Rocco (2017, esp. 112–44).
84 Small (2000, 221–34).

The Pros and Cons of allies – a quick discussion

In antiquity, allied armies were understandably not uncommon. Bringing together a greater number of troops was often desirable for several reasons, including social and political as well as the obvious military considerations, although these armies did usually suffer from a common weakness: how to deploy these troops in a cohesive fashion. The joints between allied factions were usually a weak point, as at Chaeronea, where the gap between the Athenian and Theban contingents was exploited by the Macedonian horse led by Alexander. Or at Plataea, where the Greeks were victorious in spite of various *poleis* acting independently. Or at Cunaxa, where the Greek contingents on the far right of Cyrus' army acted separately. Only rarely, as at Leuctra and Mantinea where the Thebans on the left wing deliberately left their allied contingents on their right and targeted the Spartans opposite them, was this division exploited for benefit. Typically, it was a point of weakness, particularly in armies which contained large blocks of troops.

The Romans – and, indeed, perhaps other Italian armies which featured large numbers of allies[85] – avoided this issue by using two key features. First, their army did not feature one or two large breaks which could be exploited, but rather tens of smaller ones. Each of these smaller breaks could obviously be attacked, although the risk (with an attacking force being surrounded on three sides) may not have justified the possible gain. Second, the Romans seem to have been happy to keep each unit tactically independent, able to act on its own and defend itself – famously as at Cynoscephalae.[86] Emerging from a native process of integration and association, which saw the Romans slowly bring together the various clans, families, and communities of Rome, eventually Latium, and ultimately Italy under a single banner, the Romans seem to have utilized a more detached approach which allowed each group to maintain its independent identity while still working for, and alongside, the group.

It must be noted that the obvious counter to this model is the question of how an allied army – which was so fragmented and seemingly chaotic – could ever have been as successful as the sources suggest. The logical conclusions are that either the record of success is overstated or the opposition was weaker, despite the Roman fragmentation. Both are possible options. First, on Roman success, it is worth noting that the Roman army was not uniformly victorious during the fourth and third centuries, and indeed regularly suffered sometimes catastrophic defeats.[87] The evidence does not suggest that, given equal forces, the Romans would always (or indeed regularly) win. But this does not mean the Romans were not substantially stronger than their opponents. Taking the Pyrrhic war as an example, the Roman strategy

---

85 Kent (2012a, 75–77).
86 Polyb. 18.21–26.
87 See Helm (2017) for discussion.

in this period does not seem to be to attempt to deploy a superior tactical force on the battlefield, but rather to rely on greater numbers – ideally in a specific battle, but more importantly in the long term. Rome's amalgamative system may have sometimes (perhaps often) resulted in an inferior tactical force, but its long-term strategic benefits, and the ability to simply and easily plug new groups and populations into its military structures, meant that it usually had the advantage in the long term. This model also proved generally successful socially, economically, and politically, with the Romans developing a series of relationship and networks throughout Italy which would survive the pressures of the Punic Wars and only really be tested by the changing nature of Roman imperialism in the late second and first centuries.

Conclusions

Separating out the Roman army of the middle Republic and labelling it the "manipular legion" may be something of a misnomer. Although it did seem to feature *manipuli*, these represented only one of its many divisions, and are surprisingly undiagnostic in their nature – looking very similar, in many ways, to the clan-based warbands which both predated it and still raided and waged war across central Italy in the fourth century. Additionally, focusing on and defining the legion by its *manipuli* of *hastati* and *principes* gives undue definition and meaning to these amorphous units. Despite the modern scholarly tendency to divide these units up into equal, rigid blocks of uniform troops on a battlefield, the evidence does not support this sort of model. What defined the army of this period was not the specific nature or equipment of its units, but the fact that it could field so many different units at once – and from such varied backgrounds and areas. Interestingly though, despite the success of this fragmented army in winning Rome its Mediterranean empire, from the third century onwards the Romans seem to have been keen on defragmenting their forces and moving to larger units – most notably the cohort. Perhaps because it allowed for greater control on the battlefield, greater glory for the generals commanding the armies, or greater cohesion among the troops, Rome's forces seem to have experienced a gradual movement toward greater, larger, and more structured units – albeit in fits and jolts, often spurred by a particular catalyst, and not continuously. And yet, even by the second century, this process was likely incomplete. As Jefferson recently argued about Cato's *Origines*, "The strong note of collectivity in the *Origines* almost paradoxically indicates an awareness of the diversity of Italy in Cato's time."[88] As subsequent chapters in this volume will discuss, the diverse, often nested, identities of Rome's soldiers continued to create tension within Rome's military system down through the Social and Civil wars.

88 Jefferson (2012) 326.

6 Poor man's war – rich man's fight

Military integration in Republican Rome*

Marian Helm

Introduction

The rise of Rome remains one of the most impressive feats in history. Over the course of two centuries, the *res publica Romana* managed to subjugate the Italian peninsula and subsequently the whole of the Mediterranean. Annual military operations became the norm after the settlement of the Struggle of the Orders in the fourth century, and contributed greatly to the solidity of the political system by forming a central aspect of elite competition and legitimation – as well as contributing to social harmony through the distribution of plundered resources.[1] Every citizen was supposed to serve several years in the armies of the Republic, which conducted an annual levy of at least four legions (a total of 18,000 men) – and considerably more in times of crisis. This manpower reservoir, and the resulting potential for recruiting overwhelming numbers, could not be matched by any other power in the ancient world.

Studies on the Roman military have duly acknowledged the great importance of warfare for the *res publica*, especially in regard to its political order, processes of integration and identity within Italy, and the distribution of land and spoils.[2] Influential studies have also pointed out the diverse factors encouraging military aggression, for example, the distinctive structure of the Roman family, elite competition based on military command, and, of course, economic objectives.[3] However, the individual soldier's readiness to

* I would like to thank Jeremy Armstrong, Michael Fronda, and the anonymous reviewers for their helpful remarks. All dates are BC unless otherwise noted.

1 Harris (1979, 9–67); Hölkeskamp (1993, 31–33); Rich (1993); Cornell (1995, 353–79); Bleckmann (2002, 57–112); Beck (2011b, 237–40); Drogula (2015, 182–93); Armstrong (2016c, 182–231, 2016b, 114–18); Hölkeskamp (2017, 130–39); Linke (2017, 393–97).

2 Jehne and Pfeilschifter (2006); Erdkamp (2007b); Coudry and Humm (2009); Roth (2009); Roselaar (2010); Roselaar (2012); Rosenstein (2012a, 2012b, 24–35); Fronda and Gauthier (2017). See also above.

3 Family: Thomas (1996, 322–26); Linke (2014, 82–86). Elite competition: Hölkeskamp (1987, 170–203); Rüpke (1995, 216–19); Beck (2005, 32–51); Rosenstein (2006). Economic objectives: Harris (1979, 54–104); Roselaar (2010); Rosenstein (2011); Kay (2014, 21–86); see also, especially, the articles in Beck, Jehne, and Serrati (2016).

DOI: 10.4324/9781351063500-6

This chapter has been made available under a CC-BY-NC-ND license.

fight and die in the Republic's wars remains elusive, as our sources provide neither the "little guy's" point of view nor his motivations.[4]

This chapter will argue that this problem can be at least partially circumvented by focusing on the specific position and role of different types of Roman citizen-soldiers in the legion. The social heterogeneity of the marshalled troops allows for an analysis of the factors contributing to the "vertical" and "horizontal" cohesion of each of the different units and offers potential insights into the respective groups' functions, as well as their particular interests in and their perception of military service.[5] Consequently, this chapter will first analyze the troop composition of the Roman legion in the mid-Republic to determine the approximate degree of horizontal cohesion among its troops. A second step will consider the vertical cohesion within the legion by looking at the "officer corps," especially the *tribuni militum*. The time-frame under consideration stretches from 311 to the middle of the second century and draws on Polybius' description of the Roman army.[6] Although his report might not precisely reflect the make-up of the legions at the beginning of the third century, it nevertheless represents a vital piece of information concerning the gradual evolution of the army and likely conserved some aspects of the system of the earlier period. Although, as explored in previous chapters in this volume, the interpretation is never easy or straightforward.

Let us take, for example, the *dilectus* of the Roman legions. The rotating selection of individual recruits from each *tribus* by the military tribunes, as reported by Polybius, consciously avoided the levying of regionally cohesive troops, creating motley collections of strangers.[7] This system was evidently designed to break down existing bonds, presumably shared by those within the tribes, and ensure that Rome's legions were made up of men whose only real connection was a shared Roman citizenship. And indeed, before 241, the *tribus* seemed to have constituted distinct regional areas, but afterwards rapidly lost their cohesion by successive and disjointed enlargement. Therefore, the procedure would have been superfluous in the second century, which suggests one of two things. First, as argued by Armstrong in the preceding chapter, that the Polybian description of the *dilectus* may be an idealized version of the second century system, or second, that in this

---

4 Nicolet (1980, 89–109); cf. Stietencron (1995, 41–48) on necessary strategies for motivating fighters. Rome's massive manpower advantage also relied on the successful integration and mobilization of allied troops into the Roman military, see Brunt (1971a); Baronowski (1993); Pfeilschifter (2007); Kent (2012b, 84–99).

5 Pohl and Roock (2011, 47); Armstrong (2016b).

6 The year 311 saw the introduction of the election of the *tribuni militum* by the *comitia tributa*, and probably also the formal enlargement of the consular army to two legions each. See: Cornell (1989, 373); Forsythe (2007, 36); Clark (2016, 289–94); cf. Sage (2013, 222–24).

7 Polyb. 6.19–20. For the avoidance of regional levies, see Nicolet (1980, 96–101) and especially Jehne (2006, 250–58) with references. Cf. Armstrong in this volume for a different view.

description Polybius is referring to the *dilectus* as it occurred in earlier periods, possibly the late fourth to early third centuries – or at least before 241. At the beginning, Rome's legions probably consisted of locally recruited groups (*gentes, tribus*, etc.) but, in my opinion, the disintegration of these groups had almost certainly set in with the reorganization, reform, and expansion of the Roman military around 311. A powerful argument for this is the fact that the popular election of the *tribuni militum* implemented in this year ensured that Roman troops were led in line with the interests of the *res publica Romana*, not individual strong men.[8] The popular election effectively reduced the influence of such local strong men because they would have needed to achieve a majority of *tribus* in the tribal assembly, requiring a more inclusive approach instead of relying on regional ties and loyalties.[9] Since the elections were held before the campaign season, potential volunteers, regardless of their social and economic status, had the chance to see and shape the "officer corps" that would lead the army during their time of service. This development can be tentatively described as a "democratization" of the military structures of the legion; a process that coincided with the completion of the Republican order at Rome during the time of the Samnite Wars.[10] The assertion and fixation of the political rights of the citizens, represented by the *lex Hortensia*, was mirrored by a similar self-assertion of control over the military hierarchy. Despite the justified criticism of the "Polybian lens" expressed in other contributions of this volume, there is in my opinion nevertheless little reason to doubt that the election of military tribunes and the *triplex acies* that he described were in place by the beginning of the third century.[11]

8 See Clark (2016, 289–94).
9 Livy (7.5.9, 9.30.3) reports that the election took place in the plebeian assembly.
10 Hölkeskamp (1987, 114–69); Hölkeskamp (1993, 31–33); Forsythe (2007, 35–37). The scale of the military reforms connected to the year 311 tends to be underestimated despite the stark contrast between the success of Roman arms before and after their implementation in the context of the Second Samnite War, *contra* Scopacasa (2015, 129–45). Chaotic mobilization has its limits, and it might be necessary to differentiate more strongly between troops composed of Roman *cives optimo iure* and others, especially at the beginning of the Roman alliance. See Helm (2017).
11 It seems to have been in place by 223 at the latest, since Polybius (2.33.4) reports that the frontline maniples of the *hastati* were handed the spears of the *triarii* to ward off the initial Celtic onslaught. Since Polyb. 2.14–2.35 refers in great detail to the wars against the Celts before the Second Punic War, it has been assumed that he used the reliable report of Fabius Pictor, who was an active witness of these conflicts: see Walbank (1957, 184); Timpe (1972, 938); Beck and Walter (2001, 56). See also *FRHist* 1 F29–30; Walter (2004, 205–7). Furthermore, Dionysius (*Ant. Rom.* 20.11.2) reports the employment of spears for the *principes* in the Pyrrhic War but implies that less effective troops engaged the enemy first: "Those who fight in close combat with cavalry spears grasped by the middle with both hands and who usually save the day in battles are called *principes* by the Romans" (τοὺς τοῖς ἱππικοῖς δόρασιν ἐκ διαλαβῆς ἀμφοτέραις ταῖς χερσὶ κρατουμένους μαχομένους συστάδην καὶ τὰ πολλὰ κατορθοῦντας ἐν ταῖς μάχαις Πρίγκιπας Ῥωμαῖοι καλοῦσιν). For further discussion see Daly (2002, 64–65).

Thus, the described Roman legion displayed a stark, heterogeneous composition, and it will be argued that this organization not only represented, and reproduced, the social and political order of the Republic but also served to distribute profits and burdens accordingly among its component parts. Pointedly, the legion accommodated socio-political inequalities and potential tensions among the recruited citizens and harnessed them into an efficient fighting machine.[12] The background and context of individuals is therefore important, since not every citizen fighting for Rome did so for the same reasons or was treated the same way – a fact that is most obvious in the First Punic War, when the deaths of tens of thousands of *proletarii* manning the fleets did not elicit the same reactions as the loss of legionary forces.[13]

Why fight? Mobilizing the citizens

When trying to approach the Roman citizen-soldier, particularly in the Middle Republic, it is noteworthy that the sources do not depict him as an individual warrior par excellence, but instead emphasize superb organization and discipline of the overall legion and system.[14] The rapid conquest of the Mediterranean was not attributed to the individual martial skills of Roman soldiers, but rather to a collective effort in overcoming any opposition.[15] Both the limitations of Roman soldiers – and indeed of the whole military – and the strength of Rome's collective effort and will emerge most clearly in the Second Punic War, featuring an astonishing will to persist

---

12 Armstrong (2016c, 260–69); see also the concept of "organized chaos" developed by Armstrong in this volume.

13 Linke (2016, 165–73). Only when P. Claudius Pulcher's outrageous conduct led to another defeat at Drepana in 249 did Roman public opinion shift. See also Rich (2012, 102–3).

14 Polybius (6.19–42) explains the organization of the army in detail but does not once mention its order of battle. Also note the idealized story of T. Manlius Torquatus which focuses on discipline and was widely remembered. T. Manlius followed orders by the letters and explicitly asked to engage a Gaul in combat, a lesson that was reinforced by the execution of his own son who had disobeyed orders. *Imperia Manliana* became a proverb for harsh military discipline. On the episode see: Livy 8.7–8.2; Cic. *Fin.* 2.105, *Off.* 3.112, Sall. *Cat.* 52.31–32; Dion. Hal. *Ant. Rom.* 8.79.2; Val. Max. 2.7.6, 5.8.3, 6.9.1, 9.3.4–5; Frontin. *Str.* 4.1.40–41; Flor. 1.9; App. *Sam.* 3; Gell. *NA* 9.13.20.

15 Cato *FRHist* 5 F114a (= Cic. *Sen.* 75): *sed legiones nostras, quod scripsi in Originibus, in eum locum saepe profectas alacri animo et erecto, unde se redituras numquam arbitrarentur* ("...but our legions, as I have written in the *Origins*, who have often set out with an eager and resolute spirit to that place whence they thought they would never return."). Cf. the fictitious discussion between Hannibal and Scipio in 193: Livy 35.14.5–12; App. *Syr.* 10; Plut. *Flam.* 21. Hannibal's claim that he would have named himself first among the finest generals of the time, had he not been defeated, pays tribute to the determined resistance of the Roman people rather than their generals, while Polyb. 2.24 and Livy 9.19 place emphasis on quantity and quality. The argument of Rosenstein (1990, 93–113) that the blame for defeats usually fell on the soldiers has recently been challenged by Rich (2012, 88–94). On the Roman ability to absorb losses, see: Rosenstein (2004, 107–40, 191–93); Erdkamp (2011a, 65–67); Linke (2016, 164–68).

despite devastating reverses and staggering casualties. However, this clearly exceptional case should not be assumed to represent the general attitude in all of the Republic's wars.[16] In fact, lack of enthusiasm or outright refusal to serve was as old as the Republic itself, and can be seen throughout the Struggle of the Orders in the fifth and fourth centuries, when the plebeians reportedly refused to serve in the army until their political and economic demands were met.[17] Asynchronous interests and political frictions could seriously impact both the military operations and the political order of the *res publica*, attested, for example, by the *seditio* of 342, the eccentric consulship of L. Postumius Megellus, the curious mutiny of the so-called *legio Campana*, the legislation of Fundanius Fundulus prohibiting naval warfare, and several cases relating to service in Macedonia and Spain in the second century.[18] I do not mean to imply that Roman citizens were in general opposed to military service, but a lack of identification with a war's aims could severely dampen their enthusiasm to take up arms. Certainly, military service and political participation were inextricably intertwined in the vast majority of ancient city-states, but this does not mean that the citizens became mute cannon fodder. On the contrary, they had to be convinced that supporting the war effort was in their own best interest.[19]

The Pyrrhic War provides perhaps the most compelling case. After Pyrrhus' professional, Hellenistic force outmatched them at Heraclea in 280, the Romans appear to have earnestly considered concluding a peace treaty with the Epirote king, which was averted only by the uncompromising stance of Ap. Claudius Caecus.[20] After these negotiations had failed, Pyrrhus triumphed once more over the Romans at Ausculum in 279, but did not follow-up his advantage and instead departed for Sicily.[21] Although Roman armies subsequently mustered to attack his former allies in southern Italy, their fighting spirit evidently collapsed upon his return. When the consul M'. Curius Dentatus tried to hold the levy, the first citizen to be called up in the *dilectus* refused to respond. As a reprimand, the consul declared his property forfeit, but far from "toeing the line," the man instead

---

16 See Linke (2018) for a more nuanced view on Roman aggression.

17 Cornell (1995, 242–71).

18 *Seditio* of 342: Livy 7.38.5–7.42; Dion. Hal. *Ant. Rom.* 15.3; Frontin. *Str.* 1.9.1; Zon. 7.25. Postumius Megellus: Gabrielli (2003) with references. Fundanius Fundulus: Linke (2016, 167–68) with references. Second Macedonian War: Burton (2017, 27–29). Spain in 153: Polyb. 35.4.1–6; cf. Livy, *Per.* 48. Even the rather lop-sided Third Macedonian War witnessed difficulties due to a lack of enthusiasm: Livy 43.14.2–15.1; cf. Kromayer and Veith (1963, 332–26, 414–17). See also Brice in this volume on military indiscipline in the Late Republic.

19 Rich (2012, 110) on the conviction of commanders who had jeopardized their forces in exceptionally culpable ways.

20 Battle of Heraclea: Livy, *Per.* 13; Dion. Hal. *Ant. Rom.* 19.9–12; Frontin. *Str.* 2.4.9; 4.1.24; Plut. *Pyrrh.* 16–18; Flor. 1.13.7–8; Just. *Epit.* 18.1.4–9; Eutrop. 2.11; Zon. 8.3–4. Negotiations: Plut. *Pyrrh.* 18.1–19.5; Just. *Epit.* 18.2.6–10; Cic. *Brut.* 61; Sen. *Ep.* 19.5.13; Tac. *Dial.* 18.

21 Dion. Hal. *Ant. Rom.* 20.1–4; Plut. *Pyrrh.* 21.5–22; Frontin. *Str.* 2.3.21; Polyb. 18.28.10; Cornell (1995, 363–65).

sought to challenge this verdict and Curius responded by selling him into slavery. Despite this rather harsh reported action (which may or may not be true), the reluctance to take up arms seems to have been real and arguably widespread, since it is further reported that poor citizens of the *capite censi* were armed by the state for the first time – presumably due to the lack of *assidui* presenting themselves for the *dilectus*.[22] Since the senate had already expressed doubts about Rome's ability to deal with Pyrrhus, an opinion emphasized promptly by the defeat at Ausculum, it is hardly surprising that the new recruits for the armies of 275 were reluctant to put their lives on the line in what must have appeared to be another forlorn hope. Given the clear and present risk, the benefits – both avoiding the penalties *for not* enrolling and the partaking in the rewards *for* enrolling – would need to be significant.

It is no coincidence, then, that these issues with recruitment were overcome by Curius Dentatus, a *homo novus* whose continued popular support and political success were based on his military victories over the Sabines and Samnites in his first consulship in 290 and the subsequent sharing out of large tracts of the conquered territory to his soldiers.[23] His earlier exploits may have served to emphasize the power of Roman arms, and the potential riches to be gained, while his arming of *proletarii* demonstrated that he was willing to resort to extraordinary measures to bring the legions up to strength.[24] The significantly increased costs caused by manning the legions with *proletarii* would have been shouldered by the whole citizen body. This, in turn, may have encouraged enlistment, as some would have volunteered for military service in order to avoid paying the *tributum*.[25] Furthermore, Dentatus' arming, equipping, and paying of poor recruits threatened to upset the political status quo, as it would have promoted some poorer citizens into the middle classes of the *comitia centuriata*, who then subsequently shared the same privileged position with those

---

22 Val. Max. 6.3.3–4, on the unwilling recruits. The episode is also mentioned in Livy, *Per.* 14, which suggests that Livy described it in considerable length. See also Gell. *NA* 16.4.5. See Jehne (2002, 69–72) and Beck (2005, 197–98) for discussion of the measures initiated by Dentatus. The difficulties in mustering sufficient forces against Pyrrhus probably resulted from the high casualties in past battles. Livy (*Per.* 13–14) reports a drop of 16,000 citizens from 279 to 271.

23 Livy, *Per.* 11; Columella, *Rust. praef.* 14; Frontin. *Str.* 1.8.4; Flor. 1.10.2–3; Cass. Dio. fr. 37.1; Eutrop. 2.9.3. Curius' reputation was further enhanced by the fact that he reportedly refused to accept more land than his soldiers: Frontin. *Str.* 4.3.12; Val. Max. 4.3.5; Columella, *Rust.* 1.3.10; Plin. *HN* 18.18. Cf. Forni (1953); Berrendonner (2001, 100–5); Beck (2005, 188–202).

24 Ennius depicted Curius Dentatus as a people's hero *quem nemo ferro potuit superare nec auro*, Enn. *Ann.* 456 (Skutsch); see also Cic. *Cato* 55. Fabius Pictor *FRHist* 1 F24 (= Strab. 5.3.1) comments on the wealth of the Sabines when first conquered by Rome: "Fabius the historian says that the Romans first perceived wealth at the time when they became masters of this people" (φησὶ δ᾽ ὁ συγγραφεὺς Φάβιος Ῥωμαίους αἰσθέσθαι τοῦ πλούτου τότε πρῶτον, ὅτε τοῦ ἔθνους τούτου κατέστησαν κύριοι).

25 Rosenstein (2016a, 92–96); Coudry (2009, 37–44); Enn. *Ann.* 170 Loeb (= Gell. *NA* 16.10.1): *proletarius publicitus scutisque feroque ornatur ferro* ("the proletariat at public cost with shields and savage sword was armed"). Cf. Cassius Hemina *FRHist* 6 F24; Oros. 4.1.2–4; Cic. *Rep.* 2.40. See also Tan in this volume.

citizens who had refused to respond to the *dilectus*. Even if the confrontation was less severe, the arming of *proletarii* had a distinct political dimension and should not be seen in purely military terms.[26]

A poor man's war

The disruptive effect of mobilizing *proletarii* becomes clear when taking the Roman practice of marshalling its legions into account. The composition of the legion was extraordinary, since it was designed to make the most of the differing social and economic status and capability of its soldiers. A typical (if idealized) legion of the Middle Republic featured 1,200 *hastati*, 1,200 *principes*, and 600 *triarii*, supported by 1,200 light infantry and 300 equites.[27] The panoplies of these units differed markedly, with the *equites* providing the most expensive equipment.[28] On the opposite end of the scale were the light infantry, who, according to Polybius, were recruited from the youngest and poorest citizens. The three lines of *hastati*, *principes*, and *triarii* were drawn from successively wealthier classes of the populace, with correspondingly heavier and better equipment.[29] How far Polybius' description reflects the situation at the beginning of the third century is a matter of debate. Still, the reported categorization by equipment and age suggests established guidelines for organizing units in the context of the *dilectus*.

26 Dentatus had already demonstrated a remarkable streak of independence by settling his soldiers on captured Sabine land in 290. See Beck (2005, 192–94) for discussion and sources as well as *MRR* 1.183–84. His employment of *proletarii* (Val. Max. 2.3.1; Gell. *NA* 16.10.10–13; Livy 1.43.8; Dion. Hal. *Ant. Rom.* 4.18, 6.59) would have had far-reaching political consequences by elevating them into the ranks of the *assidui*. See also Rathbone (1993, 144–47).
27 Although the legion was prone to transformative change, the general ratio of light and heavy infantry as well as their deployment probably remained roughly the same between 311 and the time of Polybius. Kromayer and Veith (1963, 261–67). The exact definition of the light infantry is disputed. They were also called *rorarii*: Livy 8.8.8; Festus, *Gloss. Lat.* 13L, 323L; Varro, *Ling.* 7.58. Lucil. 7.323 (Loeb) appears to use the terms *rorarius* and *veles* synonymously. See also: Daly (2002, 70–73); Rawlings (2007, 56); Dobson (2008, 48–49); Armstrong (2016c, 266–67).
28 Although the *equites* were amply compensated in economic and political terms: Stemmler (1997, 176–224); McCall (2002, 26–52).
29 According to Varro (*Ling.* 5.116), the more expensive ring-mail was adopted from the Celts. Polybius (6.23.15) specifies that *loricae* were worn by those men valued at more than 10,000 drachmas among the *hastati*, however, οἱ πολλοί wore the pectorale, 6.23.14, see also Livy 1.43; Dion. Hal. *Ant. Rom.* 4.16–18. Cf. Livy 28.45.16–17 on the equipment provided to Scipio in 205: "Arretium promised 3,000 shields, an equal number of helmets; and that they would furnish a total of 50,000 javelins, short spears and lances, with an equal proportion of each type" (*Arretini tria milia scutorum, galeas totidem, pila gaesa hastas longas, milium quinquaginta summam pari cuiusque generis numero expleturos*). Note the lack of armor. Polyb. 6.21.8 also implies that the units were distinguishable by equipment; the major difference between *hastati* and *principes* in this regard can only have been the *lorica*. Cf. Gell. *NA* 6.13. Rathbone (1993, 146–47) suggests that the poorest *assidui* at first made up the light infantry, with the fifth census class created in the early third century in order to exempt affluent *assidui* from service with the light infantry.

Roman battle order integrated these various groups, with their different combat capabilities, into a combined fighting machine. The engagement began with the light skirmishers, followed by the *hastati*, who were usually expected to soften up the enemy before being relieved by the *principes*.[30] Surprisingly, troops whom Polybius identifies as the most experienced and well-equipped, the *triarii*, were rarely expected to engage at all.[31] It is usually thought that the different battle lines relied on being relieved by the next line. Thus, we might expect to observe military structures aimed at generating greater cohesion among the troops.[32] However, the Romans – at least by the second century – seem to have consciously avoided the creation of locally levied and marshalled units by mixing soldiers from different regions.[33] If this mixing was not achieved by the Polybian *dilectus*, it would have been achieved, as already mentioned, by the reorganization and splitting apart of the tribes after 241. While this sacrificed regional cohesion within the units, the pronounced demarcation of troop types in camp, where the *hastati* were quite isolated due to their placement opposite the allied cavalry, while the "heavies," the *triarii* and *principes*, were in close contact, prevented close interaction between them.[34]

In my opinion, these conditions engineered a strong sense of horizontal cohesion within each individual unit and their respective battle line but also a fragmented overall horizontal cohesion across the legion as a whole, with marked divisions between the lines. Additionally, the limited interaction between the various troop types would also, to some extent, have masked the different tasks and risks which each group had to face. For example, nearly 57% of a legion's infantry was dedicated to the missile fight carried out by the *antesignani* – skirmishers and *hastati* – who were expected to engage the

---

30 Koon (2011, 87–93); Taylor (2014, 318–20).

31 Livy 8.8.11: "If the principes, too, were unsuccessful in their fight, they fell back slowly from the battle-line on the triarii. From this arose the adage, 'to have come to the triarii,' when things are going badly" (*Si apud principes quoque haud satis prospere esset pugnatum a prima acie ad triarios se sensim referebant; inde rem ad triarios redisse, cum laboratur, proverbio increbruit*). More scathing was Plautus' remark reported by Varro, *Ling.* 5.89 (modified Loeb trans.): "Come now, all of you sit by as the *triarii* are wont" (*Agite nunc, subsidite omnes quasi solent triarii*). See Campbell (2013, 427–30). Note that the number of *triarii* was never raised beyond 600, even if overall legion strength was increased: Polyb. 6.21.10. The role of the *triarii* contrasts starkly with the practice of Greek armies to make the most of their best fighters: see van Wees (2004, 177–97) and Lee (2013, 147–57). This does not mean that the *triarii* were superfluous as they provided a powerful and fresh reserve which could serve as an efficient rear-guard in case of defeat (Dion. Hal. *Ant. Rom.* 8.86.4) or provide the final punch in an indecisive battle. Overall, it seems that these troops did not regularly join battle like the other lines and might have been more akin in their deployment to elite units like the Napoleonic Guard.

32 See McCall in this volume on the killing zone and unit cohesion.

33 Polyb. 6.20.

34 Polyb. 6.33.10. Dobson (2008, 68–100), esp. 84; Jehne (2006, 252–57). Rosenstein (2012a, 93–99) argues for some interaction among the soldiers in contrast to Pfeilschifter (2007). Hesberg (2008, 71–73) on the hierarchization of space through streets.

enemy initially and absorb the first enemy shock, either in form of a charge favored by the Celts or of a sustained missile-fight preferred by the peoples of central Italy, while the *principes*, a scant 28% of the total infantry strength, supposedly only engaged at a later point, if indeed the battle progressed to that next stage.[35] Consequently, more than half of the legion – the less well equipped half at that – was employed in meeting the first advance of the enemy, while the "best" troops – or at least those with the best equipment – remained in reserve. The *antesignani* were drawn from the segment of the *assidui* below the *prima classis* and were required to possess only minor protection apart from shields and helmets, as full body armor was obligatory for the *prima classis* only.[36] As a result, while one could debate the relative risks of skirmishing to sustained hand-to-hand combat, given their numbers, position in the frontline, and their medium equipment, casualties would surely have been common and expected in these units. However, not all casualties are equal. For the state, at least, losses amongst the *antesignani* might have been more acceptable since these units were composed of largely young, unmarried sons from medium- or lower-income families, for whom replacements could be easily obtained.[37] Nonetheless, these units needed to display high morale in order to put up a spirited and disciplined defense and thus maximize the effect and impact of the second line's advance.[38]

35 Livy 8.8.5–8; Enn. *Ann.* 266 suggests that the *hastati* were primarily identified with missile fighting and thus the opening of the battle; Forsythe (2007, 32). The combination of light infantry and *hastati* indicates that the *prima acies* was basically a skirmisher line with enough backbone to slow down a determined enemy advance. The importance of missile fighting has been stressed repeatedly in recent years: Sabin (2000); Zhmodikov (2000); Quesada-Sanz (2006); Anders (2015); Armstrong (2016c, 262–68).

36 Rawlings (2007, 55–58). Note Livy's speech (8.11.7) of the Latin praetor Numisius, claiming that the Roman victory at the Veseris in 340 had been extremely bloody: "their whole army had been cut to pieces, their first and second lines had been massacred, and the slaughter had extended from the troops before the standards to those behind them; finally the veterans had restored the day" (*trucidatum exercitum omnem, caesos hastatos principesque, stragem et ante signa et post signa factam; triarios postremo rem restituisse*). Livy's fictitious speech reflects the general idea of a bloody battle, characterized by heavy casualties even among the *principes*. Casualties among the *hastati* seem to have been high in general: Livy (30.34.10–12) and Polybius (15.13) report heavy losses among them at Zama. Cf. Livy 42.7.9–10. Rosenstein (2004, 63–81) on the ability of Roman families to absorb such casualties of mostly unmarried sons.

37 Gabba (1976, 5–6) has argued, based on Livy 26.4.4–7.9, that the property qualifications for light infantry were reduced from 11,000 (Livy 1.43.7) to 4,000 asses (Polyb. 6.19.2) between 214 and 212, which would also suggest that "poor" light infantry troops were recruited to fill gaps. Although Rich (1983, 294–95, 305–12) refutes this, he nevertheless allows for exceptional recruitment in crisis situations; Keppie (1984a, 33) also assumes that this measure facilitated recruitment.

38 Note the Roman emphasis on staying in formation and holding ground rather than daredevil aggressiveness: Polyb. 6.24.8–9 on *centuriones* and their expected ability to keep their troops together under pressure. Cato *FRHist* 5 F82 (= Gell. *NA* 11.1.6) is more specific on individual troops, who were fined for breaking ranks. See also: Val. Max. 2.7.6; Livy 8.7, 34.15.4.

Balanced prospects – attempting a risk/gain calculation

Potentially high casualty rates and the necessity of maintaining morale seem to be mutually exclusive on first sight, raising the question: why would a large portion of the army have accepted a role which seems to have carried a proportionally greater personal risk? Although not exactly resembling "cannon fodder," the front lines bore the brunt of the fighting and presumably incurred heavier casualties than others in the same legion. Armstrong has recently offered a solution to this problem by pointing out that Roman armies in archaic times likely relied heavily on task-based social cohesion, that is, they were unified by the pursuit and distribution of material rewards, often acquired by raiding and plundering. References to the earlier triumphs, which mention donatives to each soldier irrespective of the unit type, corroborate this theory.[39]

However, it is necessary to conduct a "cost-benefit analysis" and to consider the individual motivation, combat burden, and possible rewards for the different groups, since a member of the *triarii* or *principes* would presumably have experienced a considerably lower net gain in relation to his investment in equipment than a skirmisher.[40] A first point to note is that any variability in risk and reward for Roman soldiers would have likely been masked by the segmented structure of the legion. On the one hand, the separation of unit types might have hidden potentially higher casualties among the light troops from the rest of the army. On the other hand, any booty or spoils acquired by members of the light infantry would have had a greater relative impact, as they required a smaller return to balance out their personal investment in terms of equipment and to increase their social standing and economic prosperity. Pointedly, the *"velites"* were in a position of small investment (even if, relatively speaking, it was quite a bit for them) coupled with high personal risk, in return for potentially high relative reward. The same holds true, albeit to a lesser extent, for the *hastati*, thus enabling a great number of less wealthy individuals and "middle class" *filii familias* to participate in the acquisition of plunder – and ultimately social advancement in the *comitia centuriata* as well as in the military sphere, where they would have risen into the more "cushioned" ranks of the *principes*.[41] Unlike the members of the

---

39 Armstrong (2016b, 117–19).
40 Churchill (1999, 91–93) and Coudry (2009, 22–28), for the division of booty on the spot between soldiers and the general, as representative of the state. Furthermore, the soldiers could expect another cash-out in the context of a triumph. See also Rosenstein (2011, 144–48).
41 Rathbone (1993, 147); Kay (2014, 31). On the impact of rewards and booty on morale, see Lee (1996, 205–6). Gellius (*NA* 16.4.2) explicitly states that soldiers could keep plunder amounting to one sesterce a day; the rest had to be handed over to the consul. Soldiers could have doubled their daily pay and still benefited from the later distribution of the accumulated booty. See also: Kay (2014, 25–35); Erdkamp (2011a, 61–63). Rosenstein (2016a, 84–97) also draws attention to the fact that serving soldiers were exempt from paying the

principes and *triarii*, risking a larger share of their economic wealth in return for a proportionally modest profit, these groups therefore had much to gain from military service, even though the hazard was comparatively greater.[42] Conversely, the comparatively lower interest in (or access to) plunder by the *triarii* and *principes* was balanced by a more privileged and sheltered role in the army, since the *triarii* not only enjoyed a protected position as the army's reserve but were also freed from camp duties. Apart from purely economic considerations, all the soldiers will have profited from the social prestige resulting from successful military operations, which was further strengthened by the honoring of individuals in front of the whole army as well as the granting of awards and decorations to individual soldiers.[43] While these decorations will have been equally sought-after, it is noteworthy that Polybius mentions them specifically with regard to younger and lightly armed troops, which suggests that they presented a unique chance for self-promotion for soldiers from a less prestigious social background.[44]

In summary, willingness to participate in military operations was dependent on very different "cost-benefit" or "risk-gain" calculations for the soldiers. The annual mobilization of the legions not only succeeded in distributing economic resources among its participants but also provided an institutionalized vehicle for regulated social mobility to less wealthy citizens, whose successful advancement legitimized and reinforced the existing social order. Referring to Armstrong's argument, the motivating force of the Roman legions could be styled, somewhat cumbersomely, as "segmented task-based horizontal cohesion."[45]

---

tributum. Colonization was arguably another boon but occurred only irregularly and has therefore been omitted.

42 These troops were at the forefront of logistics, that is, plundering and foraging, in line with Cato's famous comment: *bellum se ipsum alet* (Livy 34.9.12). According to Polybius, the *velites* were free from camp service except for guarding the gates and palisade, from which it can be deduced that they were also responsible for supplying the army. Nonius (552) quotes the satirist C. Lucilius (323 Loeb) on light infantry as carrying five javelins and a golden belt, which might refer to their individual enrichment during campaigns. Erdkamp (1998, 122–40) on foraging. All soldiers of course "invested" their lives, but when beating a hasty retreat discarded equipment would constitute a greater economic loss to the heavy infantry. On material rewards, see Nicolet (1980, 115–22).

43 See Ward (2016, 307–20) for decorations awarded for breaking formation in order to retrieve a weapon, to save a comrade, and to strike an enemy.

44 Polyb. 6.39 on decorations in general, while 6.39.1–2 specifies their great value for motivating young soldiers in particular. See also Polyb. 6.22.3 on *velites* wearing wolf-skins over their helmets to ensure that their deeds were recognized. Ward (2016, 305–6) points out that military laurels could be most easily won and witnessed among the loosely fighting *antesignani.* Decorations apparently included material awards as well, for example, Livy 7.37.1–3; Plin. *HN* 16.11–14; Festus, *Gloss. Lat.* 208L. If the deed of Decius in 343 was modelled on a similar story of the First Punic War reported by Cato (*FRHist* 5 F76 = Gell. *NA* 3.7.1–19) this would only serve to reinforce the argument that decorations were coming hand in hand with material rewards.

45 Armstrong (2016b), and also Armstrong in this volume. See also Brice in this volume.

Connecting the lines – the legion's "officer corps"

The manipular legion thus provided a highly integrative interface for the inclusion of diverse social groups and military units, but the clear divisions between the lines and the types of troops which filled them made it vital to create common points of reference. While the individual lines could rely on a certain amount of horizontal cohesion – based on similar activities, risks, age, status, etc. – the overall legion was arguably at risk of fractures appearing along the lines of these larger divisions. In this context, the command structure of the legion was responsible for linking the different "battle lines" and troop types through vertical cohesion, provided by a network of officers and relationships. Within the *manipuli*, the *centuriones* connected the individual *manipuli* to the wider command structure, in which they represented and integrated their men.[46] In contrast, officers responsible for and connected to the whole legion are surprisingly rare, but it is telling that the most important among these, the *tribuni militum*, were not elected by the *comitia centuriata* but by the more egalitarian *comitia tributa* and (in theory) were thus representatives of the whole citizen body within the military structure.

The role of these lower officers and their military service tends to be underestimated due to their low visibility, but it put 24 young *nobiles* in a vital position within the army each year, offering considerable room for communication and interaction.[47] They were the first to face each and every soldier in the *dilectus* and they administered the *sacramentum*.[48] Polybius also describes them as being in daily contact with the rank and file, since they were in charge of setting up the camp with their personal tents forming the point of orientation for the layout of the legions – indeed, the area in front of the tents of the *tribuni* constituted the public place for soldiers to meet if encamped for an extended period of time (Polyb. 6.33.4).[49] Camp routine also imposed and allowed further interactions at different levels: as noted above, the tribunes took an oath from every member of the army before setting up camp to uphold discipline, and they also came into close contact with the maniples of the *hastati*

46　Polyb. 6.34.5–6. The *centurio primi pili* was also a permanent member of the army council: Polyb. 6.24.2. In critical situations, all of the *centuriones* were addressed: Caes. *BGall.* 1.40–41 for a vivid, if late, illustration.

47　Livy (9.30.3) mentions 16 tribunes for the year 311, while the number reached 24 for the regular four legions by 207 (Livy 27.36.14). See Hölkeskamp (1993, 32); Suolahti (1955, 38–51). Rosenstein (2007) provides a succinct overview on the importance of service in the army for the political career. Note that attrition among the 24 *tribuni militum* and the *praefecti sociorum* still remained a factor, even in rather inconsequential campaigns, for example, Livy 33.36.4–5, 35.5.14 (Cisalpine Gaul) and Livy 39.31.15 (Spain). Cf. Suolahti (1955, 57–145) on the social origins of the *tribuni militum*. The nobility supplied the majority of tribunes until the end of the second century.

48　Polyb. 6.21.2; Livy 3.20.3–6; Dion. Hal. *Ant. Rom.* 10.18, 11.43.

49　Hesberg (2008, 73–76) on the layout of excavated Roman camps, which while not entirely corresponding to the neat description of Polybius did feature the strict social hierarchy and divisions of space described by him.

and *principes* through the distribution and carrying out of camp duties – the *triarii* being exempted from camp duties except for providing a guard for the *equites* (Polyb. 6.33.10).[50] Two maniples were ordered to take care of the area in front of the tents while the remaining 18 maniples were assigned by lot to each of the six tribunes, who thus rotated through close contact with three maniples every day.[51] They thus constituted a powerful vertical link in the legion, bringing at least the *hastati* and *principes* together and into contact with the command structure in the camp environment, and were thus able to help bridge the gap between the important first and second battle line. This effect would have been even higher if the tribunes also attached themselves proportionally to the ten maniples of each battle line, since this would have been a strong symbolic message – that even though some soldiers had to face greater dangers than others, their officers shared these without discrimination.[52]

The unifying and collective role of the tribunes is in line with the fact that the tribunes were also tasked with safeguarding the citizen-soldiers from potential abuse, reflected in their election by the *comitia tributa* and their accountability to the Roman people.[53] Unlike the *centuriones*, they derived their authority not from the soldiers or the commanding general, but from the whole citizen body. They thus guaranteed the fair treatment of the soldiers under military discipline and presided over any punishment that was meted out. Polybius explicitly states that the *fustuarium* was only administered if a court-martial of the legion's six tribunes condemned the culprit. Soldiers were not subjected to individual arbitrariness, but judged by the people's representatives.[54] Pointedly, the *tribuni militum* represented the Republican order within the military, as it was their task to maintain discipline but also to maintain the rights and the well-being of the troops, and, by extension, their willingness to serve. Their obligations consisted of care and command, which reinforced the legitimation of the political elite to command, on the condition of exemplary conduct and adherence to Republican ideals of leadership.[55]

---

50 Jehne (2006, 254–58) is skeptical regarding the degree of integration within the Roman army camp, although his main criticism is reserved for interactions between allied and Roman troops. In contrast, Rosenstein (2012a, 93–103) allows for a greater integrative function of the army.

51 Polyb. 6.33. See *Dig.* 49.16.12.2 on the tribunes' responsibilities, which included taking care of the food supply and the sick and responding to complaints by the soldiers; see also: Livy 8.36.6; Veg. *Mil.* 2.12, 3.2.6. The tribunes thus offered vertical cohesion, *contra* Suohlati (1955, 45–47) and Armstrong (2016b, 110–14).

52 Rosenstein (2007, 136).

53 Dion. Hal. *Ant. Rom.* 16.4; Val. Max. 6.1.11; Livy 40.41.8. The tribunes also rotated command among themselves with two tribunes taking precedence over the others on command decisions, cf. Livy 40.41.7: "Marcus Fulvius Nobilior, who was a military tribune of the second legion, disbanded the legion during his months of command" (*M. Fulvius Nobilior secundae legionis tribunus militum is erat, mensibus suis dimisit legionem*).

54 Polyb. 6.37.1–6; see also Cic. *Leg.* 3.6.

55 Suolahti (1955, 170–71); Rosenstein (2007, 136–38). The most prominent examples are T. Manlius Torquatus ([Aur. Vict.] *De vir. ill.* 28; Zon. 7.24; Livy 7.10.1–2), Q. Caedicius (Cato

The military tribunes also provide another point of investigation for explaining high mobilization rates. To put it simply, the crucial early period of a political career cannot be overestimated, since the obligatory ten years of military service exposed aspiring young *nobiles* to a wide cross-section of the Roman electorate and presented a chance for direct and unfiltered interaction between citizens and aristocrats.[56] Perhaps conveniently as well, it would also have thinned out the ranks of potential candidates through battlefield casualties and the other risks (disease, injury, etc.) associated with military service.[57] The military tribunate usually seems to have been held during this time of service, which allowed junior *nobiles* to display their worth to their fellow citizens in the ranks. For example, T. Manlius Torquatus was *tribunus militum* in 361, when he won renown for his response to the challenge of a massive Celtic warrior, whom he then slew in single combat.[58] Whether historic or not, the episode clearly illustrates the Romans' ideal image of how a *tribunus militum* was supposed to act. Furthermore, the elites' desire to participate in the fighting, and presumably to be seen doing so, was not limited to the military tribunes and is attested throughout the history of Roman expansion. Plutarch offers two striking examples for the mid-second century: his description of the battle of Pydna portrays M. Porcius Cato Licinianus and P. Cornelius Scipio Aemilianus both in the thick of the melee, with young Cato desperately trying to force his way across the battlefield in search of his lost sword and Scipio Aemilianus returning blood soaked from the battlefield in the evening.[59] Such bravery in battle was expected from *nobiles*, as Oakley's list of 20 formal duels between 400 and the Social War alone – without figuring in ordinary fighting and attrition – attests.[60] Pliny's account of the *laudatio funebris* for L. Caecilius Metellus (cos. 251 and 247) corroborates the need for martial displays, as the list of aristocratic qualities is headed by being an eminent warrior (*primarius bellator*), followed by oratory and military command skills.[61]

FRHist 5 F76; Livy, *Per.* 17; Livy 22.60.11; Plin. *HN* 22.11; Flor. 1.18.13–14; [Aur. Vict.] *De vir. ill.* 39; Oros. 4.8.2; Zon. 8.12), and Scipio Aemilianus (Vell. 1.12.4; 149: Plin. *HN.* 22.13). See Rawson (1971a, 14–15) on the likelihood that some kind of military manual existed.

56   On the ten-year service obligation, see Polyb. 6.19.3. Cato the Elder, born in 234 (Cic. *Brut.* 61, 80; Cic. *Sen.* 10.14.32), started to serve at age 17 (Plut. *Cat. Mai.* 1.6) and became quaestor at age 30 (Livy 29.25.10).

57   Rosenstein (2006, 365–68); Rosenstein (2007, 136–39).

58   Claudius Quadrigarius *FRHist* 24 F6 (= Gell. *NA* 9.13.4–19); Cic. *Fin.* 1.23, Livy 7.9.6–10.14; Val. Max. 3.2.6. See also Oakley (1985, 393–94); Clark (2016, 284). *Contra* Goldsworthy (1996, 149–63) who argues that elite members tried to avoid single combat in the late Republic, but this seems hardly to have been the case in earlier times, see Pliny below.

59   Plut. *Aem.* 21–22, *Cat. Mai.* 20.7–8; Livy 44.44.1–3; Polyb. 29.18. Another example is the young Ti. Sempronius Gracchus who was the first to scale the walls of Carthage, Fannius *FRHist* 12 F4 (= Plut. *Ti. Gracch.* 4.5–6).

60   Oakley (1985, 393–96); see McCall (2002, 69–72) on cavalry combat.

61   Plin. *HN* 7.140: "for he had made it his aim to be a first-class warrior, a supreme orator and a very brave commander" (*voluisse enim primarium bellatorem esse, optimum oratorem,*

Although the mentioned spectacular cases probably occurred only rarely, it is nevertheless clear that the aspiring scions of the *gentes* did not enjoy the option of leading from the rear.[62] The presence of young aristocrats among the rank and file – as comrades and officers – may have demonstrated to the ordinary soldiers that the dangers were equally shared by all citizens, reinforcing the legitimation of the elite and of army service in general. It is noteworthy in this regard that the most vividly memorized and heroic examples refer to the infantry, suggesting that the military service of young aristocrats was not exclusively restricted to the cavalry.[63] From a political perspective, service with either the cavalry or the infantry would have provided the chance to gain valuable experience and to make a favorable impression on fellow citizens. Considering the internal diversity of a legion, these political benefits for *nobiles* serving with the army become even more pronounced. By the third century, campaigns lasted for at least six months and thus presented a rare opportunity to engage with fellow citizens on an unmatched scale.[64] In this regard, each legion provided a captive audience to the conduct of potential future candidates and could serve as a huge multiplier due to their regional diversity. This "news" aspect should not be underestimated, since the next stations of the *cursus honorum* were decided in the *comitia tributa*.[65] It was all the more important due to the massive expansion of the *ager Romanus* since the Samnite Wars, which meant that the city of Rome was no longer regularly visited by a majority of the citizens.

---

fortissimum imperatorem). While *orator* and *imperator* refer to the political and military aspects of the *cursus honorum*, *bellator* claims precedence, making it clear that the performance in the line of battle was a necessary precondition, Polyb. 6.54. See also Lendon (2007, 509–12) and Ward (2016, 302–4). The price of this display of martial prowess could be high as Livy's casualty list for the battle of Cannae confirms, which numbers more than half of the 48 military tribunes and an additional 80 senators or men eligible for elevation to the Senate among the fallen (Livy 22.49.16–17). See also Barber in this volume.

62 Lendon (2007, 512–15). Also note the prevalence of heroic combat in Roman myths: for example, M. Valerius Volosus at Lake Regillus (Livy 2.20.1–3); Dion. Hal. *Ant. Rom.* 6.12; *Inscr. Ital.* 13.1.64); Horatius Cocles (Polyb. 6.55; Livy 2.10.2–11); the *spolia opima* of Cornelius Cossus (Livy 4.17–20) and Marcellus (*MRR* 1.233 for references).

63 Polybius does not equate the cavalry with the senatorial elite (6.19.1–3). See also Stemmler (1997, 100). Compare McCall (2002, 1–12), who emphasizes the importance of cavalry service for elite status. A middle course might be indicated by the famous speech of Servilius Pullex showing off the wounds sustained in single combat, while his service in the cavalry is separately mentioned (Livy 45.39.16–19), which suggests that elite status did not preclude service in the infantry. See also Plut. *Cat. Mai.* 20.4 on the early military training of a young *nobilis*, which aimed at preparing for both infantry and cavalry combat.

64 Rosenstein (2007, 136–38) and Clark (2016, 276–77). Appius Claudius Caecus served three campaigns as *tribunus militum*, *Inscr. Ital.* 13.3, No. 12. Beck (2005, 168–69) considers these to have been instrumental in paving the ground for his remarkable career.

65 Cato's election in 191 also suggests that political benefits could be reaped from holding this office, since he had already been consul at that point, see Cic. *Sen.* 32. Similarly, M. Valerius Corvus served as *tribunus militum* in 297, even though he had already been consul in 312, see Livy 10.14.10.

In this context, the legion offered a setting for wooing potential voters and for accessing new networks and areas of support but also the chance for the citizens to engage with potential patrons, whom they could not reach by the traditional means, for example, the *salutatio*, due to increasing distances.[66] This probably served to ensure a respectful interaction with the soldiers who might have become important in future elections.

In sum, this system – with the tribunes acting as the vital links between the legion's varied and discrete unit types and battle lines – ensured that discipline was maintained while respecting the citizen-soldiers' rights and opinions at the same time. Correspondingly, discipline and morale seem to have usually been high in the legion because soldiers were inherently moti-vated to cooperate in its enforcement to guarantee the success and safety of the army from which they themselves benefitted the most.

Conclusions

Without a doubt, Rome's mid-Republican legions were effective military institutions, but their full potential was dependent on the willingness of Ro-man citizens to fill its ranks. The tactic of exposing the weakest troops to "tire out" the best enemy troops before the commitment of the Roman heavy infantry generally worked quite well, but only on the condition that the sol-diers making up the different battle lines were willing to accept its inherent risks to achieve victory. In this regard, this chapter has argued that the setup of the legion promised an attractive risk-gain equation custom-built for the heterogeneous troops: the less affluent soldiers had to endure a greater risk in return for potentially greater economic gains (at least relative to their investment and initial worth), while the wealthier heavy-infantrymen faced a less intense battlefield role but also had relatively less to gain from plunder and pay. While the material benefits held different degrees of importance for soldiers from differing social backgrounds, all of them were motivated by the chance to gain social prestige by exemplary performance of duty, which promised enhanced status and advancement within the military and civic order – if they survived. Furthermore, the presence of *nobiles* in the army, especially in the form of the elected general and the 24 *tribuni militum*, em-phasized that the burdens and benefits of army service were equally shared.

This being noted, it is worth emphasizing that the legion was no great social equalizer, but rather served to reinforce the dividing lines of various status groups. Nevertheless, it harnessed the social and economic distinc-tions which existed between the individual groups and used these to create a fragmented, but horizontally cohesive, series of formations which were motivated by the promise of social advancement and recognition through

---

66 Plutarch (Plut. *Ti. Gracch.* 7), reports that the young quaestor Ti. Sempronius Gracchus saved the soldiers from slaughter at Numantia and was later able to rely on their support in the political arena. See also Beck (2015).

material rewards. In political, as well as military terms this was unproblematic, since the potential social climbers presumably had little incentive to question the established structures, thereby maintaining the status quo of the social order of the Republic.

This brings us back to the case of M'. Curius Dentatus and the reluctant recruits. Their refusal to serve and the corresponding arming of *proletarii* put the whole staggered system of losses, rewards, and advancement into question, since the latter's recruitment created a cohesive body of troops with similar interests, as opposed to the segmented horizontal cohesion resulting of the typical heterogeneous legions. Over time, such a practice held the risk of creating a military special interest group and was only considered under the direst of circumstances until the end of the second century.[67]

---

67 See Brice in this volume on the dangers posed by too much cohesion among troops.

7 "Take the sword away from that girl!" Combat, gender, and vengeance in the middle Republic

John Serrati

Introduction

As warfare evolved during the early and middle Republic, so too did religion, the concept of the divine, and role of the gods in Roman life.[1] As warfare came to encompass and influence every aspect of life at Rome, not only did war deities seem to rise in importance but also many divinities not traditionally associated with conflict – such as Neptune, Quirinus, and Janus – developed martial aspects as part of their characters. Moreover, deities for whom war had traditionally formed only part of their role saw non-martial aspects of their personas diminished to the point where they only came to represent war. Just as Roman armies began to spread throughout and dominate Italy, warfare seemed to spread throughout and dominate Roman society.

In this regard, Mars is particularly instructive. In earlier eras of Roman history, Mars seems to have had clear agricultural functions to go along with his role as the most important Roman war god; he is one of the most ancient gods on the Italian peninsula and, originally, a multifaceted deity. In his war guise, as Mars Gradivus, he was a god of battle who symbolized the *militia*, the area outside the *pomerium*, the sacred boundary of the city, where enemies lay and where the legions operated. Specifically, Mars represented warfare as a masculine pursuit – something undertaken by citizen males. Undoubtedly, the majority of ancients, from soldiers and generals to historians, saw warfare as an exclusively masculine pursuit. And this has transferred over to the present, as modern scholars have traditionally (and almost exclusively) examined ancient warfare through the lens of the male fighters. More often than not, women have been seen as passive

---

1 This paper was originally presented at McGill University and at University College Dublin; I am grateful for the excellent comments and suggestions which I received from the audiences. I am equally indebted to Prof. Alison Keith, who generously read over a draft; the peer reviewers; and most of all the editors of this volume. Finally, I am grateful to my former graduate student, Ms Meghan Poplecean, for some of the initial ideas within this paper. Any errors or omissions remain my own. All dates are BC unless otherwise noted. All translations are my own.

DOI: 10.4324/9781351063500-7

This chapter has been made available under a CC-BY-NC-ND license.

actors – victims to whom warfare happened. In mid-Republican Rome, however, as with most martial societies, the feminine played a relevant and necessary role in the annual rhythm of the campaign season. On a practical level, women gave birth to Roman warriors, and thus had a degree of efficacy as wives and mothers. This likely contributed to their greater freedom of movement in comparison to contemporary Mediterranean cultures. To the Romans, *domus* referred to a sphere of influence rather than an actual physical structure: this was most often the space of the home itself, but it could also refer to the city of Rome within the *pomerium*. Women by and large had freedom of movement within the wider sphere of the *domus* and were not consigned and confined to the private space of the physical house. On the other hand, the sphere of the *militia*, which existed outside the *pomerium*, was largely forbidden to females. Thus, while women at Rome were present in the public sphere, this was only accepted if they operated within their assigned gender roles.

That being noted, alongside Mars the Romans had another major divinity associated with conflict: the goddess Bellona. As nearly everything at Rome associated with war tended to be male, that she was a female deity in a highly patriarchal society is significant, and why the Romans had a female deity representing battle will be one of the main questions addressed in this chapter. At first blush, Bellona appears as a primordial Roman war deity. She is associated with strife and chaos, and, as with Ares in a Greek context, appears to represent conflict as an inversion of civic harmony. Yet by the mid-Republic, warfare was arguably no longer an inversion of civic harmony, and instead part of the normal cycle of life at Rome. Thus, like Mars, Bellona evolved, and she emerges in the first century, though in all likelihood much earlier, as goddess of battle. As much as Mars seems to have represented masculinity, duty, and citizenship, Bellona represented the chaos, fear, and blood involved with combat. Her rites involved a cacophony of noise and, in equal measure, she could fortify courage and evoke terror. Furthermore, her temple came to be the *locus* of the fetial rite for declaring war, and throughout the Republic it was also the place where returning generals petitioned the *patres* for a triumph. Thus, Bellona was present for, and presided over, the ritual opening and closing of conflict. In the end, this chapter aims to show that Bellona's femininity was not happenstance; as a woman, she brought specific elements to the intersection of religion and warfare at Rome, and these elements serve to provide us with insights into both the Roman concept of gender, as well as the role of women in warfare at Rome.

Mars, *virtus*, and Roman warfare

Although the worship of Jupiter was of undeniable importance to the Romans, Mars commanded often equal and at times even greater significance. And, as the prevalence and frequency of warfare increased at Rome, so too did Mars' role in the city's religious life. While other deities certainly had warlike associations, Mars was the preeminent god of the martial culture which had developed by the middle Republic. Of the archaic Roman triad of Jupiter, Mars, and Quirinus, only Mars has a month named after and dedicated to him; he is the only god with a full "season" of festivals. In his role as both a god of war and of agriculture, Mars touched on virtually every aspect of Rome life, and was therefore a personification of the annual Roman rhythm of farming, political activity, and war.[2]

In the guise of a war god, Mars was known as Mars Gradivus. He had his own college of priests, the 12 *Salii Palatini*, who spent the month of March engaged in war dances around the Palatine that formed part of the rituals for the opening of the annual campaign season. The priesthood was undoubtedly very ancient, and individual members of this college were referred to as *sodales*, the term used for the bands of fighters who formed the private forces of a warlord in archaic Rome.[3] Both Livy and Vitruvius speak of a temple to Mars outside the *pomerium*, with the former stating that it was founded in 388. A statement by Servius confirms that this was the temple specifically to Mars Gradivus.[4] This identification is germane as Gradivus is the god Mars when he is representative solely of war. The word *gradivus* may refer to marching, an apt etymology for a temple from which armies departed on campaigns. It can also be translated as "rampaging," signifying the power of a Roman army and the effect of warfare on the country. Equally, however, *gradivus* may refer to the act of physically stepping outside the *pomerium*, furthering the idea of the *pomerium* as the dividing line between *domi militiaeque*.[5] The temple stood on the outside the *Porta Capena* on the *Via Appia*, between the first and second milestones from Rome [see Map 1]. On account of the temple, this stretch of the *Via Appia* came to be referred to as the *Clivus Martis*, and the temple was sometimes called "Mars in Clivo." The site served as the muster point for Roman armies who were about to set out on campaign.[6] A temple of this sort dedicated in 388 fits well with Roman history, as in the years and

---

2 This is especially true if one considers Mars and Quirinus to be two aspects of the same divinity, with the former representing the male citizenry assembled for war and the latter the male citizenry assembled for politics. See Beard, North, and Price (1998a, 14–18); Belier (1991, 79–100, 135–38); Cooley (2006, 231); Dumézil (1970, 1.141–280, esp. 154–61, 165–75); Forsythe (2012, 28); Lajoye (2010); Lipka (2009, 3–4, 17–18, 59–61, 169); Momigliano (1983); Scheid (1983); Scheid (1985, 74–94); Scheid (2016, 6–7, 128); Scheid (2011, 34–38).

3 *CIL* 1².2832a (the "Lapis Satricanus"); Livy 1.20; Polyb. 21.13.11–13. See also: Diom. *CGL* 1.476; Cornell (1995, 74–76); Habinek (2005, 20); Lipka (2014, 59–60); de Vaan (2008, 570).

4 Livy 6.5.8; Vitr. *De Arch.* 1.7; Serv. 1.292.

5 See: Dion. Hal. *Ant. Rom.* 2.48.2; Festus, *Gloss. Lat.* 115.6–12L; Livy 22.1.12; Ov. *Fast.* 6.191–2; Serv. 6.860; Beard, North, and Price (1998b, 370); Scheid (2011, 62–66); de Vaan (2008, 268–69); Ziolkowski (1992, 101–4, 238).

6 Livy 7.23.3.

decades immediately after the Gallic sack of the city, the senate appears to have taken greater control of the city's military forces, slowly eliminating the private, clan-based forces which dominated beforehand. Thus, the temple of Mars outside the *Porta Capena* served as the muster point for armies which were genuinely Roman in character, raised via the *dilectus* and commanded by elected magistrates.[7]

The sources from the middle Republic illustrate not only Rome's martial culture but also Roman ideas about gender. Plautus certainly reflected contemporary attitudes, and his plays, written for mass consumption, must have resonated with audiences.[8] As such, allusions to war feature prominently in his work. Moreover, in Plautine Latin, *virtus* primarily meant martial courage. Cato and Ennius, along with the fragments of several others, although perhaps written for more boutique audiences, nonetheless must have reflected a number of contemporary attitudes as well, and their frequent allusions to war and Roman manliness would presumably have been familiar to readers. In the middle Republic, Mars was synonymous with *virtus* as well as the ability to endure the hardships of combat, including death. The evidence for martial courage as the main, though certainly not the only, meaning of *virtus* in the mid-Republic is very strong. Dying in an act of *virtus* meant that a man would live on in the collective memory of his comrades and family.[9] This concept likely had its origins much earlier in Roman history, as the Twelve Tables of the mid-fifth century mention a *corona virtute* ("a crown won by *virtus*") for those who died in war.[10] M. Claudius Marcellus' (cos. 222) vow to construct a temple to Honos and Virtus, as well as his dedicatory inscription to Mars, are two examples that highlight the strong association between *virtus*, Mars, and warfare.[11] A temple to Virtus was also dedicated by Marius in 101, while Scipio Aemilianus had previously dedicated a shrine to Virtus in 133.[12] Additionally, the term *virtus* plays a prominent role in a

---

7 See Serrati (2011, 20–21).
8 For the presence and role of non-elites in Plautine audiences: Beacham (1991, 101–16); Fantham (2005, 222–24); Leigh (2004, 22–39, 51–56, 85–96); Moore (1998, 40–43, 197–99); Richlin (2017, 125–48).
9 See the sepulchral inscription for Publius Cornelius Scipio (son of Africanus; died c.170) from the Tomb of the Scipios (*CIL* 1^2.10); Enn. *Ann.* 382 Skutsch; Naev. *Ex inc. ab., ex com.* 1–3 *ROL* (= Gell. *NA* 8.8.5); Plaut. *Capt.* 690; *Ex inc. inc. trad.* 126 *ROL*; Earl (1960) 240; Harris (1979) 20.
10 Twelve Tables 10.1–8 Courtney (=10.5–7 Loeb); see Courtney (1999, 15–16, 22–26).
11 Temple: Cic. *Nat. D.* 2.61; Livy 27.25.7–9, 29.11.13. The temple was first vowed by Marcellus in 222 at the Battle of Clastidium, but he did not begin construction until 208, when he planned to fill it with the spoils of his Sicilian campaign. The temple was finally dedicated by his son in 205: see *CIL* 6.3735; Cic. *Rep.* 1.21, *Verr.* 2.4.121; Livy 25.40.1–3, 26.32.4; Plut. *Marc.* 28.1, *Mor.* 318D-E; Val. Max. 1.1.8; Orlin (2002, 131–32, 136); Ziolkowski (1992, 58–60, 252–55). Inscription: "To Mars, dedicated by Marcus Claudius, son of Marcus" (*CIL* 1^2.609).
12 Marius: *CIL* 11.1831. Festus, *Gloss. Lat.* 468.1–3L; Val. Max. 2.5.6; Vitr. *Arch. praef.* 17, 3.2.5. Scipio Aemilianus: Plut. *Mor.* 318D–E.

number of inscriptions from the Tomb of the Scipios, beginning with that of Scipio Barbatus himself from the first half of the third century.[13]

There seems to be no question that our earliest Latin authors use the term *virtus* to mean martial courage: Livius Andronicus employs the term in a lament for a dead hero, while Naevius calls upon young men to defend their homes by means of *virtus*.[14] The term also appears in this context in a speech by Q. Caecilius Metellus from 201.[15] Cato used the term differently on a work-by-work basis; while *virtus* never has a military sense when used in his *De Agricultura*, it refers exclusively to martial bravery when he employs it on every other occasion. The use of ἀρετή by Plutarch when quoting Cato is likely a translation of *virtus*.[16] The word is used to mean physical courage on the majority of occasions when it is employed by Plautus; of these, 23 refer specifically to battlefield courage.[17] In some of these examples, *virtus* is employed ironically or even metaphorically, but the martial overtones are clear. While it often refers to actual soldiers in the field, it can equally refer to civilians or slaves who take on the characters of soldiers for a time. Terence does not use the term nearly as much as Plautus, but does to refer to battlefield courage on at least two occasions.[18] Ennius uses the word in a clearly martial context six times, Lucilius five times, and Accius four times.[19] *Virtus* appears three times in a martial context in a collection of fragments from the mid-Republic which are not attributed to any one author.[20] Additionally, Polybius often uses ἀρετή as the equivalent of the Latin *virtus* when he is describing courageous battlefield action.[21] *Virtus*, like ἀρετή, could be passed from father to son.[22] Although attitudes toward warfare had changed in the late Republic and early Imperial period, the sources of this time were in many ways meant to act as repositories of collective memory for Rome's more martial past, and thus also attempted, albeit imperfectly, to reflect the world of the mid-Republic.[23]

13 The term appears on four inscriptions from the Tomb of the Scipios. On three of these, ranging in date from the first half of the third century to the first half of the second century, it very likely means battlefield courage (*CIL* 1².7, 10–11).

14 Liv. Andron. *Ajax Mast.* 16–17 *ROL*; Naev. *Taren.* 90–91 *ROL*. On the use of *virtus* in the mid-Republic, see Balmaceda (2017, 14–42); Courtney (1999, 42–43, 74–78, 85); McDonnell (2006b, 17–24, 29–38, 44–71); Rosenstein (1990, 95–111); de Vaan (2008, 681).

15 Quintus Caecilius Metellus *ORF* 6.3.

16 Cato *FRHist* 5 *Orig.* 4 F76.19 (twice), unassigned F113, *ORF* 8.58, 141, 146. Plutarch's quotation of Cato: *Orig.* F135 (*Cat. Mai.* 10.4).

17 Plaut. *Amph.* 75, 191, 212, 260, 354, 648–53, *Asin.* 556–58, *Cas.* 88, *Cist.* 198, *Curc.* 179, *Epid.* 106, 381, 442, 445, *Mil.* 12, 32, 57, 1027, 1327, *Pseud.* 532, 581, *Truc.* 106. On the use of *virtus* by Plautus, see Lodge (1962, 2.878–9).

18 Ter. *Eun.* 778, 1090.

19 Enn *Ann.* 6.188–9, 10.326, *Inc. Ann.* 562 605 Skutsch, *Hec. Lyt.* 62 *FRL*, *Phoen.* 109 *FRL*; Lucil. 5.245, 30.1013–4, *inc.* 1131, 1177; Acc. *Arm. Jud.* 108, 123, *Neop.* 482, *Nyct.* 493 *ROL*.

20 *Ex inc. inc. trag.* 98,102, 126 *ROL*.

21 Polyb. 3.84.7–10, 6.39.1–10, 24.9, 52.11, 54.2; see Cic. *Tusc.* 2.43.

22 See Eur. *Arch.* F 232 (Collard and Cropp) for a Greek example; Plaut. *Pseud.* 578–92 for a Roman example.

23 See Serrati (2013, 166–67).

There is some evidence to suggest that during the later fourth century, Greek ideas about the pairing of wisdom with battlefield courage, personified by the goddess Athena, had begun to take hold at Rome.[24] To this end, the early second-century Latin tragedian Marcus Pacuvius appears to have written a play about *virtus* and *sapientia*.[25] These notions were likely influential in the development of Minerva, who perhaps now took on more warlike traits, as a mid-Republican dramatic fragment of unknown provenance which refers to a marriage between Minerva and a warlike man (*bellicosus*) speaks to this idea.[26] Nonetheless, *virtus,* as battlefield courage, seems to have been an exclusively masculine ideal, and thus the joining of wisdom with martial bravery had an even greater effect on Mars. The god easily fits into this mold, as there is no question that by the late third century, if not much earlier, he had become synonymous with both battlefield courage and generalship. Furthermore, war and wisdom were at some point twinned on the first day of the Quinquatrus festival, 19 March, which had rites for both Mars and Minerva.[27]

As a war god, and in contrast to the Greek Ares, Mars seems to have had no negative aspects to his persona; like *virtus*, he was always viewed as a positive force whose role was to aid Rome's armies and generals.[28] In the middle Republic, Mars was not only a god of war but specifically the god of Roman victory, and the latter was brought about by *virtus*. This can most clearly be seen in an allusion to the First and Second Punic Wars by Plautus:

> Conquer through true *virtus*, as you have done before. Protect your allies (*socii*), old and new, increase your auxiliary forces through just laws (*iustis legibus*), destroy your enemies, and earn praise and laurels, so that the Carthaginians, whom you have conquered, may be punished.[29]

In this period, auxiliaries for the Roman army came from allied states, and thus the former could increase if the latter were protected via just laws. Conquest, on the other hand, was clearly achieved through *virtus*. And to display *virtus* on the battlefield was the ultimate demonstration, not only of one's manhood but also of an individual male's status as a citizen: "But it is a man's duty to live a life inspired by true *virtus*, to stand fast with irreproachable bravery in the face of enemies. True freedom belongs to the

---

24 Plin. *HN* 36.28; see Courtney (1995, 40–43, 223–28); Lipka (2009, 34–35); Zevi (1968–9).

25 Pacuvius, *Antiopa* 1 *ROL*.

26 *Inc. Trad. Com.* 154 *ROL*.

27 Festus. *Gloss. Lat.* 446.29–448.4, 480.25–9L; Varro, *Ling.* 6.14.

28 For example: Enn. *Ann.* 10.326–8 Skutsch. While *virtus* was viewed as a positive character trait, it could nonetheless be employed for less than noble purposes; for example, Enn. *Hec. Lyt.* 62 *FRL*.

29 Plaut. *Cist.* 197–202: *vincite virtute vera, quod fecistis antidhac; servate vostros socios, veteres et novos, augete auxilia vostra iustis legibus, perdite perduellis, parite laudem et lauream,ut vobis victi Poeni poenas sufferant.*

man who conducts himself with purity and steadiness."[30] Similarly, *"Virtus, without doubt, comes before everything; it is what preserves our liberty, safety, life, property, parents, country, and children. Virtus has everything in itself. The man who shows virtus possesses all that is good."*[31] Given the early second-century time frame, when warfare remained the primary concern of most *nobiles*, such sentiments are unlikely to refer to anything other than martial prowess. Thus, in essence, all that is good in Rome depends upon the *virtus* of its male citizen warriors. The statement is quintessentially Roman, and there can be little question that it would have resonated greatly with Roman audiences in the period immediately following the Second Punic War and during Rome's first forays into the Eastern Mediterranean. It is therefore almost certainly original to Plautus and not borrowed from the unknown Greek work upon which he based the *Amphitryon*.[32] As Rome's main martial deity, Mars was associated with the aspects of war which the Romans most valued in the middle Republic, and chief among these was *virtus*. In fact, as the god of courage, battlefield aggression, and conquest, Mars was the very personification of Roman *virtus*, and was thus representative of both citizenship and manhood at Rome during the third and second centuries, if not earlier. Thus, the connection between Mars, *virtus*, and masculinity is clear, and altogether, enough evidence exists to draw similar conclusions about the roles of gender within the martial culture of mid-Republican Rome, as well as to make a general statement about the Roman perception of the feminine at the time.

Bellona, the feminine, and Roman warfare

Although warfare was traditionally associated with men and masculine characteristics in Rome, the complex nature of both warfare and gender in Roman society defy easy divisions and hierarchies. Given the rigid social hierarchies which governed Roman life in the middle Republic, gender roles were likely clearly defined and understood. Moreover, in the middle Republic, a man's *virtus* was primarily displayed on the battlefield, not necessarily over women within the domestic sphere.[33] Roman women were the mothers of the male warriors who possessed *virtus*, and therefore, as opposed to their Greek counterparts, could more easily be conceived of as having both strength and individual aegis. As such, the primary duty of Roman women was motherhood, and through this role, they facilitated Rome's culture of

---

30 Enn. *Phoen.* 109 *FRL* (= Gell. *NA* 6.17.10): *sed virum vera virtute vivere animatum addecet fortiterque innoxium astare adversum adversarios. ea libertas est: qui pectus purum et firmum gestitat.*
31 Plaut. *Amph.* 648–53: *virtus praemium est optumum; virtus omnibus rebus anteit profecto: libertas, salus, vita, res et parentes, patria et prognatitutantur, servantur:virtus omnia in sese habet, omnia assuntbona quem penest virtus.*
32 Baier (1999, 27–28); McDonnell (2006b, 32–33); Segal (1975, 254).
33 Goldstein (2001, 306–17).

militarized hyper-masculinity. In theory, Roman women acted as societal arbiters of male martial courage, as can be seen in legendary women like Lucretia and Cloelia, who demonstrated that (at least part of) a women's role was to spur a man to do his duty and to fight.[34] It is true that their primary space was within the domestic sphere; this was not necessarily limited to their physical dwelling, as women were known to exercise a degree of freedom at Rome because the city itself was a *domus* writ large.[35] Outside of the *domus*, however, they were significantly more restricted, and although the evidence is scant, references would seem to indicate that women were discouraged from accompanying a Roman army on campaign, and could have been banned from camps altogether by certain generals. Scipio Aemilianus famously expelled all the female camp followers in order to reimpose order among the legions besieging Numantia in 134. Metellus Numidicus acted similarly in 109, ejecting the camp followers in order to restore discipline. Although the gender and status of these people are not mentioned on this occasion, it is highly likely that they included female prostitutes. Even the idea of having women in a military camp could be used as ammunition against a political opponent, as Catiline's fellow conspirators were derided by Cicero in 63 over the prospect of taking women on campaign with them.[36] Servius implies that women were not to be regularly found in military camps, and indicates through wordplay of *castra* (camp), *casta* (chaste), and *castro* (castrate) that Roman military installations were meant to retain their purity specifically by the exclusion of women: "But the camp (*castra*) is called chaste (*casta*), because in this place, lust is castrated (*castraretur*) since a woman has never been present there."[37] Later, he states more overtly that "female camp followers were regarded as shameful by our ancestors."[38] Propertius also speaks of Roman camps as being wholly off limits to women. Although composed in 16, Propertius' fourth elegy likely reflects the state of the late Republican army, as it was certainly written before the formal ban on the marriage of Roman soldiers – something which may have been instituted by Augustus in 13, but equally could have been significantly later.[39]

Rome therefore seemed to feature a strict separation of the domestic (female) and military (male) spheres. This sharp distinction featured

---

34 Lucretia: Livy 1.58.5–11; Cloelia: Livy 2.13.6–11). On the role of women as mothers during the Republic, see Tac. *Dial.* 28. For women and gender in martial cultures, see De Pauw (1998, 1–25); Elshtain (1995, 181); Goldstein (2001, 301–22). For Rome specifically, see Culham (2014, 133–38); De Pauw (1998, 67–69); Fantham et al. (1994, 220–30); Goldstein (2001, 306–17); Rawson (2006).

35 Milne (2012, 25).

36 Scipio: App. *Hisp.* 14.85; Frontin. *Str.* 4.1.1; Livy, *Per.* 57; Plut. *Mor.* 201B; Val. Max. 2.7.1. Metellus: Frontin. *Str.* 4.1.2; Sall. *Catil.* 45.2; Val. Max. 2.7.2. Catiline: Cic. *Cat.* 2.23.

37 Serv. *Aen.* 3.519: *Dicta autem castra quasi casta, vel quod illic castraretur libido, nam numquam his intererat mulier.*

38 Serv. *Aen.* 8.688: *Mulier castra sequebatur quod ingenti turpitudine apud maiores fuit.*

39 Prop. 4.3.45. See: Campbell (1978, 153–54); Phang (2001, 124–29).

prominently within the mid-Republican conceptualization of space, which the contemporary Romans described as *domi militiaeque*: at home and where the legions operated. On 1 March, the men of Rome celebrated the *Feriae Martis* which likely involved a sacrifice at the Altar of Mars; this structure was in the Campus Martius and was thus outside the *pomerium* and in the *militia*. Meanwhile, the women observed the *Matronalia* festival for female fertility from within their own homes.[40] Ovid himself draws a very direct parallel between the two rituals, as the poet has Mars explain that while war and the *militia* is the domain of men, motherhood and the *domus* is the equivalent for Roman women.[41] And even though the camp itself might be perceived as a *domus* in some ways, it was nonetheless within the *militia*, and significantly, although the *Lares* and the *Penates*, who were the male domestic deities, accompanied the army, the goddess Vesta was conspicuously absent from Roman military camps.[42] Thus, the Roman conceptualization of space was itself inherently gendered, as women may have had considerable freedom within the *domus*, but were forbidden from entering the masculine *militia*, where their presence in a military installation was thought to cause a degradation of discipline or even an outright emasculation of the soldiers themselves.

Such gender roles are equally present in the Roman martial pantheon of the middle Republic. While Mars personified the idea of *virtus* as manliness as well as martial courage, Bellona stands out as playing a significant role in the interplay between warfare and gender from the mid-fourth to the early first centuries. Although her status waned during the Imperial period, in the middle Roman Republic, evidence appears to show Bellona as a major Roman war deity, perhaps second only to Mars. Having Bellona as a major female war deity alongside Mars appears to have been completely acceptable in contemporary Rome. Unlike Athena, who is often spoken of in masculine terms,[43] Bellona is no mere *virago* ("woman with masculine or warlike qualities"). She is a war goddess who was fully female; she represented the chaos of the battlefield itself, as well as the idea of Roman warfare as vengeance.

While Bellona is often interpreted as simply the personification of war or as emblematic of the chaos of conflict, evidence points to a far more nuanced deity who was an important part of the Roman pantheon during the Republic. First, Bellona features prominently as the fifth

40 Altar of Mars: Fest. 204.15–7L; Livy 35.10.12, 40.45.8; Richardson (1992, 245). Matronalia: Mart. 5.84.10–11; Ov., *Fast*. 3.167–248; Plut. *Rom*. 21.1; Holland (2012, 212); Takács (2008, 40–41).

41 Ov. *Fast*. 3.167–248.

42 *Lares* and *Penates* accompanying the army: Livy 44.39.5; Serv. *Aen*. 2.296, 3.12; Tib. 1.10.15–25; Verg. *Aen*. 8.678–81. In general: Jung (1982, 334); Langlands (2006, 265–75); Milne (2012); Rudán and Brandl (2008, 4–6); Rüpke (1990, 66); Wintjes (2012, 42–43).

43 For example, Aesch. *Eum*. 734–43.

deity mentioned in the *devotio* of P. Decius Mus at the Battle of Veseris (Vesuvius) against the Latins in 340, where she follows Mars Pater and Quirinus:

> Janus, Jupiter, Mars Pater, Quirinus, Bellona, Lares, new gods (*Novensiles*), native gods (*Indigites*), you gods who hold both us and our enemies in your power, and you, divine Manes, I invoke and worship you, I beseech and beg your favour, that you fortify the might and bring about the victory of the Roman people, the Quirites, and visit fear, weakness, and death upon our foes. As I have pronounced these words, on behalf of the *res publica* of the Roman people, the Quirites, on behalf of the army, the legions, and the auxiliaries, I devote the legions and auxiliaries of the enemy, together with myself, to the divine Manes and to the earth.[44]

With these words, Decius charged headlong into the enemy lines, sacrificing himself to the gods. The battle supposedly turned on this action and the Romans won the day. Decades later, at the high point of a battle in Etruria between a Roman army and a force of Etruscans and Samnites in 296, the Roman commander, Appius Claudius Caecus, made the following vow:

> 'Bellona, if you grant us victory today, then I hereby vow a temple to you.' Having pronounced this, he began to match his colleague [Lucius Volumnius Flamma Violens] in courage, and then the army began to match his, as though the goddess were inspiring him.[45]

As with Decius, this turned the battle and the Romans emerged victorious. Fulfilling the vow, Caecus dedicated a temple to Bellona on 3 June 293.[46] Epigraphically, Bellona also features among eight other Roman deities whose names are individually inscribed on a series of dishes which have been dated to the mid-third century. These were likely used for libations. Bellona's dish also contains the earliest, and only mid-Republican, representation of her.[47]

---

44 Livy 8.9.6–8: *Iane Iuppiter Mars pater Quirine Bellona Lares Divi Novensiles Di Indigetes Divi quorum est potestas nostrorum hostiumque Dique Manes, vos precor veneror veniam peto oroque uti populo Romano Quiritium vim victoriam prosperetis, hostesque populi Romani Quiritium terrore formidine 8morteque adficiatis. Sicut verbis nuncupavi, ita pro re publica populi Romani Quiritium, exercitu legionibus auxiliis populi Romani Quiritium, legiones auxiliaque hostium mecum Deis Manibus Tellurique devoveo.*
See Ando (2009, 181–85); Roth (2009, 62); Versnel (1976).
45 Livy 10.19.17: *"Bellona, si hodie nobis victoriam duis, ast ego tibi templum voveo." Haec precatus, velut instigante dea, et ipse collegae et exercitus virtutem aequavit ducis.*
46 On temple foundations resulting from warfare, see Fronda in this volume.
47 Musée du Louvre, Collection de Campanie, Inventory number K 614. The inscription reads BELOLAI POCOLOM (for the inscription: *CIL* 1².441).

Finally, in the early second century, she appears in a list of deities within the prologue to the *Amphitryon* (43) by Plautus. Here, she comes after Neptune, Victoria, and Mars, three other deities associated with war (Neptune was a god of victory at sea) and who did not seem to have negative connotations. Clearly, Bellona was connected to war, but if she was regarded as a purely malevolent force in the mid-Republic, she would have struck Plautus' audience as a very strange inclusion in a list of deities who were seen as representative of Rome's martial prowess. That she features prominently on the aforementioned mid-third century libation dishes would seem to greatly further this point.

Moreover, Bellona's name itself likely reflects the idea that warfare itself was viewed with a degree of positivity at Rome during the middle Republic. The older form of her name, Duellona, as well as the archaic Latin word for war, *duellum*, had shifted to *bellum* and the corresponding Bellona over the course of the mid-third to the mid-second centuries.[48] *Bellum* was associated with the description of warfare as *bella acta*, "good or valorous deeds." This perhaps came about because warfare had by this time come to be seen as the best role for the citizen, something reflected by the contemporary works of Cato the Elder, particularly his preface to the *De Agricultura*.[49] While battle could be negative, war itself – and its corresponding personification in Bellona – could be viewed as a positive undertaking, both on the level of the individual and for society as a whole. Therefore, considering this evidence, it is difficult to believe that the goddess was seen as a purely malevolent force or even a simple personification of war itself. In every respect, she appears to be a deity who could and did help her adherents, and her powers in war were not to be taken lightly.

Bellona is perhaps best known for the temple dedicated by Caecus, which stood prominently at the north-western end of the Forum Holitorium, next to that of Apollo (see Map 1).[50] Here, the senate greeted returning generals and listened to their petitions for a triumph. As such, there is no question that the temple was *extra pomerium*, and likely close to the very beginning of the Roman triumphal route.[51] As with the aforementioned temple to Mars Gradivus outside the *Porta Capena*, the positioning beyond the *pomerium* was normal for war gods; they specifically operated in the *militia* rather than

---

48 See Varro, *Ant. Hum. Div.* F189 Cardauns, *Ling.* 5.73. Plautus perhaps uses *duellum* as an archaism (*Amph.* 189, *Asin.* 559, *Epid.* 450; *Capt.* 68, *Truc.* 483). See also Lodge (1962, 1.210). Ovid uses the term only once (*Fast.* 6.201) and very purposefully: when mentioning the Third Samnite War and the vow of a temple to Bellona by Appius Claudius Caecus. In using the archaic term, Ovid was likely highlighting both the antiquity of the war as well as the goddess.

49 Astin (1978, 189–203); Courtney (1999, 50–53); Gratwick (2002); McDonnell (2006b, 57–59); Pinault (1987); de Vaan (2008, 70).

50 Ov. *Fast.* 6.205, 209.

51 Beard (2009, 92–105, 201, 206); Champion (2017, 130–42); Versnel (1970, 132–63); Ziolkowski (1992, 292–95). On the granting of triumphs by the senate, see Fronda in this volume.

the *domus*, and any magistrate requesting a triumph would have to do so before crossing the *pomerium* and surrendering their *imperium*.[52] Indeed, no temple for a deity exercising a function related to war stood within the *pomerium* until that of Mars Ultor (dedicated in 2).[53] After 280, the *Columna Bellica* in front of Bellona's temple was home to the fetial rite. As described by Livy, once the Romans determined that a foreign entity had committed an offence against Rome or one of its allies, the priestly college of the *fetiales* undertook an ambassadorial mission to the potential enemy and demanded recompense, returning to Rome afterwards. If none were forthcoming, after 33 days the fetials returned to the border of Roman or allied territory and there called upon the gods to witness that the Romans had been unjustly treated. The priests thus declared the conflict to be *ius* and ceremonially cast a special spear onto the foreign soil. This rite served as the formal declaration of war on the part of the Roman people.[54] However, according to Servius, the *fetiales* were unable to perform their ritual declaration of war at the border of enemy territory for the Pyrrhic War in 280; in consequence, they acquired a patch of land by the *Columna Bellica*, in front of the temple to Bellona, and henceforth declared that the spear-throwing rite was now happening "as though in enemy territory" (*quasi in hostili loco*).[55]

Spears were the most ancient symbols of warfare at Rome; not only were they used in the ancient fetial rite but also the weapon was sacred to Mars as well as Bellona.[56] The *hastae Martis*, along with the *ancilia* or sacred shields, were displayed in the Regia and were said to vibrate when there was impending war or a grave danger to the state.[57] Moreover, the *triarii*, the most experienced and distinguished troops during the mid-Republican period,

---

52 Livy 26.21.1, 28.9.5, 38.2, 30.21.12, 31.47.7; Ando (2009, 116 n. 82); Scheid (2011, 62–66).

53 Augustus possibly dedicated an earlier temple to Mars Ultor on the Capitoline in 20 (Dio Cass. 54.8.2). However, this is most often referred to as a *templum* rather than an *aedes*, and so the extent to which it was an actual building and not just a consecrated space remains unknown. Therefore, the first war temple within the *pomerium* which can be securely dated is that of Mars Ultor in the Forum of Augustus; *contra* Ziolkowski (1992, 266–68), who argues that several temples associated with war were *intra pomerium*.

54 Livy 1.32.5–14. See: Cic. *Rep.* 2.31; Dion. Hal. *Ant. Rom.* 2.72.6–8; Plin. *HN* 22.5; Serv. 9.52; Varro, *Vita Pop. Rom.* F93 Riposati. On the *fetiales* in general, see Livy 1.24, 32.6–14, 31.8.3, 36.3.7–12. See also: Serv. 9.52; Beard, North, and Price (1998a, 26–27, 132–33); Ferrary (1995); Rich (1976, 56–60, 109); Rich (2011, 2013, 559–64); Rüpke (1990, 97–117, 2016, 106–10); Santangelo (2008); Warrior (2006, 58–59); Wiedemann (1986); Zack (2001, 1–73); Zollschan (2011).

55 Serv. 9.52. See: Festus, *Gloss. Lat.* 30.14–6L; Livy 1.32.6–14; Ov. *Fast.* 6.203–8; Suet. *Claud.* 25.5; Ando (2009, 115–16). On the *Columna Bellica* see Ov. *Fast.* 6.205–8, Serv. 9.52; Ziolkowski (1992, 18–19, 47–49).

56 Who is often portrayed holding one, for example, Stat. *Theb.* 2.719, 4.6–7. There is likely a link between Bellona and the *hastiferi*, but evidence for the existence of this college is entirely imperial.

57 Arn. *Adv. Nat.* 6.11; Gell. *NA* 4.6.1–2; Livy 21.62.4, 40.19.2; Obseq. 36, 44, 47, 50; Plut. *Rom.* 29.1; Varro, *Ant. Div.* F254 Cardauns; Rüpke, (1990, 133–36).

also used spears. The spear is likewise the weapon of choice for heroes in Greek and Latin epic, and is used by both Livius Andronicus and Ennius for dramatic effect, even when the latter is writing of historical episodes.[58] Thus, through the fetial spear as well as her temple, we see that Bellona is ritualistically present at both the initiation and at the termination of conflict. As the fetial spear symbolically crossed from Roman territory onto foreign soil, so too did a victorious general and his army cross the *pomerium* to return to the *domus*. More than the mere personification of war, Bellona, like Janus, also represents the division between war and peace.

That said, Bellona remained the only female deity in the mid-Republican pantheon with clear links to warfare. The aforementioned fetial rite with which she was involved provides a clue as to why this was. In framing themselves as victims and in demanding recompense, we can see that the Roman concept of warfare is essentially one of revenge. There is no question that, on a practical level, warfare at Rome was about social prestige and personal enrichment, but on a conceptual level – as illustrated by the speech Livy puts into the mouth of the Samnite general Herennius Pontius after the defeat of a Roman army at the Caudine Forks in 321 ("The Romans... shall not rest until they have wreaked manifold vengeance on your heads")[59] – war, for the Romans, was seen as the primary method for righting wrongs, bringing about justice, and, as stated by Varro, restoring Concordia.[60] Furthermore, the idea of a lack of recompense as being the cause of Roman conflict is embedded within the Latin language itself, as the archaic verb *hostio* means both "to recompense" and "to fight against." Ennius even uses the term as pun, with Achilles promising to recompense his enemies via his weapons ("You weapons, my sword and my spear, in close combat some recompense will come from my own hand").[61] Pacuvius employs the term similarly, where the character Telamon promises to return (*hostio*) the violence of any enemy in kind.[62] Moreover, the idea that the purpose of warfare was to restore the *pax deorum*, as well as the notion that all conflict was rooted in vengeance, can be seen in the words related to *hostio*, namely *hostis* (enemy) and *hostia* (sacrificial victim). The latter were not merely offerings to the gods out of piety, but were given in exchange for something tangible within the human *cosmos*. Thus, an enemy was anyone who had not provided Rome with the deserved remuneration for the wrong it had suffered, and enemies who died on the battlefield were themselves offerings which were given to the gods in exchange for a Roman victory.[63]

---

58 Enn. *Ann.* 8.266, 11.355–6, 15.391–8, 16.413, 421–22 Skutsch; Liv. Andron. *Od.* 21 F42, 22 F43–4 *ROL*.

59 Livy 9.3.13: *nec eos [Romanos] ante multiplices poenas expetitas a vobis quiescere sinet.*

60 Varro, *Ling.* 5.73.

61 Enn. *Hec. Lyt.* 56 *FRL*: *Quae mea comminus machaera atque hasta hostibitis manu.*

62 Pacuv. *Teuc.* 377–78 *ROL*.

63 *Hostio* as "to recompense": Plaut. *Asin.* 377; as "to fight against": Laev. *Erot.* F1 Courtney; in employing the term as a pun, Enn. *Hec. Lyt.* 56 *FRL* and Pac. *Teuc.* 377–78 *ROL*

In this regard, one has to look no further than Lucretia's demand for justice, Dido's curses, or the burning of Aeneas' ships to see that women are portrayed as more vengeful in Roman literature.[64] And female lamentation rituals have likewise been interpreted as essentially calls for vengeance.[65] Therefore, vengeance in Roman conflict fits well with the notion of a female war deity. Vergil even picks up on the idea as he associates Bellona with the Furies, three beings whose primary duties were vengeance, justice, and the righting of wrongs.[66] Since the Romans had conceived of warfare as a form of ritualized revenge since at least the fourth century, Vergil may very well have been drawing on an earlier tradition, which itself would have been highly influenced by Hellenic and Near Eastern contexts where the idea of vengeance as a significant trait for female deities was even more developed.[67] Thus, the pursuit of justice through warfare may have been Bellona's primary role as a goddess in the mid-Republic; her wrath on the battlefield was itself the physical manifestation of the fetial rite and, like the Furies themselves, she was the personification of Roman vengeance.

Bellona on the battlefield

Of course, in between the beginning and the ending of a conflict, there is the actual fighting. Beyond her temple, combat is perhaps the aspect with which Bellona is most associated. Although she could bring her adherents victory and glory, she likewise represented the pains and negativities of war. Despite the multitude of modern, popular works that refer to the Roman army of the Republic as some sort of "war machine," killing is not an innate behavior and comes unnaturally to the majority of humans. This rings especially true for ancient warfare, where the bloodshed was up close and very personal, and where battlefields would have been horribly sonorous, fear-inducing places. In spite of the reputation of Roman soldiers for stoic toughness and discipline, the psychological trauma of the battlefield must have been very real, and was likely felt even by seasoned veterans.[68] Given these realities, that the Romans represented battle with a deity other than Mars, the positive personification of *virtus* and battlefield courage, is hardly surprising. The Romans understood that warfare could not be so cleanly

---

capture both senses. See: Fest. 91.7–8, 334.16–9L; Eichner (2002); Lodge (1962, 1.723–4); de Vaan (2008, 291). On the *pax deorum* to explain military defeat, see Rosenstein (1990).

64 Lucretia: Livy 1.58.5–11; Dido: Verg. *Aen.* 4.584–629; Aeneas' ships: Hellan. *FGrH* 4 F84.

65 Alexiou (2002, 21–22), who discusses the idea in a Greek context; Keith (2016, 164, 171, 178 n. 33).

66 Verg. *Aen.* 8.700–3; cf. Stat. *Theb.* 1.46–87.

67 For the Roman traditions: Cic. *Nat. D.* 3.46; Festus, *Gloss. Lat.* 74.11L; Serv. 4.609. For the Greek and Near Eastern, traditions: Breitenberger (2004, 23–24); Budin (2002); Marcovich (1996); Pryke (2017, 139–40, 152–53, 160–82).

68 On the physical and psychological stresses of ancient hand-to-hand combat, see Grossman and Christensen (2008, 30–99, 214–16); Grossman (2009, 131–54); see also Milne in this volume.

and easily defined. The association between Bellona and the maelstrom of combat is perhaps responsible for the goddess' disheveled hair in the only portrait of her from the mid-Republic.[69] In later sources, she is strongly connected to the chaos, fury, and especially the sound of battle. As with actual soldiers in an actual war, when she enters the field of battle, she is preceded by trumpets, her armor and weapons clang, and she shouts with the force of many men.[70] Ovid has her accompanied by "the clash of arms, the groans of fallen [and]... a sea of blood."[71] In fact, most of Bellona's imagery has both her and her weapons permanently caked in blood. As opposed to the aforementioned opening and closing of wars, this aspect sees the goddess in perpetual combat.

In both Greek and Latin literature, the washing of the body, as well as one's weapons, is a highly symbolic gesture which purifies the fighter and removes him from the chaotic realm of war so that he may return, even temporarily, to the everyday world. In the *Iliad*, Andromache prepares a bath for Hector as she awaits his return, and Ares bathes and puts on fresh clothes after combat.[72] Achilles, on the other hand, refuses to bathe following the death of Patroclus, and is thus himself suspended in combat.[73] Similarly, Bellona does not bathe, and therefore remained in a similar state. A historical illustration of this is the Roman dedication of captured enemy arms after a victory, which were traditionally burnt and offered to a deity. These were normally dedicated to Vulcan, and on more than one occasion were offered to Mars.[74] Yet, despite being a goddess of the battlefield itself, Bellona is never recorded as having received such a dedication. This is likely because she was viewed as permanently on the field of war. While Mars might represent war as a communal undertaking by the civic community, winning these wars required actual fighting, and Bellona was combat itself.

Bellona's association with combat and the battlefield is likewise specific to her gender. A male citizen, performing his military duty and displaying his *virtus*, was seen as orderly and in harmony with society at large. The battlefield, as a place of extreme immediacy and *audacia*, devoid of logic and reason, runs contrary to this notion. Although the term became wholly negative in the first century, in the middle Republic, *audacia* was often used in reference to battle and the battlefield, and carried the same ambiguities as the English "audacity."[75] It could signify boldness, decisiveness, and conviction;

---

69   See above n. 47.
70   Hor. *Sat.* 2.3.223; Juv. 11.5; Mart. 12.57; Stat. *Theb.* 2.719, 4.9; Tib. 1.6.45-54; Val. Fl. 3.60; see Scheid (2016) 128.
71   Ov. *Met.* 5.154-6.
72   Andromache: Hom. *Il.* 22.437-46; Ares: Hom. *Il.* 5.905.
73   Hom. *Il.* 23.40-6; see Grethlein (2007). I am grateful to Ms. Meghan Poplacean for the references to washing in Homer.
74   App. *Pun.* 133; Livy 45.33.1-2.
75   Cato *FRHist* 5 *Orig.* F76.12; Corn. Sis. *FRHist* 26 F81; Plaut. *Mil.* 464; cf. Cato *ORF* 8.22, who uses the term to describe his political enemies.

but equally, the word could signify rashness, indiscipline, and recklessness. There is no question that *virtus* was regarded as an exclusive masculine trait; the only time the term is used in reference to a woman is when Plautus has the character Alcumena play a man in a comedic sexual role reversal.[76] In fact, a lack of *virtus* was seen as characteristic of the feminine in the work of Plautus: "Without *virtus*, an eloquent citizen is, in my opinion, like a wailing women."[77] *Audacia*, on the other hand, could equally apply to men and women. The term is often used to describe women in the dramatic sources of the middle Republic, where it refers to a lack of rational thought before an action, regardless of whether that action is positive or negative. The early second-century comedic playwright Caecilius Statius employed the term to describe a crafty prostitute in a fragment of unknown provenance, whereas the mid-second-century tragedian Lucius Accius used it to describe a bold woman in his *Tereus*.[78] And the connection between women and *audacia* is particularly prominent in Terence's *Eunuchus*, where the term is first used to describe a courtesan named Thais, and it is later applied collectively to all the women of Thais' household: "Look at the audacity of those tramps!" (*Audaciam meretricum specta*).[79] The association of *audacia* to the feminine is made even more explicit in the *Miles Gloriosus* of Plautus. When the character Palaestro describes a young woman named Philocomasium, *audacia* features prominently in his list of her "feminine ways" (*ingenium muliebris*).[80] Perhaps the strongest example, however, comes from Plautus' *Casina*, where the handmaiden Pardalisca warns her master Lysidamus and his wife Cleostrata about the female servant Casina:

> Inside the house just now, I've seen strange things done in strange ways and with new and unheard-of audacity. Be on your guard, Cleostrata! Stay away from [Casina], please, so that in her rage she doesn't harm you. Take the sword away from that girl! She has lost all reason.[81]

The reference to the sword makes the allusion explicit: when a weapon is placed in the hand of a woman, there is a high potential for chaos. The final

---

76 Plaut. *Amph.* 925.

77 Plaut. *Truc.* 495: *sine virtute argutum civem mihi habeam pro praefica.*

78 Caecil. Stat. *Ex inc. fab.* 267 *ROL* (though the fragment is admittedly very unclear); Acc. *Ter.* 655 *ROL.*

79 Ter. *Eun.* 525, 994; cf. Ter. *And.* 217.

80 Plaut. *Mil.* 185a, 189.

81 Plaut. *Cas.* 625–29: *Tanta factu modo mira miris modis intus vidi, novam atque integram audaciam. Cave tibi, Cleostrata, apscede ista, opsecro, ne quid in te mali faxit ira percita. Eripite isti gladium, quae sui est impos animi.* These events are, of course, part of a comedic ruse being perpetrated on Lysidamus by his wife Cleostrata, and they are not happening in actual fact. Pardalisca is in on the joke. As well, the reference to a sword foreshadows the episode where a character named Olympio is tricked into seducing a male slave dressed as Casina, and finds a hilt (*capulus*) under the man's clothes (907–10). Nevertheless, Lysidamus believes the ruse, and thus the idea that a female would cause chaos with a weapon appears secure. On the use of *audacia* by Plautus, see Lodge (1962, 1.185–6).

word of the passage, *animus*, can also be used to denote boldness based on emotion and lacking in discipline. These sentiments also seem to reinforce the aforementioned idea that a female presence around a Roman army would result in a diminution of discipline.

Such references hint at why the Romans may have represented the frenzy of battle as feminine. Moreover, the prefix *au-* in *audacia* alludes to something which is perceived by the senses, and particularly refers to sound. This furthers the notion of the battlefield as a highly sonorous place, and of Bellona being deeply associated with loud noise. The *audacia* of Bellona, as well as the battlefield, was something which could be heard as well as seen. In this sense, Bellona's gender is no accident of history or borrowing from an older people; as the sources make clear, to the Romans of the middle Republic, a female could more easily be seen as chaotic, frenzied, visceral, and existing on pure emotion. If Mars represented *virtus* and the embodiment of duty, discipline, and courage, then Bellona was *audacia*, personifying the disorder of combat, and the bold and swift actions on the battlefield which could bring about victory or ruin.[82]

Conclusions

Although warfare was inexorably intertwined with Roman order in the middle Republic, this did not mean that the Romans were unaware of the chaos it typically involved, particularly on the battlefield itself. Similar to the Roman view of femininity, battle could be illogical and highly emotional; yet if the violence were properly controlled, the community could reap significant benefit. Through the vehicle of the *domus*, Roman women in theory were supposed to act as the social arbiters of the masculine *virtus*, and thus they had a strong role to play in the martial culture of the middle Republic. While they may have been viewed as an emasculating force within the sphere of *militia*, femininity itself nonetheless was present on battlefields through Bellona as the personification of both Roman vengeance and combat. These two aspects, vengeance and combat, intersect by means of the feminine; in both Greek and Latin literature, a primary purpose of female characters whose relatives had died fighting was to demand revenge. Women, like Bellona herself, were seen as the main motivators behind the idea that vengeance for past wrongs, including the death of loved ones, was obtained through war.[83] Within these conflicts, however, discipline as well as daring, *virtus* as well as *audacia*, were required to bring about victory. Discipline and duty are all well and good for an army as a whole, but the Romans equally rewarded men for individual acts of bravery. Mars and

---

82 Sallust (*Catil.* 3.3) directly contrasts *virtus* with *audacia*, but this comes from a time when the definition of the latter had shifted towards being entirely negative. See: Balmaceda (2017, 52–53); McDonnell (2006b, 59–61).

83 Alexiou (2002, 171–86); Foley (2001, 153–59).

Bellona were necessary in equal measure for success in ancient combat. While clearly existing alongside him, Bellona is in some ways Mars' direct opposite, where he is strategy and she is chaos. Like Fortuna, her presence was unavoidable, and she could be cruel and helpful in equal measure. Female deities were often more uncontrollable and unpredictable than their male counterparts, but without question, they were also seen as necessary for the maintenance of society.

Bellona's role as one of Rome's main war deities lasted until Sulla in the early first century; never having experienced a civil war before, the conflict between Sulla and the Marians (88–7, 83–2) must have shattered the Roman sense of social, political, and cosmological order. Bellona was a particular favorite of Sulla, and when she became heavily associated with him, she likewise became representative of civil war itself. This association destroyed the idea that Bellona was a goddess primarily representative of proper Roman vengeance. Divorced from this element, only the aspect of battlefield chaos remained. When this came to be combined with civil war, she began to be seen in an almost purely negative light. Moreover, Sulla himself associated her with the Eastern goddess Ma, a deity whose rites involved self-laceration and human blood, things deeply foreign to Roman ritual at the time.[84] Augustus and Vergil both picked up on these new themes, and during the imperial period she was largely seen as emblematic of Romans spilling Roman blood.[85] That said, it is possible to postulate an earlier wane in her worship in that almost all our evidence for Bellona as a major war deity antedates the Second Punic War, when the Romans began to employ genuine strategy and generalship to warfare. Battle in the preceding period must have been a significantly more rampageous affair, with the Romans emerging victorious through greater experience, discipline, and numbers.[86] The time afterwards, however, was an era when the Romans began to study Greek tactical manuals and to apply these principals of grand strategy to campaigns as well as individual battles.[87] This was more the domain of Mars. These changes may very well have resulted in a decline in Bellona's importance from the early second century onward. She was still a goddess of some prominence, and her temple remained a strong focal point for the senate, but certainly, by the time we reach the late Republic, Mars is effectively Rome's dominant war god. Nevertheless, prior to the gradual professionalization of Roman generalship and the army, when the legions were still drawn from the *assidui*, the pantheon of Roman war divinities was significantly more diverse, and serves to illustrate the role of religion and gender within the martial society at Rome in the middle Republic.

---

84 Lennon (2013, 92, 109–14).
85 For example, Verg. *Aen.* 7.17–20, 8.700–3.
86 See Armstrong in this volume.
87 Though see McCall in this volume.

8 The middle Republican soldier and systems of social distinction*

Kathryn H. Milne

Introduction

In 214, the four legions of Ti. Sempronius Gracchus met a Carthaginian army close to Beneventum. The Romans, in this case, were an unusual group, for many of Gracchus' soldiers were men scraped together hastily during the manpower shortage caused by Rome's disastrous defeats in the preceding years of the Second Punic War. Some of the men were hardly men at all, but boys who had been enlisted while they wore the *toga praetexta*, a garment put aside at the age of 17.[1] Six thousand were criminals, debtors, and those convicted of capital crimes.[2] About 8,000 were slaves, who had been asked to join with promises of freedom.[3] Only a year or so under arms, this was their first pitched battle. Trained in legionary maneuvers at Liturnum, they had participated in the ambush, or perhaps slaughter, of a Punic-allied Campanian army in its camp, and they had been besieged by Hannibal's army at Cumae, including making a successful sortie from the gates.[4] They had not stepped onto a proper battlefield though, and had self-selected into the legions only in the very loosest sense, since for many their alternative had been slavery, imprisonment, or execution. On the battlefield at Beneventum, they slowly began to divide by temperament. Gracchus had hoped to spur them on by offering freedom to anyone who brought back the head of an enemy, but this had proved a miscalculation; the bravest, Livy writes, were occupied in hacking at the body of the first man they had slain, leaving the thrust of the battle to the slack and fearful.[5]

This incident, as recounted here by Livy, forms a neat and didactic story. The horrors of the battlefield adjudicated who succeeded as a soldier and

---

* All dates are BC unless otherwise noted.
1 Val. Max. 7.6.1b; Livy 22.57.9.
2 Val. Max. 7.6.1b; Livy 23.14.3; Zonar. 9.2.
3 Livy 22.57.11; Macrob. *Sat.* 1.11.30–1. On the *volones* in general, see Rouland (1977, 45–58).
4 Training: Livy 23.35.6. The Campanians: Livy 23.35.10–19. At Cumae: Livy 23.37.1–9. For the context of these army movements in Campania, see Fronda (2010, 100–47, esp. 124–25, 134–35).
5 Gracchus' offer: Livy 24.14.6–9. The distribution of troops: Livy 24.15.3–4.

DOI: 10.4324/9781351063500-8

This chapter has been made available under a CC–BY–NC–ND license.

who failed, separating – in dramatic fashion – those who could launch themselves into the fray and kill, and those who hung back. Although Livy perhaps presented this story with a mind to its literary effectiveness and poignancy, the underlying principle is sound: combat and danger are dividers of men. This applies not to slaughter or ambush but what is called "complex war," defined as involving self-sacrificial practices, in which soldiers, by their very participation, take high risks with their lives.[6] This type of violence relies on an underlying social framework of positivity toward the military and its actions. This modern framework of anthropological discourse is remarkably similar to the assessment of Polybius, who attributed the Romans' military success to a superior social system that provided distinction to those who excelled in battle and thus encouraged emulation of that excellence.[7] This system is described in Polybius' sixth book, where he concludes his description of the army's practices by linking Rome's military success directly to the importance of rewards and punishments (τιμὰς καὶ τιμωρίας: 6.39.11). Polybius specifically tells us that he observed what we might now call the societal ennoblement of the warrior figure. He states (Polyb. 6.39.8) that rewards had an impact beyond the immediate effect on their recipient, since they incentivized both soldiers and civilians to emulate that honored soldier's deeds.

This chapter will consider some of the ways in which the Romans assigned value to particularly proactive military activity, and elevated and ennobled aggressive and violent acts, thus encouraging military participation. I will focus on two types of geographical space. The first is in the city of Rome itself, how representations of war contributed to the ennoblement of warriorlike activity. Here, the Roman triumph is arguably the most ostentatious event which took place in the city and directly contributed to validation and elevation of the soldiers' role. The second space is in the cities and towns outside of Rome where the soldiers normally lived, where they returned after war, and where, as Polybius writes (6.39.8–10), tokens of honor earned during service gained the soldier a positive reputation at home.

Natural born killers?

Polybius, who was present at the siege of New Carthage, described some soldiers as throwing themselves off the scaling ladders when met with even

---

6 As distinguished from raiding, ambushes, feuding, and other actives which involve easy killing of rivals with little risk, found in primitive societies and even animal groups – see Wrangham and Glowacki (2012). Keegan (1993, 89–94) calls this concept the "military horizon," a term borrowed from the anthropologist Turney-High's book *Primitive War* (1949). The atrocious and unnecessarily violent sacks of cities in the ancient world fall into this category, and are excluded here. There are many involving the Romans (most notoriously, the sack of New Carthage: Polyb. 10.14–15), but equally other cultures of the ancient Mediterranean, see Eckstein (2006, 203–5).

7 Eckstein (1997).

slight resistance, whereas others charged eagerly up to take their places (Polyb. 10.13.6–10). That some men are distinguished by an ability to throw themselves forward into danger is also the key principle of Polybius' description of the young Scipio Africanus' actions during the battle of Ticinus in 218. Quoting Gaius Laelius' firsthand account, Polybius writes that when Scipio saw his father injured and surrounded by the enemy, he urged his comrades to help him intervene, but none were willing to do so, and he charged to the rescue alone.[8]

In Livy's description of Beneventum, he labeled the men who hung back as *timidi*, "fearful" or "cowardly" (Livy 24.15.4), and the same implication is intended to distinguish Scipio from his peers in Polybius' story. Contemporary scholars of battlefield dynamics would perhaps be less critical. In his work *Men Under Fire*, the U.S Army Officer S.L.A. Marshall famously made the claim that in the Second World War, only a quarter of all US citizen soldiers fired their weapons.[9] After the Battle of Gettysburg in the American Civil War, 90% of the muskets found were loaded, thousands of them for multiple times, indicating that men were repeatedly only pretending to fire and hiding their behavior by loading their weapons again and again.[10] Grossman, comparing the psychological impact of different means of violence in war, called the job of the Roman soldiers the hardest of all. They had a difficult, gory, and intimate task that required two acts that are generally hard for humans to do: to kill at the extremely close range of edged weapons, and to deliver the Roman signature piercing thrust straight into flesh, something which is psychologically much harder than slashing at an opponent.[11]

Although Roman soldiers' moments of fighting at close quarters would have been limited, the results of encounters with the Romans and their swords are shocking, and there are plenty of soldiers who performed the kind of brutal job to which Grossman refers.[12] Polybius describes Celtic soldiers

---

8 Polyb. 10.31.3–7. Gaius Laelius, the consul of 190, personally met and told the story to Polybius, and his account should be considered trustworthy: Walbank (1967, 198–99).

9 Marshall (2000, 50–63). For the controversy over Marshall's figures, see Chambers (2003, esp. 120 n. 3) for a bibliography of supporters and detractors. It is, perhaps, a fair suggestion that having been a killer in a martial context in the ancient world would not have carried the same stigma or power to intimidate as it does in the modern. Lintott (1999, 36) remarks that the culture of the ancient Mediterranean "did not place such a high value on human existence in itself as ours do now."

10 Grossman (2009, 23–26).

11 See Grossman (2009, 120–30), on the historical reluctance of soldiers to use piercing weapons. He posits that even Roman soldiers frequently did prefer to slash (esp. 340 n. 7). For the signature Roman thrust, see Veg. *Mil.* 1.12. For examples of the use of this technique previous to 214: Livy 7.10.9–10; Gell. *NA* 9.13.17; Polyb. 2.33.6. Carter (2006, 155 n. 10, 159) notes the technique was in use before the introduction of the *gladius hispaniensis*. This may have been the case at Beneventum, even though many of the soldiers were armed with Gallic and other arms taken from Rome's temples: Livy 22.57.10–11; Val. Max. 7.6.1b.

12 Sword fighting was not required of all Roman soldiers with unrelenting frequency, since the *velites* were skirmishers and *triarii* fought primarily with the heavier spear called the *hasta*. Less than 60% of a legionary's role might involve fighting with a sword, see Quesada

as being "cut up" (διακοπτόμενοι) at the battle of Telamon in 225 (2.30.7). Polybius also mentions the Romans using their swords as thrusting weapons in the subsequent fight with the Insubres (2.33.6). Here, the expression for the Romans' actions (ἐκ διαλήψεως) means to thrust at an enemy at close quarters with the tip of the sword.[13] Polybius writes of swords as both a cutting and a thrusting weapon when he compares the Roman and Macedonian armies (18.30.7). After taking New Carthage, Scipio's army trained there, including practice in sword fighting (Polyb. 10.20.1–4). Polybius is definite about the battle coming down to a sword fight at Zama, where initially the professional mercenaries were more skilled than the Romans (15.13.1–4). The sword used was likely the *gladius Hispaniensis*, carried by both cavalry and infantry.[14] Livy tells us that the damage wreaked by this weapon by the Roman cavalry on the Macedonian army in 200 resulted in the horrific sight of corpses with arms torn off the shoulder, or headless, or with the innards laid bare.[15]

Just as in Livy and Polybius' accounts, today we recognize stark variations in the temperaments of soldiers. It is observed that a small minority of individuals are not only willing to participate in battle, but eager and enthusiastic killers. Two psychiatrists of the Second World War, Swank and Marchand, identified 2% of American soldiers whom they termed "aggressive psychopaths."[16] These men were notable in that they suffered no remorse for their killings at all. Pierson called such men "natural killers," who possess certain traits of birth and behavior in common, and so could be identified and used effectively.[17] Natural killers have drawbacks, Pierson cautioned, but advantages too:

> They will personally kill the enemy in droves. They are natural leaders who will motivate other soldiers to kill. They are also fiercely competitive and will aggressively pursue victory. In a battle of attrition, the natural killer can single-handedly tip the scales.[18]

Sanz (2006, 2). For those who did, a normal ancient battle probably consisted of repeated incidents of small groups of soldiers who dared to dart forward to attack, see Sabin (1996); Sabin (2000); Zhmodikov (2000); Quesada Sanz (2006). Sword encounters were therefore limited, see Goldsworthy (1996, 224 with n. 149 and n. 150), citing Clausewitz and Major General Fuller, who from their own field experience estimated hand-to-hand fighting lasted around 20 and 15 minutes, respectively. Goldsworthy writes that this limitation was due to the "physical and emotional strain" of such fighting.

13 Schweighäuser (1795, ad loc.) points out that Polybius' other two uses of this term are his description of Philopoemen striking an already crippled Machanidas with the butt of his spear (11.18.4) and the citizens of Abydos who, having thrown down their Macedonian opponents, struck at their faces as they lay on the ground (16.33.3).

14 On the origin and development of the *gladius Hispaniensis*, see Quesada Sanz (1997) and Bishop (2016).

15 Livy 31.34.4–5. Quesada Sanz (1997, 253–54) identifies the sword used in this encounter as the *gladius*.

16 Swank and Marchand (1946, 180).

17 Pierson (1999, 61–65).

18 Pierson (1999, 64).

It is probable that a small number of such men has always existed. Indeed, reading Pierson's description, it is difficult *not* to think of the legendary (or semi-legendary) L. Siccius Dentatus, whose military decorations numbered in the hundreds, and who, according to Valerius Maximus, "always seemed to take the greater part of the victory."[19] Although L. Siccius Dentatus himself may never have existed, he may be based on a real legate of the 140s, Q. Occius Achilles.[20]

Not everyone is a Dentatus. Nor even, according to Polybius, were the Romans the most naturally warlike of peoples: they were upstaged by the Macedonians, who were made ferocious by poverty and proximity to barbarians.[21] When faced with the Macedonians, other nations fled (Polyb. 4.69.6; 5.100.6). But this was not the case with the Romans, and it is precisely because the Romans were no more bellicose than any other people of their time that we may be able to see in their military culture a mixture of the factors which create killers: training, experience, the presence of comrades, close supervision by superiors, and, as is the focus of this chapter, the widespread societal ennoblement of violent acts.[22] The legend of the idealized Dentatus and his extensive decorations show that the acts for which he won those awards were considered positive and valuable. By contrast, if a soldier's activities in war are considered undesirable or even feared by the civilian population, rather than elevated and held in esteem, service can become unpopular, and soldiers' views of their own experiences become tainted by others' negative evaluations, as happened in twentieth-century Britain and in the United States after the Vietnam War.[23] Behind an enthusiastically fielded army of willing combatants lies a wider social system in which the act of soldiering is endowed with positive value. Societies supportive of war must have, in some way, specifically ennobled the activities involved in

19 Val. Max. 3.2.24: *ut maiorem semper victoriae partem traxisse videretur.* See Plin. *HN* 7.101–2; Gell. *NA* 2.11; Festus, *Gloss. Lat.* 208L s.v. *obsidionalis corona*; Amm. Marc. 25.3.14, 27.10.16. On the origins of the story, see Forsythe (2005, 208–9). Oakley (1985, 393 and n. 9) notes that he need not be real to exemplify attitudes of later generations.

20 The story of Occius is given in Val. Max. 3.2.21 and Livy, *Epit. Oxyrh.* 53–54; See also Oakley (1985, 396 and n. 33) and Forsythe (2005, 208) for the suggestion that he influenced the figure of Dentatus.

21 Eckstein (2006, 200–16, esp. 202–3) on the Macedonians.

22 For a breakdown of all the factors which contribute to encouraging soldiers to actively participate in combat, see Grossman (2009, 141–94).

23 In Britain in particular, soldiers returning from war were feared for "their chosen calling and their continued potential to use force": see Reese (1992, 5). The same was true of returned soldiers from Vietnam, who were suspected of unjustified violence. The psychiatrist Jonathan Shay, who treated veterans of the Vietnam War, wrote that for those at home, the veteran "carries the taint of a killer, of blood pollution": Shay (2002, 152). Another modern study observed that upon their return, soldiers look for gestures of acceptance, a societal endorsement that their actions had been legitimate and justifiable: see Fontana and Rosenheck (1994, 683).

organized violence.[24] In particularly war-positive cultures as, for example, the Yanomamö of the Amazon, military achievements are valued and those who can claim them are held in high esteem.[25] When the behaviors of a warrior or a soldier are generally viewed as positive, these positive values have the effect of perpetuating the military career, offering others a pathway to seek esteem through military means. Once ennobled, soldiering becomes an attractive option because it offers an avenue to social distinction, and distinction, then as now, is an innate desire.[26] "No-one" wrote Valerius Maximus, "is so humble that he is not touched by the sweetness of glory."[27]

The Roman triumph

From antiquity to the present, scholars have typically seen the triumph as for and about the general and his career, as well as the elite political landscape, with soldiers as mere accessories of little interest or regard.[28] Polybius wrote that a triumph was a means to display the achievement of a general to his fellow citizens, and a tool of the senate to either amplify or obscure individuals (Polyb. 6.15.7–8). There are no extant artworks that show the soldiers marching in their section of the parade.[29] When soldiers do turn up in accounts of triumphs, they are often viewed via their relation to the commander. Livy reports that Gaius Aurelius Cotta, the consul of 200, complained to the senate that they had authorized a triumph without hearing the testimony of the officers and soldiers. However, he interprets the physical bodies of these men as evidence – living attestations, not to their own efforts, but to the commander's achievements. They should have been allowed to walk in the triumph, he wrote, so that the Roman people could see the witnesses to the triumphator's deeds.[30] Modern authors have deviated little from this approach and have generally examined the triumph

24 Goldschmidt (1986, 8–9): "If a society is to have the advantages of military personnel, the motivations for warriorhood must be established. It is a matter of great significance that these must be created."
25 Chagnon (1988, 990).
26 Frey (2007, 7–8).
27 Val. Max. 8.14.5: *Nulla est ergo tanta humilitas quae dulcedine gloriae non tangatur.*
28 See Beard (2007, 241–42).
29 Beard (2007, 242) calls this phenomenon "a striking testimony to the selective gaze of Roman visual culture." Soldiers do sometimes appear in artwork in roles related to other parts of the triumph, such as carrying spoils.
30 Livy 31.49.10: *Maiores ideo instituisse ut legati, tribuni, centuriones, milites denique triumpho adessent, ut testes rerum gestarum eius cui tantus honos haberetur, populus Romanus videret* ("Their ancestors had ordained that legates, tribunes, centurions, and even the common soldiers should participate in a triumph, and their purpose was to let the Roman people see the men who had witnessed the feats of the person to whom that great honor was being accorded"). Of the two surviving manuscripts, one has a gap between *ut* and *rerum* where apparently the scribe could not read the word in the text he copied, and the other has *virtus*.

as a tool of elite power, authority, and influence, or as a religious ritual with the conquering *Imperator* at its center.[31] The theory that the songs sung by the soldiers had an apotropaic function similarly robs them of their agency in expressing their thoughts and opinions, and positions them as ritual protectors of their general against supernatural forces.[32]

Without consistently considering the soldiers' perspective, triumphs can look as if they did little in the interests of soldiers. The soldiers present were sometimes few, or even entirely absent.[33] A stipulation that the triumphing commander must return his army to Rome seems to have been little more than an infrequently applied instrument of senatorial control.[34] Despite this, some ancient sources are quite insistent that early iterations of the triumph were intended to highlight the soldiers' bravery, and that the emphasis on the role of the soldiers had originally been greater, then diminished over time. In Dionysius of Halicarnassus' version of the triumph of Romulus, the triumph is the same moment when the soldiers re-enter the

---

The emendation of *testes* is Madvig's and is generally considered correct, see Briscoe (1973, 162–63). On the dispute about the triumph, see Pittinger (2008, 168–80).

31 Richardson (1975) investigated how much control the senate exercised over triumphal procedure. Warren (1970, 49) equates non-combatant family members of the general and the soldiers, "the triumph's chief importance lay in the *auctoritas* and consequent political power it bestowed upon the victorious general, and the honour it brought his family and his troops." Östenberg (2009, 12) writes comprehensively of the triumph but still reflects this commander-centric default: "As the victorious general approached the city, Rome herself, her senators, magistrates, and people, went out to welcome him." Stroup's (2007, 29–30) well-written and dramatized description of a typical triumph unfortunately includes every part of the procession apart from the soldiers. More relevant to the perspective of soldiers is Pittinger (2008) who considers their role in permitting triumphs within the framework of Livy's histories. Rüpke has written extensively on how religion in a military focus centers the commander (1990, 243–45), and especially the triumph, Rüpke (2008) and notably (2006, 255): "The whole procedure was supposed to honour the triumphing general."

32 Rüpke (1990, 230–33); Versnel (1970, 380). Richlin (1992, 10) sees this, with the *Ludi Florales* and the Saturnalia, as a reversal of the normal moral and social order, the common man protecting the prominent from evil.

33 Q. Fulvius Flaccus triumphed with only the soldiers who had been selected for release from the province of Spain in 180, either because they had served their time or had been discharged for bravery: Livy 40.40.14–15, 40.43.4–7, 42.34.9–10. In 195, Q. Minucius Thermus celebrated a triumph over the Spanish, despite having turned his troops over to his successor when he left the province: see Richardson (1975, 61), citing Livy 31.10.6–7 (Minucius' triumph) and 31.17.1 (the army handover). M. Claudius Marcellus was allowed to celebrate an ovation in 211, despite having handed over his troops to his provincial successor (Livy 26.21.4; Plut. *Marc.* 22.1), and in 200 the praetor Lucius Furius Purpureo triumphed with no soldiers at all (Livy 31.49.1–3).

34 On the *deportatio exercitus* stipulation, see Richardson (1975, 61–62). It is attested first in Marcellus' case in 211, where it is cited as a reason for the ovation instead of the triumph. Although applied to the case of C. Claudius Nero in 207, the rule seems to drop out of consideration soon afterwards, was revived only when it became useful to stem the tide of triumphators from the protracted wars in Spain in 185.

city, while male citizens with their wives and children line the road welcoming the soldiers back and congratulating them, before ushering them into the city to partake of food and wine.[35] He complains that the triumph had grown more luxurious over time, and was in his day "rather an exhibition of wealth than an appreciation of valor."[36] Plutarch says that the turning point came at the *ovatio* of M. Claudius Marcellus over Syracuse in 211, and he describes the landscape of the city as having a warlike appearance at that date, "quite full of barbarian arms and bloody spoils, encircled with memorials of triumphs and trophies."[37] Livy blames Cn. Manlius Vulso's triumph in 187 for the introduction of luxury goods. For him, Manlius' over-indulged troops heralded the beginning of the softening of the Roman character, and then found expression both in the luxury goods of the triumph and an enthusiasm for elaborate banqueting (Livy 39.6.6–9).

The attempt to isolate the watershed moment that Rome was tipped into self-indulgence is a literary trope, and in fact, statuary and paintings captured in Rome's wars were likely a familiar sight in Rome from the fourth century.[38] The idea of a Roman culture that was previously simple and austere must be an overstatement.[39] The crucial point at issue here, however, is not the introduction of luxury *per se*, but the fact that the Dionysius, Plutarch, and Livy all claim a past in which the focus of a triumph had been on soldiers and martial prowess. They allege that the change occurred sometime around the end of the third century and beginning of the second century. Although all suffer from the anachronistic assumption that armies in very early periods were state-run and controlled – a situation much more in keeping with the second century than the sixth and fifth centuries – there is evidence that both the general and community had strong incentives to celebrate soldiers during early triumphs.[40] These soldiers had a permanent, or semi-permanent, relationship with their general. In Rome's earliest periods, the distinction between public and private military enterprise was blurred, exemplified by the war of the *gens* Fabia and their retainers against Veii in 479–77.[41] Members of the Italian elite and their followers were able to move between communities, and we hear of Etruscan, Sabine, and Roman individuals who migrated into and out of Rome.[42] Dubbed *condottieri*, after similar warlords in Italy in the Middle Ages, these elites led members of their own *gens* but also clients, freedmen, and the more mysterious *soldales*

35 Dion. Hal. *Ant. Rom.* 2.34.2, see also 5.17.2.
36 Dion. Hal. *Ant. Rom.* 2.34.3: εἰς πλούτου μᾶλλον ἐπίδειξιν ἢ δόκησιν ἀρετῆς.
37 Plut. *Marc.* 21.2: ὅπλων δὲ βαρβαρικῶν καὶ λαφύρων ἐναίμων ἀνάπλεως οὖσα καὶ περιεστεφανωμένη θριάμβων ὑπομνήμασι καὶ τροπαίοις.
38 Gruen (1992, 84–113, esp. 107).
39 McDonnell (2006a, 78) makes this point but argues that the general idea of a Rome which admitted more luxury over time need not be dismissed out of hand.
40 On the state control of armies, see Drogula (2015, 20–21).
41 Rawlings (1999, 102); Armstrong (2016a, 48–50).
42 Rawlings (1999, 103–6) provides an extensive list; see also Rich (2007a, 15–16).

or "sword-brothers" who fought with the *gens* but did not seem to belong to it.[43] Armstrong has pointed out that the middle and late Republican practice of distributing donatives to the soldiers outside the city gates prior to the triumphal procession, without the presence of city magistrates, is a reflection of the early private relationship between general and soldiers.[44] On the occasion of a triumph, the warlord would benefit from giving donatives to his troops, recognizing bravery, courting the loyalty of his soldiers, and advertising his military strength to the wider community. These early triumphs were a means by which a community could stake a claim in otherwise private victories and were crucial to the societal acceptance of warfare.[45]

The early war bands could be a source of safety and prosperity to an otherwise vulnerable community. The site of Rome was, like all communities in archaic Italy, a small place on a bellicose peninsula characterized by constant low-level raiding. By the fifth century, Rome's new civic recruitment system was still dominated by aristocrats and their clans. When these semi-private armies were fighting close by, or repelling attacks from other states, there would have been an immediate sense that these men had put themselves in danger for the lives and properties of members of the community. It is worth comparing this to Polybius' account of the aftermath of the battle of Zama, where he writes that the Roman people were overjoyed to be freed from fear, and the contents of the triumph reminded them vividly of their former peril (Polyb. 16.23.4–5). Set in the fifth and fourth centuries are also tales of ordinary soldiers doing extraordinary deeds, and this reflects the idea that violence in the military context had become a legitimate and desirable source of individual distinction.[46] The number of recorded single combats in particular suggests that this was a real practice from an early period, or at least regularly attributed.[47]

Exactly how and when Rome's military endeavors came completely under the control of centralized government is not easy to discern from the available evidence.[48] It seems that generals and magistrates were not necessarily one and the same until after the Licinio-Sextian Rogations in 367, and the practice of having two consuls – who were also army leaders – sometime after that.[49] Around the end of the fourth and the beginning of the third centuries, the merging of the two offices likely started to change the character of the triumph. When the magistrate-general's triumph marked an end to a campaign, it was often the end of his military career too, and so he had no need to build military strength and following, engender personal loyalty, or advertise that he had the best warriors and the most lucrative opportunities,

43 Armstrong (2016a, 31–36).
44 Armstrong (2013a, 16–17).
45 Armstrong (2013a, 13–16).
46 Rich (2007a, 20).
47 Oakley (1985).
48 See Drogula in this volume for one suggestion.
49 Drogula (2015, 33–45).

as the earliest generals had done. A triumphing general needed to prioritize translating his victory into political power so that a one-time event could raise the level of his own personal influence and maximize his opportunities in the perpetually competitive political landscape.[50] It is more than coincidental that the earliest date given for larger and more ostentatious triumphs is offered by Florus (1.13.27), who points to the parade of Greek art in a triumph in 275 or 272 as a deviation from the normal display of purely military spoils.[51]

Just as the practice of greeting the triumphing soldiers with honeyed wine is attested in Plautus, confirming that this was still in place in the early second century (Plaut. *Bacch.* 1074), there are other ways in which the triumph had – and continued to convey – "an appreciation of valor." War creates a discrepancy between the soldier's experience and the civilian's, especially because it can (and often does) involve acts of extreme violence. Veterans returning from Vietnam feared what other people would do or think if they knew what they had done abroad.[52] The Romans offer a stark contrast in their approach to negotiating that gap in experience. Rather than hide its details, they sought to tell that story, with pomp and ceremony, with detail, and pageantry, and celebration. In triumphs, they brought forth representations of the towns, cities, and places where the soldiers had been; the standards, equipment, and even the people whom they had fought. Many soldiers bore military decorations which attested to exact deeds that they had done, and these are perhaps the symbols of status and distinction that accord least to our own perspectives on the world. They celebrate what so many modern societies have found unpalatable: individuals engaged in specific, proactive, enthusiastic participation in military activities that either confirm or imply lethal violence. The soldiers with prizes were conspicuous, and the public and celebratory occasion ennobled these activities and conveyed, in no uncertain terms, that they honored and distinguished their recipients. The whole endeavor facilitated collective and individual narratives of soldiers' experiences in a detailed and exacting manner.

The physical context of the soldiers' experiences, or the places where they had been, was displayed in the form of models. Representations of towns were carried in the procession, which depicted the towns the Romans had conquered. These were three-dimensional models rendered in ivory, wood,

---

50 The concept of transforming one-time victories to political power comes from Hölscher (2006), who addresses structural, institutional political power on the level of the state and society, and frames triumphs as a means to accomplish this by involving citizens in a ritual of victory.

51 On exactly where this triumph sits in the history of Greek art at Rome see McDonnell (2006a). For Florus' confusion of triumphs in 275 and 272, *ibid.* 74. For the increasing aggrandizement of triumphs and especially the rise in public buildings in the second century: see Fronda in this volume.

52 Shay (2002, 152).

and other materials.[53] In Latin, these are *simulacra oppidorum*, literally "representations of towns." From Strabo and Appian, the Greek word *pyrgoi* gives us "towers."[54] Östenberg suggests that this means the emphasis of the representations was on city walls, a feature which would parallel the representation of walls and battlements on the *coronae muralis*, the award given to soldiers who were the first to scale the walls of an enemy fortification. Certainly, the image of a city as a set of walls may have been a more accurate impression of a soldier's memories of the campaign than a representation of the city's inner map. Sieges could take weeks, months, or even years. The siege of Lilybaeum in 250 lasted nine years. Syracuse, in 213, lasted for two years. The Third Punic War was, in almost its entirety, a siege of Carthage, which lasted three years and culminated in six days of street fighting and destruction within the walls.[55] It could take weeks just to construct towers and ramps, and while this is a small part of the typical time frame of ancient warfare, it is a rather long time for a human being to sit under a city's fortifications.[56] Just as a postcard suggests the sense of a travel experience to someone who stayed at home, the model depicted the soldier's viewpoint. Against this backdrop was implied action or drama, since its inclusion in the triumph indicated that the Romans had succeeded in taking that particular town or city. The fact that city walls were depicted on *coronae muralis* hints at a parallelism between the symbols of a collective achievement and those honoring distinguished individuals.[57]

Other military actors, instruments, and weapons were included in a triumph, which were real rather than representations if possible. Thus, we find ballistae, catapults, and "all the other instruments of war" (*alia omnia instrumenta belli*) at Marcellus' *ovatio* over Syracuse (Livy 26.21.7). M. Fulvius Nobilior had the same in his triumph of 187 (Livy 39.5.15). There were also the enemy people, and their standards and arms. Captured enemy standards opened the parade.[58] Plutarch describes the weapons and armor displayed in the triumph of Aemilius Paullus in 167 as the choicest pieces of the captured weaponry, freshly cleaned, although their display

---

53　These models are described in triumphs which took place from 201 to the Imperial period. See Östenberg (2009, 199–203).

54　App. *Pun.* 66, in reference to the triumph of Scipio Africanus in 201. Strab. 3.4.13, quoting Posidonius in such a manner as to confirm that both Strabo and Posidonius knew the word and its use for representations of cities in triumphs. For discussion, see Östenberg (2009, 199–205).

55　The siege: App. *Pun.* 94–126; the duration of the sack: App. *Pun.* 130.

56　For the length of time to construct siege works, see Levithan (2013, 128–29). For an example of a soldier who seems to spend much of his time staring at walls, see the pseudo-caesarian author of the *Bellum Hispaniense*, who mentions walls 14 times in his short account (*BHisp.* 3, 13, 15, 19, 29, 34, 35).

57　Östenberg (2009, 204–5) elaborates on this parallel, but sees it as between the triumphator's achievements (the *simulacra oppidorum*) and the soldiers' (the *coronae muralis*) rather than collective and individual soldiers.

58　Östenberg (2009, 38–41).

had been carefully stage-managed to gesture at authenticity. He relates that the helmets, shields, swords, greaves, and spears taken from the enemy were heaped as if at random, and as they were rolled along on their wagons clashed against each other with such an alarming sound that to witness them was chilling (Plut. *Aem.* 32.5–7). The intimidating presence of this equipment would have invited the audience to reflect upon the context in which it was used, and to imagine sights and sounds: the armor clanking and ringing on the bodies of Rome's enemies, swords and spears struck against shields in a fearsome pre-battle din. Nor were those enemies lacking, for the triumph included captured prisoners. These were preferably chiefs, kings, and their families, although there were non-elites as well – like the 50 survivors Scipio Aemilianus took from the siege of Numantia to walk in his triumph, or the Celts sent to Rome by L. Aemilius Papus in 225 for the same purpose.[59] Some middle Republican triumphs included horses and elephants from enemy armies.[60] The enemy did not have to be imagined, for those captured and taken to Rome were a small group of the exact people the Roman soldiers had encountered, and their clothes, hair, faces, and statures offered a close paradigm for those who had not been selected for the triumph, and had instead perished in battle or been sold as slaves. It is clear that the point of these captives was to detail actual situations and sights as accurately as possible. Florus (1.37.5) records that in 120 the Gallic king Bituitus appeared in the triumph of Fabius Maximus over the Allobroges, dressed in his arms on his war chariot, "just as he had fought" (*qualis pugnaverat*).

Actions and awards

Audiences evidently liked to see a high number of soldiers with military decorations and rewards in the triumphal processions. Livy (10.46.2–3) writes about the remarkable triumph of L. Papirius Cursor over the Samnites, where the first reason given for its distinction was that both infantry and cavalry were highly decorated. Those soldiers who had particularly distinguished themselves during the campaign were marked with various forms of awards, some for specific, named deeds, and others for acts of bravery observed by the general or his officers. Awards, of course, signal strongly the kinds of behaviors and acts that are desirable to the donor or the donor institution.[61] The set of behaviors that are rewarded create concrete examples to help shape the ideas of others in the group about what they are supposed to be doing.[62] Just as Roller observed that Rome's legendary stories had a strong proscriptive quality, intended to encourage emulation

---

59 App. *Hisp.* 98; Polyb. 2.31.1–6.
60 Östenberg (2009, 171–84); Östenberg (2014a).
61 Marinova, Moon, and Dyne (2010, 1471); Frey and Neckermann (2008, 199).
62 Mickel and Barron (2008, 335).

of the heroic acts depicted in them, soldiers who were decorated in triumphs showed behaviors that were wanted, needed, and valued while serving in the legions.[63]

The behaviors which gained awards in the middle Republican legions relate not simply to duty but eager, proactive engagement in military activity. They were decorations that rewarded the habitual behaviors of the "natural killers" – or the *Dentati* of this world – that most people will not do instinctively but can be induced to, given the right motivation. Polybius gives us a list of various articles of war given as awards and the set feats necessary to win them. Soldiers who had wounded an enemy earned the *hasta pura*, a type of spear. Those who killed and stripped an enemy earned *phalera* (horse-trappings) if he was in the cavalry and *patella* (a shallow dish) if he were infantry.[64] Polybius is clear that these acts would not win decorations if done in the normal course of a battle, but were given to a man "who, faced with no compulsion, in skirmishing or any other opportunities voluntarily and deliberately endangered himself."[65] Similarly, there was the *corona muralis*, already mentioned, and its sister award the *corona vallaris*, for the first man over the wall of an enemy camp. The winner of these awards had to have been an impatient and competitive soldier who had rushed to the very front of the fighting, with the intent of an aggressive attack on the defenders inside the fortification. Even the decorations which relate to defending or preserving life could be achieved only with the application of proactivity and initiative. The *corona civica* was awarded to a soldier who saved the life of a fellow citizen in battle, and then held the ground where the deed had happened for the rest of the day. The implication of this latter stipulation was the halt of a reversal. The awards as a whole reflect an ideal soldier who was at the front when others were charging, and holding his ground when others were falling back. Other awards, such as the *corona aurea,* were at the discretion of the commander and awarded for miscellaneous acts of bravery. Mentions of these awards are unfortunately rarely accompanied by detail about the deeds for which they were won, although hints can be found in the idealized versions. Siccius Dentatus, for instance, claimed eight golden crowns, and Dionysius of Halicarnassus has him detail the exploits that led to one of them. It was during a rout of a Roman cohort that Dentatus wrestled their standards back from the enemy, "taking on danger on behalf of all,"[66] thus saving the centurions of the cohort from disgrace. The key point is the reversal of fortune of a cohort that had been defeated and was in retreat.

---

63 Roller (2004, 2009). That Roman and allied soldiers had similar access to awards and decorations, see Rosenstein (2012a, 102–3); see also Fronda in this volume.

64 Polyb. 6.39.3. For the identification of the objects with their Latin names, see Maxfield (1981, 61–62).

65 Polyb. 6.39.4: ἐὰν ἐν ἀκροβολισμοῖς ἤ τισιν ἄλλοις τοιούτοις καιροῖς, ἐν οἷς μηδεμιᾶς ἀνάγκης οὔσης κατ᾽ ἄνδρα κινδυνεύειν αὐτοί τινες ἑκουσίως καὶ κατὰ προαίρεσιν αὐτοὺς εἰς τοῦτο διδόασι.

66 Dion. Hal. *Ant. Rom.* 10.36.5: μόνος ἐγὼ τὸν ὑπὲρ ἁπάντων κίνδυνον ἀράμενος.

Since discretionary awards had no set rules, it must have been difficult to enforce consistency across a middle Republican system which continually fielded consular legions that were raised as an entirely new collection and configuration of men each year. There was no standing institution of the "army," no physical base, no permanently serving officers.[67] In order to remain meaningful incentives for emulation in the way that Polybius described, the decorations and spoils won by various soldiers needed to represent broadly the same quality of achievements with the same degree of difficulty. The value of awards depends on them being perceived as being fair. The army, in other words, needed to maintain a broad level of "fairness" in order to maintain the perception across the community that it was a legitimate arbiter of distinction. Unreliable, unfair, or overly liberal distribution of awards would decrease their exclusivity and hence their value. This is what is behind the Elder Cato's scorn of Fulvius Nobilior, whom, according to Aulus Gellius, Cato criticized for giving awards in mundane circumstances: not for actions at the capture of a town or the destruction of an enemy camp, but for industry in building ramparts or digging wells (Gell. *NA* 5.6.24–6). Noticeably, these are not aggressive or violent acts directed at an enemy.

In the same way, the value of an award also depends on the prestige of the donor. Commanders and officers needed to be generally similar in their judgments in order to uphold the value of prizes in general but also because their own prestige, and the perception of others about their fairness and discernment, depended on how they distributed awards.[68] This mattered for the soldiers, too, for an award for bravery from an extremely well-respected and discerning commander would automatically convey more prestige than from one who had a reputation for mistakes or terrible judgment. This dynamic means that awards create a psychological bond between awardee and donor, since the recipient becomes personally invested in upholding the prestige of their source.[69] In our case, a decorated soldier would become invested in the reputation of the commander who made the award and the officer whose recommendation it had been, as well as the idea of the army as a legitimate path for distinction. The act of accepting an award means that the recipient is incentivized to protect or even boost the donor's reputation in order to protect the value of his prize, since "[recipients] benefit from speaking in favourable terms, and acting in support of, the award giving institution."[70] This brings us back to the soldiers in the triumphal parade, who had good reason to be positive about their leaders in order to uphold

---

67 On the general abstractness of the middle Republican "army," see James (2002, 38–39).

68 Frey (2007, 11) illustrates this principle using the church's process of beatification and sanctification, which is scrupulous because of how badly an unworthy recipient would impact the church's reputation.

69 Frey (2006, 382).

70 Frey (2007, 8).

the legitimacy of their own distinctions. The triumph was the perfect opportunity to do so, for as they processed, they sang songs, most famously of the ribald sort as in the triumph of Julius Caesar (Suet. *Iul.* 49, 51). Scholarly approaches to this practice have seen the verses as apotropaic, intended to ward off evil from the general celebrating success, or to allow him to reintegrate into the community by diminishing the standing he had acquired by his great victory.[71] The songs, however, were not exclusively mocking, and even those making fun conveyed a positive impression of the general: Caesar claimed to be delighted in his soldiers' frankness because it showed their trust that he would not become angry (Cass. Dio 43.20.4). If there is truth in this, then enduring his soldiers' mockery could be a rich opportunity for the commander to demonstrate his even-handedness. Certainly, soldiers who really thought their commander unfair did not reveal it in jesting song: at his triumph in 177, C. Claudius Pulcher gave his allied troops half the amount of money that he gave citizens, and they registered their protest not in verses but by following his chariot in silence (Livy 41.13.8). Many triumphal songs were simply complementary of the commander, such as those sung in the triumphs of M. Claudius Marcellus in 222, Cn. Manlius Vulso in 187, and L. Aemilius Paullus in 168.[72] Neither were all the songs about the generals. Some honored other officers. Appian says that at the triumph of Scipio in 201 the soldiers referenced several of their leaders in both complementary and derogatory ways (App. *Pun.* 66). Livy notes that the tribune P. Decius Mus stood out in the triumphal parade over the Samnites in 343 because of his awards and because of the songs the soldiers sang about him, and at the triumph of M. Livius Salinator in 207, the cavalry used their verses to distinguish two legates.[73] The soldiers' praise of their commander and officers helped protect the value of their service and especially the awards they had won.

The soldiers' songs also amounted to a collective review of their leaders. The words used to describe them imply that they were not particularly polished. Dionysius of Halicarnassus says that the mockery was once performed in prose, but had changed to verse by his time, and he calls such works "offhand" or "improvised."[74] Livy refers to them as *inconditus*, "rude" but also literally "unbuilt" or "unformed."[75] This suggests that they were composed around the occasion of the triumph, perhaps during the assembling, while the soldiers waited to process, or on the spot as they marched to the tune of a well-known song or rhyme. Another possibility is that they had been made

---

71 Versnel (1970, 70); Warren (1970, 65); O'Neill (2003, 3–4).
72 Plut. *Marc.* 8.2; Livy 39.7.3; Plut. *Aem.* 34.7. Cf. on the triumph of Romulus, Dion. Hal. *Ant. Rom.* 2.34.2.
73 Livy 7.38.3; 28.9.19.
74 Dion. Hal. *Ant. Rom.* 2.34.2: τὸν ἡγεμόνα κυδαίνουσα ποιήμασιν αὐτοσχεδίοις ("they praised their leader in improvised poems").
75 Livy 4.20.2, 5.49.7, 7.38.3.

up, wholly or in part, while the soldiers had been on their campaign – either simply for sport or with a potential triumph in mind.[76] Since the soldiers composed the songs, they also chose both their subjects and whether the verses praised, mocked, or derided those subjects. A certain level of consensus must have existed on these issues. Some officers had been popular, or capable, or brave, while others were mocked in friendly jest, and some reproached. These opinions must have been reflections of the general feeling in the camps and the subjects of soldiers' gossip. When sung in triumphs, they gave voice to their direct, collective impression of their leadership, and these positive and negative reviews were important enough that they were noted in sources and made their way into the works of our extant historians.

Monuments and memory

Not all soldiers, of course, won spoils and decorations on campaigns which ended in a triumph.[77] In the middle Republic, many soldiers would have served Rome's armies who never had occasion to triumph in Rome, or who had done so but came from communities located too far away for community and family to make the journey, or who had served in undistinguished campaigns.[78] There were, however, permanent commemorations of campaigns. Plutarch describes the city of Rome in the time of Marcellus as full of arms, spoils, memorials, and trophies, and that the sight was not for the faint of heart (Plut. *Marc.* 21.2). Similarly, Silius Italicus imagines the Curia on the eve of the Second Punic War, adorned with chariots, shields, armor, and even weapons with the blood still in evidence.[79] This display has a narrative quality, evoking past wars against the Carthaginians, Gauls, Ligurians, and others. Both Plutarch and Silius seem to take delight in the magnificently intimidating quality of such a landscape and, as before, the picture is likely to be overzealous. There is, however, plenty of evidence for displays of spoils on public buildings and in temples. In the aftermath of the battle of Cannae, the ad hoc legions raised to continue the fight were armed with the spoils taken from the Gauls by C. Flaminius, which had

---

76 See Chrissanthos (2004, 355–56), for instances of soldiers discussing matters in camps.
77 The chances of walking in a triumph varied massively in the middle Republic. The most bountiful era for triumphs was from 200–166, in which there were 41. At the bleak end of the scale, there were only six triumphs during the 18 years of the Second Punic War 218–201. For these and the frequency during other periods, see Rich (2014). At the height of hostilities, in 212–11, there were no less than 25 legions in service: see Brunt (1971a, 417–22). This means that soldiers were much more likely to be killed in battle than participate in a triumph.
78 The exception is the "triumph" of T. Quinctius Flamininus through Italy, on which see Fronda in this volume.
79 Sil. *Pun.* 6.617–29. On the dubious historicity of Silius' description, especially that spoils were hung on the Curia, see Rawson (1990, 160–1).

presumably been on display since his triumph in 223.[80] Statues and paintings taken from conquered cities stood as reminders of campaigns. Discussing the dedications made by M. Marcellus in the temples of Honos and Virtus, which were apparently tourist attractions, Orlin remarks that the dedications in the temples would have reminded visitors of the "campaign and the individual who had brought these objects to Rome."[81] In one case, a military tribune dedicated part of his share of the spoils to Mars and to Fortuna, and dedications like this could have been made by other officers.[82] The reminder of the fame of the *imperator* and a prompt to remember the efforts of the soldiers of that campaign are not mutually exclusive. For many visitors, it would have been more pertinent to remember the soldiers who had fought, especially if they were friends, relatives, community members, or young boys anticipating what their own service might be like.

It was not only weapons and armor that preserved the memory of campaigns but also the so-called "triumphal paintings." These were painted depictions of battle scenes, and were likely not, as sometimes assumed, carried in the triumph, but were made for commemorative use to be displayed on temples or public places.[83] Four examples of this practice are known from the middle Republic. Three are mentioned in Pliny's *Natural History*: M. Valerius Messala's depiction of his victory over the Carthaginians in 263, L. Scipio Asiaticus' victory over Asia, and L. Hostilius Mancinus, who was a legate in the Third Punic War and commander of the fleet, and presided over the first break into the city of Carthage (Plin. *HN* 35.7.22–3). According to Pliny, the latter stood by his painting in the forum, narrating the events depicted to an audience, and was so popular for doing so that he won the consulship in the next election. The existence of these paintings and the remark about Hostilius' stories both indicate an appetite for detail among Roman audiences. The care taken over the informative details of triumphs and the attractiveness of narrative accounts are evidence that war stories were popular and desirable. The fourth painting from the middle Republic was one in the shape of Sardinia, which Ti. Sempronius Gracchus displayed in the temple of Mater Matuta in 174. This painting of the various battles had an accompanying explanatory text, which, according to Livy, began, "Under the auspices and command of Tiberius Sempronius Gracchus, the legion and the army of the Roman people subdued Sardinia."[84]

80 Livy 23.14.4. The order from the senate is recorded at 22.57.10–11. Valerius Maximus (7.6.1b) says merely that they were taken from temples, plural.
81 Orlin (1997, 136).
82 *ILLRP* 100, 221. Degrassi identifies the military tribune as M. Furius Crassipes, praetor of 187 and 173. Poccetti (1982, 669–70) points out that this must have come from the booty personally given to him by the *imperator*, who alone had jurisdiction over the dedication of captured spoils.
83 Östenberg (2009, 192–99).
84 Livy 41.28.8: *Ti. Semproni Gracchi consulis imperio auspicioque legio exercitusque populi Romani Sardiniam subegit.*

This illustrates the dual function of such paintings rather well: Gracchus is named first as the commander, but it is clear in the subject of the sentence that those directly responsible for the action commemorated are the Roman soldiers who fought. We know of one painting, not strictly "triumphal," in which the subject was the soldiers: after Sempronius Gracchus, whose soldiers' story opened this chapter, won his battle, he commissioned a painting of the soldiers celebrating and feasting in the streets of Beneventum. Many of the slave soldiers had won their liberty that day, as Gracchus had spared and freed even those who fell back – all 4,000 of them – who later retreated away from the camp for fear of punishment. According to Livy, these men were divided in the painting of the celebration as well. Although they all wore caps of liberty, by Gracchus' order those who had fought reclined like citizens, while those who had retreated stood and waited on the others, like slaves.[85] Placed in the temple of Liberty, this painting separated the men just as the battle had done: the eager, proactive, violent men enjoying a higher level of prestige and distinction than those who had backed away in fear.

Those means of commemoration that were under the control of Rome's elite – like the triumph, "triumphal" paintings, and dedications of spoils in temples – were naturally centered in the city of Rome. Of course, not all soldiers were Roman, and even fewer were Romans who came from the city itself.[86] The display of spoils, however, was not entirely exclusive to urban Rome, as sometimes spoils from Roman victories were sent to other towns. Cicero tells us that in the middle Republic, commanders placed spoils not in their own houses, but in the city, the temples of the gods, and "all parts of Italy."[87] Some spoils, then, were fixed in public spaces in Italian towns just as in Rome. Additionally, many more soldiers would have returned to their communities without triumphing than those who did – because they had left a campaign early, or it had not been that successful, or because the Romans had been defeated – although this did not preclude them from winning distinctions, which were often awarded in army camps following the battle in which they had been earned.

The nature of these award ceremonies in the field facilitated the dissemination of heroic stories and their value in a civilian context. Polybius writes that an assembly of the soldiers was called, in which the commander brought forward the men to be decorated (Polyb. 6.39.2). Of note here is that the distinction won by a soldier had two parts: the tangible object of reward and the accompanying narrative that was spoken by the commander. The form of the award was connected to, and symbolic of, an act, as described

---

85 Livy 24.16.18. On the possible appearance of this painting and on the reliability of Livy's description, see Koortbojian (2002, esp. 36–37). It may have still been extant in Rome.
86 Brunt (1988, 253–54).
87 Cic. *Verr.* 2.1.55: *omnisque Italiae partis*; see also Livy 24.21.9, who writes that Roman spoils were sent to Syracuse. See Fronda in this volume on various "triumphal" displays before Italian audiences.

above. A narrative account of the deed itself was described to the audience of soldiers along with any other act of note which the soldier had done previously.[88] This speech aided any future use the soldier might wish to make of his achievements. The fact that the awards were made at assemblies of the soldiers served to share and legitimize the narrative, making witnesses of men who had not seen the deed itself. In turn, this would have helped to both spread and corroborate the soldier's claim to distinction at home.[89] Every soldier with army experience would have been able to identify the meaning of the decorations displayed on houses and the acts they symbolized, whether he knew its owner or not. A narrative heard by many would also have facilitated the soldier's continued service in legions. Since middle Republican armies had few sources of continuity, ordinary soldiers held their rank only for the duration of a campaign, and petitioned to be (re)appointed at particular ranks after gaps in service.[90] This practice implies that some form of record, memory, or soliciting of testimonies from others was sought to corroborate claims.

Conclusions

Polybius and modern scholars of war agree that good soldiers, for the most part, are made and not born. At the same time, it seems that in the Roman system there was the assumption that almost everyone was capable of distinguished deeds. As well as those who were present in the camp and listened as the man was given his award, those who stayed at home were also "urged on to competition and rivalry" through the display of spoils on private houses.[91] Pliny writes that the great men of the past had spoils fastened on the outside of doors and at thresholds, which it was forbidden to remove, even if the house was sold (Plin. *HN* 35.2.7). These were never repaired but left to disintegrate with the passage of time (Plut. *Mor.* 273c–d). Any soldier

88 Polyb. 6.39.2. The information (and verification) about who had done what potentially involved a number of people: Polybius says that soldiers could be punished for lying about their achievements, which implies self-reporting (6.37.10), and he states that those citizens whose lives had been saved by another crowned that man either voluntarily or under compulsion if the tribunes had adjudicated the case that way, indicating the involvement of senior officers (6.39.6). According to Livy, Scipio himself adjudicated a dispute at his assembly after the siege of New Carthage (26.48.13–4).

89 Here, with de Ligt (2007), I assume a preliminary stage to the *dilectus* in which only certain tribes were called, making it likely that a limited number of areas were represented in each levy. The exact mechanisms of the *dilectus* are, however, unknown, and the question is further informed by three chapters in this volume: the matter of which communities were contributing troops to Rome (Tan), the possibly idealized and dubiously practical Polybian account of the *dilectus* (Armstrong) and the view that the tribes had become geographically disjointed by 241 (Helm).

90 Consider the centurion Spurius Ligustinus (Livy 42.34) who says specifically that tribunes assigned rank to the men they recruited.

91 Polyb. 6.39.8: ἐκκαλοῦνται πρὸς τὴν ἐν τοῖς κινδύνοις ἄμιλλαν καὶ ζῆλον.

could display spoils that he had been awarded (Polyb. 6.39.9–10). A number of authors refer to the practice, imagining it to stretch back to the Bronze Age.[92] One of Plutarch's suggestions for why spoils were left to disintegrate in place was that men felt their reputations lasted only as long as the spoils did, and they would thus be motivated to earn fresh distinctions (Plut. *Mor.* 273c–d). This implies that the condition of spoils on a man's doorpost could be read by others to convey not just the existence of, but also the quality and consistency of, the owner's bravery; whether it had been proven once, long ago, or whether it was proved continually and consistently through renewed honors. Pliny remarks that the owner of the new house would have a powerful prompt to emulation (Plin. *HN* 35.2.8). Spoils returned to soldiers' homes did not have the context of ostentatious pageantry, a ceremony with detailed visuals, or a huge crowd whose collective enthusiasm confirmed and authenticated the distinction the man had earned. Nevertheless, they were a visual prompt for visitors to ask the homeowner their origins, an advertisement, and endorsement of the legitimacy of Roman armies to bestow distinction and an affirmation that proactive and violent military activity was a positive value to which one should aspire.

The display of spoils or decorations on one's house amounted to a declaration that the homeowner personally stood behind these principles. The soldier at home held a large stake in continuing the culture which ennobled the acts for which he had won his awards and spoils. His reputation, and the status he enjoyed from his decorations, depended on the community believing and continuing to believe that whatever deed he had done was "good" and had value, and that the military and its representatives were legitimate arbiters of that worthiness. By accepting a decoration from his commander and displaying it on his house, a soldier had invested himself in the reputation of that commander and, underlying that, the value of the military endeavor in general. The value of that award, and the social distinction he enjoyed from it, would benefit from his promotion of the military and the narrative he himself told of his achievement. The more he aggrandized his deed, explained his peril and the gory, successful result, the more exclusive the award would seem – and the same was true of describing the deeds of others. The soldier was invested also in others believing that his service had been fairly judged and that his officers were worthy men. The legions, in other words, sent soldier-ambassadors back to towns who were deeply invested in speaking well of army service as a path to greatness. The consequence of this arrangement is that Rome's biggest promoters of military service were those with the ability and the temperament to engage in proactive, enthusiastic acts of war, which either explicitly required or heavily implied brutal acts of lethal violence.

---

92 Spoils attached to doorposts appear in: Serv. 7.183; Sil. *Pun.* 6.434–5, 445–46; Livy 38.43.10; Verg. *Aen* 2.504 (at Troy); 7.183–5 (at the palace of Latinus); Prop. 3.9.26; Ov. *Met.* 8.154 (King Minos' palace at Crete).

9 Uncovering a "Lost Generation" in the senate

Demography and the Hannibalic War*

Cary Barber

Introduction

During the two years that followed Hannibal's winter invasion of Italy in 218, Roman and allied casualties swelled first into the tens of thousands, and from there perhaps even into the hundreds of thousands of lives.[1] These figures are shocking even by modern standards, and a simple calculation can reveal their true extent in better detail: Brunt estimates (rather optimistically) that there were roughly 300,000 Roman citizens at the time of Hannibal's invasion, three quarters of whom were of fighting age.[2] Against these figures, his conservative estimate of 50,000 citizen casualties from 218 to 216/5 represents the loss of 17% of the total citizen population and over 22% of Rome's potential fighting forces. The loss of one in five, or even one in four, Roman citizens – a far more approachable and appreciable figure than 50,000 – imbues the extant narratives of these events with added meaning, and helps to contextualize and explain the radical recruitment that followed, as men over 35, adolescents, and even the enslaved were levied in response to severe levels of attrition among the ranks of the *iuniores*.[3] It was not only the lower classes who suffered such appalling losses in the early stages of the war. Rather, the ruling elite may have experienced an even higher proportion of wartime mortality. For among the countless thousands who perished at the hands of natural and battlefield attrition between 220 and 216, there lay 177 citizens who accounted most within the Republic's steeply hierarchical

---

* All dates are BC unless otherwise noted.

1 Hannibal defeated a series of Roman and allied forces between 218 and 215. Brunt (1971a, 419) offers a relatively conservative estimate of 50,000 Roman citizens dead by 215. Even such a low estimate is astounding. Supposing equal casualties among citizens and allies, then combined losses could easily reach the mark of 100,000 lives lost at a minimum.

2 See Brunt (1971a, 61ff.), though this does not include his 10% for underreporting. Against the results of the most recent census for which there are figures (in 234, reportedly 270,000 citizens), these losses grow far worse. Brunt's calculations should therefore represent the "best-case scenario" for the Republic: a high figure in terms of total population, and a low figure in terms of overall casualties.

3 Men over 35 conscripted after Trasimene: Livy 22.11.8. Recruitment of young and slaves in 215: Livy 22.57.9–12.

DOI: 10.4324/9781351063500-9

This chapter has been made available under a CC-BY-NC-ND license.

social and political economies.[4] These 177 had been senators, many of whom had fought and perished based on a shared ethos of duty, honor, and service to the *res publica*.[5] Collectively, they represented 60% of the senate,[6] and thus nearly two of every three senators, a truly staggering proportion unparalleled even in the bloodiest years of the Roman Revolution.

Given the magnitude of the losses, the deaths of these 177 senators have not gone unnoticed by ancient and modern observers. Two decades ago, Feig Vishnia asserted that between the last *lectio* of 220 and M. Fabius Buteo's emergency *lectio* to refill the depleted senate in 216, "apart from the loss of many of the 'middle generation' … it must be stressed that two generations of future magistrates had been wiped out."[7] Feig Vishnia appears largely to have intuited this blunt conclusion. I agree with her position in the main, but more can be said. This chapter aims to quantify more precisely the impact of senatorial mortality in the early years of the Hannibalic War, the ramifications of which were felt for decades. By exploiting analytical tools developed in the social sciences – namely, several recent, updated population-modeling techniques based on new or improved model life tables that can more accurately reconstruct the demographic conditions of antiquity – I will demonstrate that, in actuality, nearly every senator aged 45 and younger between 220 and 216 had been killed, which brings into focus for the first time the true size and scope of a fully "Lost Generation" of senatorial *iuniores*.[8]

The effects of this "Lost Generation" on Roman politics were dramatic in the short term, and it potentially threatened both the short-term and long-term stability of aristocratic competition. In a recent article, Stein argued that senatorial losses were not so serious as to require emergency enrollment of new senators; the special *lectio* in 216 being instead some sort of political maneuver on the part of the aristocracy.[9] Stein seems to greatly underestimate just how significant the loss of 177 senators would have been in comparison to the normal rates of attrition in the senate, which would be replaced every five years. As will be demonstrated, this high level of mortality

---

4 Livy (23.23.7) provides the figure of 177 senators killed in his report of the new *patres*, whom M. Fabius Buteo adlected as replacements (*in demortuorum locum sublecturum*) in his emergency *lectio* of 216. Livy 21.59.9, for example, attests to the relative impact of elite deaths against those of ordinary citizens: "…but the loss of the Romans was out of proportion to the number slain, for it included several knights, five tribunes of the soldiers, and three praefects of the allies" (*sed maior Romanis quam pro numero iactura fuit, quia equestris ordinis aliquot et tribuni militum quinque et praefecti sociorum tres sunt interfecti*).

5 Senators and military service: Rosenstein (2006, 2007).

6 The mid-Republican senate had a normal complement of 300 senators, for which see Barber (2016, 74–174); Mommsen (1887/88, 3.2 847); *contra* Cornell (2000); Jehne (2011, esp. 222–23 and n. 53).

7 Feig Vishnia (1996, 99–104, quote at 101). See also Hölkeskamp (1999) and Beck (2003, 136), who, to varying degrees, accept this argument.

8 The last attempt at this type of study was Cavaignac (1932), based largely upon the pioneering work performed by Willems (1878) and Hofmann (1847).

9 Stein (2007).

disrupted key mechanisms of aristocratic competition, namely the funda-mental connection between the natural cycles of death and renewal within the senatorial elite on the one hand, and the stability of the Republican sys-tem that emerged from the early third century on the other. In the decade following Cannae, the "Lost Generation" meant that a clique of senior state-men who survived the years 220–216 could – and indeed, did – dominate Roman politics, with few age-rank peers to check them and without the usual challenge from rising competitors. This chapter will demonstrate the clear connection between the massive demographic disruption of 220–216 and subsequent patterns of office-holding.[10]

As indicated, the application of demographic models is central to this chapter. Indeed, such methods are necessary to allow us to go beyond the ev-idence that the literary sources can offer – evidence that, simply, is incapable of answering the necessary questions. No texts indicate, for instance, how many *iuniores* were present in the mid-Republican senate nor how many *pa-tres* were past fighting age.[11] Nor is it clear from literary sources how many *iuniores* and *seniores* would perish from natural attrition over the course of four years and therefore how many of the 177 dead could be accounted for through "natural deaths," nor how these different types of deaths might dis-tribute themselves among the various age cohorts of the senate. It is entirely uncertain then, from literary sources alone, whether the ranks of younger senators could absorb this extraordinary blow without affecting the normal political workings of the Roman state, or if such losses represented a ma-jor shock to the system. A demographic approach, however, combined with traditional historical methods, can unlock features of the mid-Republican *curia* that would otherwise remain inaccessible to modern observers.[12] In recent years, demographic analysis of the ancient world more generally, and of Rome in particular, has withstood several critiques, most importantly that we lack the statistical data necessary to draw firm conclusions about ancient populations. I do not share this pessimism. Thus, an additional goal of this chapter is to demonstrate the validity of applying demographic tech-niques to the study of ancient populations, and their absolute usefulness in helping to answer important historical questions.

Model life tables and Roman demographics: use and limitations

It must first be noted that the 177 deaths under discussion were not evenly distributed across the 300 members of the mid-Republican senate. Rather,

---

10 Jehne (2011, 226–27) has noted this connection.
11 While there is certainly no reason to assume that *patres* served up to the age of 45 under normal conditions, the Second Punic War is clearly an abnormal era that required im-mense sacrifice from its ruling class.
12 For a similar approach to the vexing question of the demography of Italy in the second century, see Rosenstein (2004).

they occurred at a vastly disproportionate rate among the *iunores*, upsetting the internal balance of the senate between young and old. To substantiate this claim, it will first be necessary to reconstruct the membership of any given age cohort, or generation, within the senate of the Second Punic War. The problems with relying entirely upon the extant sources for identifying these groups, however, as well as the reasons for turning to demography to supplement the available textual evidence, are manifest. Despite Livy's precise report of senatorial deaths between 220 and 216, there is very little evidence for who these 177 *patres* had been;[13] nor is it possible to reconstruct the entire senate of 220 and 216, as Pierre Willems did long ago for the senates of 179 and 55.[14] There is, in other words, no way to check off each of the known deceased from a list of all known *patres*, and to determine from there the resulting composition of the *curia* of 216. Nevertheless, we can begin to categorize these fallen senators demographically in order to determine analytically useful features about them such as age and cause of death.

The most important groups to consider here are the *iuniores*, who were aged 17–45 and therefore liable for conscription, as well as the subgroup of *iuniores* aged between 17–35 who, if they were anything at all like their counterparts amongst the "ordinary" ranks of the citizenry, were those most likely to have seen active combat in the field as infantry, cavalry, or officers.[15] When the number who died in battle, out of the 177 who perished from 220 to 216, are separated from those who likely died of natural causes, and when these excess casualties are then superimposed upon the *iuniores* of 220, the full extent to which the younger generation of senators suffered during Hannibal's initial advances into Italy becomes far clearer.

A critical first step in reconstructing the qualitative and quantitative impacts of Rome's curial losses is therefore an approximation of the normative age structure of the mid-Republican senate. Such a model is altogether elusive in the sources, however, though this is not solely the fault of the general dearth of textual or material evidence that imperils much of the study of the middle Republic. Key here, too, is the issue of the credibility of what ancient numerical evidence does exist, even for later periods when both textual and material sources arguably improve in quality and quantity.[16] Moreover,

---

13 Scattered references to the deaths of named *patres* can be found on rare occasions, and there are also mentions of unnamed senators who belonged to various rank groups and who died in specific engagements – most notably consuls, praetors, and other current or former curule magistrates who fell dramatically in the heat of battle at Trasimene, Cannae, and elsewhere. These are generally grouped in with other senatorial deaths, for example, at Livy 22.49.16. More rarely, individual losses of specific senators can be inferred by piecing together various clues from accounts of the era, for example, the case of C. Servilius Geminus, who was presumed dead but in fact captured by the Boii in 218.

14 Willems (1878).

15 For more on the age groups who took part in warfare: Rosenstein (2004, 84–85).

16 Inconsistent or insufficient sample size, a general class bias in terms of the epigraphic habit, and issues of age rounding (or even inaccurate age awareness), are only a few of the hurdles that can vitiate the use of funerary inscriptions, tax documents, and other epigraphic or

where some precision is made possible through an analysis of extant ancient sources, the scope of the conclusions that can be made in these cases is often limited, and extrapolation from them largely unsupported.[17] In sum, the demographic regime that characterized life in Republican Rome is obfuscated by several source problems that, when taken together, make absolute certainty in the reconstruction of any demographic aspect of the ancient world a virtual impossibility. By shifting focus away from the ancient evidence, however, and by looking instead to recent attempts to model the demographic profiles of other pre-modern societies using model life tables, it is possible to estimate the age structure of the senate within an acceptable margin of error.[18] This last caveat – within an acceptable margin of error – is critical: the results of this inquiry should be understood as heuristic indicators, and not exact representations, of the age profile of the Republican *curia*.

While this has long been acknowledged as the proper approach toward the model life tables, for some scholars at least, this admonition is no longer sufficient. Even luminaries in the field of ancient demography have leveled serious critiques on the life tables over the last two decades, to the point that their defense is once again necessary if the results of any study that employs them are to be taken seriously.[19] It should be acknowledged from the outset that the arguments against the life tables are not at all unfounded, though a series of notable rebuttals have countered some of these recent claims.[20] Nevertheless, some of these critiques have merit, particularly their illustration of the problematic nature of the Coale-Demeny tables and the data

---

literary evidence for establishing the demographic profile of Romans of any place or period. The seminal study on the difficulties of using ancient epigraphic evidence is Hopkins (1966); see also: Hopkins (1987); Brunt (1971a, 131–55); Duncan-Jones (1977); Duncan-Jones (2002); Parkin (1992, 6–8); Scheidel (1996a); de Ligt (2012, 1–39); Hin (2013, 101–9).

17 Parkin (1992, 4–41); Duncan-Jones (2002, 93–104) (on the limitations of a single source's importance, in that case the album of Canusium), *contra* Frier (1982). See also Scheidel (2001a) for a critique of Frier's analysis.

18 Essentially, these tables are designed to supplement whatever can be known about the demographic regimes that affect populations for which there is very little data, or data of indeterminate value: so Parkin (1992, 78–81). Compiled in large part from an enormous set of records of conditions in mostly eighteenth- to early-twentieth-century communities from around the world, the life tables are statistical models that offer an assortment of information about hypothetical populations, including their likely age structures, net reproduction rates, and rates of mortality at various ages. In order to cover a broad swath of possible demographic conditions, Coale and Demeny produced 25 model tables for each sex, representing 25 different life expectancies at birth ranging from 20 to 80 years of age at death. For each of these 25 tables, there are four distinct patterns of age-specific mortality that are designed to reflect the conditions of various regions around Europe and the Mediterranean.

19 For some relatively recent critiques: Engels (1984); Harris (1999); Scheidel (2001a); Sallares (2002); Earnshaw-Brown (2009).

20 Parkin (1992, 81ff.) offers cogent arguments against those who question the usefulness of the life tables for ancient historians. Hin (2013, 109ff.) has skillfully updated many of Parkin's arguments in her own invaluable study.

and techniques that were used in their creation. What emerges from a consideration of the drawbacks of these models, however, is not their general unreliability, but rather their functional limitations. These limitations do not present a fundamental impediment to the demographic reconstruction of the Roman past, so long as one remains cognizant of what the tables are reasonably capable of showing.[21]

The most forceful arguments against the Coale-Demeny life tables center upon the techniques that were used to construct their high-mortality, low life-expectancy models. Perhaps most problematically, the bulk of the data that Coale and Demeny drew upon in their calculations derived from nineteenth- and twentieth-century populations undergoing the "demographic transition."[22] For these transitioning populations, life expectancy at birth never dropped below 35 – a figure substantially higher than the traditional 20–30 years often posited for citizens in the Republic.[23] Despite the lack of data from pre-transitional communities with low life expectancies at birth, the Coale-Demeny tables nevertheless provide projections for populations of just this type. In lieu of empirical data upon which to construct these models, however, Coale and Demeny instead apply a series of mathematical regressions on low-mortality tables in order to approximate demographic conditions for pre-transitional, high-mortality groups. Through extrapolations from the better-supported tables, for which there is more substantial data, the lower life-expectancy models assume an age-specific mortality curve similar to that of their higher life-expectancy counterparts. In order to reach the requisite lower life expectancies at various ages, however, the parameters for mortality in these tables were amplified significantly, especially at the lower ranges, in order to account for high infant mortality regimes like those presumed to be prevalent in the ancient world.[24]

Parkin has suggested that this offsetting is a reasonable solution that may weaken, but not invalidate, the reliability of the resulting high-mortality tables.[25] Against this contention, several scholars have remarked that high-mortality populations are often subject to demographic pressures, such as contagious disease and high infant mortality, which target age-specific groups but that are largely absent from low-mortality communities.[26]

---

21 A similar position was sketched out nearly 30 years ago by Hopkins (1987, 116). See also Talbert (1987, 132–33), for a characteristically judicious use of the life tables.
22 As Parkin (1992, 81) notes, the potential problems stemming from this were already pointed out by Hollingsworth (1969, 343) following the publication of the first edition of the Coale-Demeny Model Life Tables. For the demographic transition, see Engels (1984) and Parkin (1992, 71).
23 Estimate of 20 to 30 years' life expectancy: Hopkins (1966); Hopkins (1983, 70ff.). This figure is supported more recently by: Parkin (1992, 84); Saller (1994, 12–25); Scheidel (2001a).
24 Coale and Demeny (1983, 24); Parkin (1992, 81ff.); Scheidel (2001b, 21–25); Woods (2007, 375–76); Hin (2013, 109).
25 Parkin (1992, 81).
26 Scheidel (2001a, 4–11); Sallares (2002); Woods (2007); Hin (2013, 109ff.).

As a result, any sort of extrapolation from low-mortality populations to those with high mortality runs the risk of masking the demographic complexities of certain age groups within high-mortality communities, especially in the earliest years of life. Complicating this problem even further, the very features of high-mortality populations that can render their age structures distinct are those features that, if found to be present in a nineteenth- or twentieth-century community, led to the exclusion of that community's demographic dataset from the Coale-Demeny tables. In order to attain uniformity in the statistics they collected for the construction of the life tables, the researchers omitted demographic data from populations suffering from epidemic outbreaks, warfare, famine, and high infant mortality.[27] As all of these are held to be common features of life in antiquity, the exclusion of modern communities experiencing these conditions potentially attenuates the use of the life tables for the study of the ancient past, for some even fatally.

Because the aim here is to reconstruct the age structure of the senate, and not the broader Roman population, the potential impact of famine can be dismissed without much hesitation.[28] While nutritional deficiencies may have afflicted the lower classes during the Republic, it is reasonably certain that the "senatorial class" enjoyed similar, if not sometimes better, access to food and general nutrition than the nineteenth- and twentieth-century populations who provided data for the Coale-Demeny life tables. Warfare, disease, and disproportionately high infant mortality, however, remain at issue, and their impacts must be tempered or dispelled lest they invalidate the use of these demographic tools.

The absence of warring communities from the Coale-Demeny datasets represents a very clear problem, to say the least. Roman elites were active participants in the campaigns that were undertaken during virtually every year of the Republic's existence, and as the 177 senatorial deaths from 220 to 216 make clear, the prospect of death during military service was a very real one for *patres*.[29] But the Hannibalic War is, by all measures, an extreme aberration from the norm. Just how much "normal" warfare might have skewed the age structure of the senate in other eras is therefore not immediately clear, and a lack of ancient sources or extensive modern scholarship on the subject means that it is best to begin by speaking about elite

---

27 Coale and Demeny (1983, 5, 11–12, 24–25).

28 Hopkins (1983, 71). Garnsey (1988, 6–39) suggests that the food supply was less of a concern to the Romans than to other societies, though in any case the "senatorial class" would not suffer as the rest of the population might. Here, I follow Rosenstein (2006, 6 n. 31) in positing a *de facto* requirement of one million sextental *asses* for inclusion in the "senatorial class," that is, the office-holding aristocracy. Holding office was a prerequisite for enrollment into the senate: Barber (2020).

29 Harris (1979). Rosenstein (2004, 3) argues military service in this era "constituted the paramount arena for the display of aristocratic *virtus* and the acquisition of prestige." See also Rosenstein (2006).

casualties in generalities.[30] Hopkins has suggested that during the Republic, the involvement in warfare of senators and their sons would have affected the age structure of the political aristocracy in some unknown, but significant way.[31] This may be true for a broader political elite that included all non-senatorial males from senatorial families, though this is not at all certain. It is nevertheless unlikely to be the case that warfare impacted the age structure of those actually in the senate, as once they had been adlected, very few senators would have seen active combat as infantrymen, prefects of the allies, or military tribunes – the positions that would expose them to danger – outside of highly unusual circumstances such as the Second Punic War. Though senators could, and apparently sometimes did, serve in these capacities during "normal" years, scholars have long associated the military tribunate and the prefecture of the allies with the careers of younger elites who had not yet been enrolled among the *patres*.[32] As a result, very few actual senators – as opposed to those who actively sought the *curia* – would have succumbed to battlefield attrition.

Epidemic disease is a more serious hurdle. In recent years, a growing body of scholarship has been dedicated to the study of communicable diseases in Roman Italy, particularly malaria and tuberculosis.[33] Within the city of Rome, these two diseases are believed to have been endemic until only very recently. Because these diseases are often highly selective in terms of age, with malaria disproportionately affecting children under five and tuberculosis targeting those between 18 and 35, there is a real possibility that the age structures of the populations in the Coale and Demeny tables varied significantly from the age structure(s) present in Rome.[34] And while there are reasons to suspect that the impact of communicable disease on mid-Republican elites was less severe than Sallares, Scheidel, and others have suggested, there can be no doubt that mid-Republican senators were exposed, throughout their lives, to a range of fatal illnesses, even if they were likely to have fared better than the *patres* of later periods who continued to live within the walls of the city.[35] Although senators were likely to

---

30 Harris (1979, 39–40).
31 Hopkins (1983, 71).
32 Hill (1952, 27ff.). See Suolahti (1955, 213ff.) on the prefects, about whom knowledge is limited. Suolahti's information on the office is built on very few cases, though he is right to equate the prestige of the position with the military tribunate. His evidence suggests that the holders of both offices were likely to be similar in terms of experience and rank, and so were mostly young men about to begin their political careers. For the holding of the military tribunate before becoming a senator: Suolahti (1955); Wiseman (1971, 143–45); Cagniart (1989, 147–49); Rosenstein (2007, 136–41). Senatorial mortality in battle: Barber (2016, 33–68).
33 Effects of diseases like malaria and tuberculosis on Roman Italy: Scheidel (1996b, 1999b); Scheidel (2001a); Sallares (2002); Scheidel (2003); Woods (2007); Hin (2013).
34 For malaria: Scheidel (2003, 164). For tuberculosis: Scheidel (1994, 157).
35 The most glaring issue for these scholars lies in the suitability of their evidence for a mid-Republican context, as most of it derives from late Republican or imperial sources that describe conditions within the city that would be anachronistic for the middle

enter the senate around the age of 30 in the middle Republic, and thus past the period when they were most vulnerable to malaria, and nearly past the age when they were most susceptible to tuberculosis, these maladies would have already taken their toll on both their life expectancies and the overall mortality curve of the senate. High infant mortality, the source of the other remaining critique of the Coale and Demeny tables, would impact the age structure of the senate similarly.

In sum, unlike famine and warfare, it would be disingenuous to claim that neither disease nor infant mortality had a measurable impact on the age structure of the senate. Even still, this is no reason to despair. In recent years, Woods and Hin have surveyed comparative data from modern populations that are characterized both by high levels of mortality and relatively diligent record-keeping of demographic data. Several of these populations present conditions that are comparable to what scholars propose for various regions and eras of the ancient world, and two of these communities in particular offer strong candidates for comparison with mid-Republican Rome. For Woods, the best model for the Roman world comes from early-twentieth-century Chile, where infectious diseases like tuberculosis, influenza, and pneumonia (though, noticeably, not malaria) were the primary causes of death.[36] Hin, on the other hand, turns to the population of Navrongo in Ghana, which offers comparable conditions in terms of the disease regime and medical treatments present in Roman Italy.[37] Both Woods' "Model South" Chilean tables and Hin's "Navrongo Model" offer actual data from well-documented populations experiencing high mortality overall, relatively high infant mortality, and disease regimes that are similar to those that have been suggested for Republican Italy. While the Coale and Demeny tables can continue to offer important insights into high-mortality populations in general, the updated tables provided by Woods and Hin can offer a better sense of the demographic profiles of Middle Republican elites, and therefore the relative age structure of the senate.

Age structure of the mid-Republican senate

Based on the age structures derived from Coale and Demeny, Woods, and Hin, it is possible to approximate the relative number of *iuniores* and *seniores* within the senate, and from there the impact on these groups of wartime

---

Republic. Rome was less populated in the middle Republic, less dense, and less of a mixture of different groups than it would later become in the late Republican iteration of the city, not to mention the sprawling imperial metropolis of the Julio-Claudian era.

36 Woods (2007, 382). Chile was relatively urbanized at this time, with nearly half of its population living in cities, and its disease regime, aside from the absence of malaria, seems to have largely matched what scholars have proposed for Roman Italy.

37 Hin (2013, 114ff.) is somewhat skeptical of Woods' comparison due to both the high rate of cardiovascular disease among Chileans and the absence of malaria. Characterized by a Sahelian climate and an altitude similar to Roman Italy's, Navrongo is inhabited by a population that is largely resistant to the use of modern medicine and that counts malaria and gastroenteritis as primary causes of death.

Table 9.1 Relative Number of *Iuniores* and *Seniores*
within the Senate after a *Lectio*

Model Life Table	# Iuniores	# Seniores
Level 3 West Female	165	135
"Navrongo"	172	128
"Woods"	170	130

casualties during the Second Punic War. Table 9.1 provides the relative number of *iuniores* and *seniores* within a "normal" mid-Republican *curia* based
on the model life tables under study.

As is immediately clear from these models, the figures for the *iuniores* in
all three reconstructions are enticingly close to the number of 177 dead. Still,
more information is required before establishing a "Lost Generation" of
mid-Republican elites, including the rate of mobilization of *iuniores* following Hannibal's invasion in 218, as well as the impact on the *patres* of "normal
attrition." For the former, one can turn back to the ancient evidence, and
particularly Livy (22.49.16–18) who claims that there had been former consuls, praetors, and aediles among the 29 *tribuni militum* who perished at Cannae. Moreover, there were found among the fallen *milites* some 80 *senatores*
or those who had held an office that granted them entry into the senate. The
number of senators on campaign from 218 to 216, and especially at Cannae,
must therefore have been staggering. The fact that at least 80 *senatores* perished among the soldiery in a single battle should reflect a near-total mobilization of the 165–75 of the senate's *iuniores* during the early years of the war,
and the presence of ex-curule magistrates among the *tribuni militum* suggests
that this may have been true across all magisterial ranks. The punishments
inflicted by the censors of 214 and 209 on *equites* who had not yet fought since
the beginning of the war (and in 214, on those who had not fought in every
year of the war thus far) rounds out this body of evidence.[38] What one finds is
the wholesale mobilization of virtually all *iuniores* within not only the entire
senatorial elite but also the entire equestrian class. Based on this evidence,
it should be assumed that all, or nearly all, of the roughly 165–75 *iuniores*
would have seen military service of some sort between the years 218 and 216,
unless they were otherwise employed on behalf of the Republic.

The total number of *iuniores* should indicate the size of the contingent of
senators who were guaranteed to be in the battle lines, and who were therefore most likely to suffer the greatest number of casualties. An obvious question, however, is how many *seniores* fought and died in the period 220–216
alongside them. While Livy twice provides evidence for the forced recruitment of men up to the age of 50 during later decades of military crisis, there
is no evidence for similar service by *seniores* during the Second Punic War.[39]

38 For the censors of 214: Livy 24.18.7. For the censors of 209: Livy 27.11.12–16.
39 For the recruitment of men aged 50 and younger: Livy 40.26.7, 42.31.4.

This *argumentum ex silentio* is admittedly problematic. Yet, we may consider Livy's report (22.11.9) that during the post-Trasimene crisis of 217, conscription for naval service was limited solely to those freedmen who were under 35, while those who were older would stay in Rome for garrison duty. While certainty on this point is impossible, it seems reasonable to assume that so long as there were *iuniores* from which to recruit, *seniores* would typically have appeared less attractive as candidates for frontline duty, while more suitable for garrison duty at Rome (and a city garrison would have been necessary in any event). Thus, while a small number of *seniores* fought as consuls, proconsuls, and *legati* from 220 to 216, virtually all *iuniores* could be found among the ranks of the *equites* and the *milites*, if not also as magistrates serving in positions within the army.

Even if *seniores* did not perish on the battlefield, however, they would have nevertheless fallen victim to the normal human lifecycle. Natural attrition among the *seniores* must therefore also be calculated using the life tables' mortality rates to determine how many of the *seniores* perished among the 177. Table 9.2 shows the anticipated number of *seniores* who would survive a "normal" four-year period between *lectiones*, assuming (again, heuristically) that none of them had fallen in war.

These impressionistic findings indicate that, out of a senate of some 123 surviving members, around 75%–85% would have been *seniores* (assuming very little mortality from this group due to combat), though under normal conditions following a *lectio*, these *seniores* would have generally made up only 30%–35% of the senate's 300 members (and an even smaller proportion in the years preceding the *lectio*'s completion). Those *patres* who were eligible for conscription, on the other hand, had seen their numbers plummet from around 165–170 in 220 to as low as 20 in 216 (and so from roughly 55%–60% of the full senate, down to 15%–20% of a much-reduced *curia*). The group of senatorial *iuniores* who survived those four long years from 220 to 216 would have therefore returned to a senate in which they were no longer a significant majority, but a considerable minority, in comparison with their elder counterparts who now dominated the senate's ranks. This finding alone is highly suggestive of a radically new comparative relationship between the various age groups of the senate in the year 216. It also reinforces what Table 9.1 seems to show at first glimpse: namely, a "Lost Generation" of younger *patres* who had fought and perished almost in their entirety between 220 and 216.

Table 9.2 Surviving Senatorial *Seniores* in 216

Model Life Table	# Seniores in 220	# Natural Deaths among Seniores (220 – 216)	Approx. # Surviving Seniores in 216 (of 123 Survivors)
Model Life Table 3 West	135	32	103
"Navrongo"	128	31	97
"Woods"	130	33	97

The "Lost Generation" and Roman politics after 216: a re-evaluation

Certain connections between this massive loss of life on the one hand, and identifiable patterns in contemporaneous Roman governance on the other, seem natural. In terms of the latter, it has long been posited, based on the consular *fasti* and other textual evidence, that a clique of elder *consulares* controlled the state after 216 through their monopoly over military commands. Among the elder elites, the three most powerful were Q. Fabius Maximus, M. Claudius Marcellus, and Q. Fulvius Flaccus.[40] This regime would endure for a decade, so this theory goes, until new candidates rose to the consulship and iteration no longer prevailed.[41]

To explain the predominance of these elder elites, some have argued for a voluntary transfer of power to the senate's elders – and a curbing of competition within the elite – for as long as the crisis loomed.[42] Once Rome had recovered, so this model contends, competition for office would have continued just as it had in the past. More plausibly, others have argued that aristocratic competition for public office in fact continued to function normally during the Hannibalic War. In this period of crisis, however, with military factors now playing a more prominent role in voters' minds, the natural result was that older, more experienced generals who were better positioned to persuade voters of their strengths in this area would win the consulship and the praetorship more often.[43] Once the acute military crisis passed, typical electoral patterns simply re-emerged. While this latter explanation is more compelling, it fails to take into account the legitimate demographic and generational factors that shaped patterns of office-holding.

Feig Vishnia's interpretation does invoke generational considerations to explain the restoration of normal electoral patterns. According to her argument, a significant degree of inter-generational anger and frustration arose in response to Rome's gerontocracy, particularly among the 177 individuals adlected in 216 whom she identifies as younger, inexperienced *equites*.[44] As the war dragged on, she argues, these younger *senatores* grew resentful of the dominance exerted by Rome's elder statesmen, and in time actively and collectively resisted oligarchic control. The scenario she envisions is thought-provoking, but ultimately unconvincing; there are fatal flaws in the foundations of her argument, which portrays neither fully nor accurately the extraordinary *lectio senatus* of 216 and the impact it would have on the senate's ordinary age- and rank-distribution.

---

40 Q. Fabius Maximus (cos. 233, 228, 215, 214, 209); M. Claudius Marcellus (pr. 224, cos. 222, pr. 216, cos. 215, 214, procos. 213–11, cos. 210, 208); Q. Fulvius Flaccus (cos. 237, 224, pr. 215–14, cos. 212, 209). These lists are not exhaustive, though they provide a good sense of the dominance of these three men during this period of great crisis.
41 Patterson (1942); Crake (1963); Rosenstein (1993).
42 Develin (1985).
43 Rosenstein (1993).
44 See Feig Vishnia (1996, 102ff.).

Here, a demographically informed, generational perspective can add nuance and help explain more fully the distortion of Roman politics in the years after 216. The loss of virtually every senator under 46-years old was indeed an obvious catalyst for the unusual patterns that appear in the magisterial *fasti* – particularly the consulship. At the same time, while the group of 177 newly-enrolled senators certainly did include many younger men who had recently held the quaestorship or the tribunate of the plebs, it would have also included very many who had been elected to public office (perhaps much) earlier, but who had never entered the senate.[45] Indeed, these older men were likely adlected *en masse* to replace the 177 dead. As Livy notes, the censor Fabius Buteo enrolled "in the place of the deceased first those who since the censorship of L. Aemilius and C. Flaminius had held a curule magistracy and had not yet been chosen into the senate, in each case in the order of his election."[46] After this, he chose those who had held lower magistracies (aediles, tribunes, and quaestors) but were not yet enrolled.[47] This would have resulted in a disproportionately greater number of older – in some cases much older – ex-magistrates entering the senate than usual.[48] This, in turn, would have led to a *curia* that resembled even less the "ideal type" found in Table 9.1, and that would have been marked instead by a further decline in the overall proportion of *iuniores* in the *curia*, thus exacerbating what Hannibal's inroads had already established, namely a sharp divergence from the senate's normal age- and rank-distribution.

This situation also means that the post-216 *curia*, which bowed to the pressure of a dominant few elder statesmen, was far more multifaceted in terms of its age- and rank-groupings than has previously been suggested. But their eventual capitulation should not be taken to imply that all of these "lesser"

45 Though their careers and entry into the *curia* had long ago stalled (whether because of some perceived moral failing, or because there were not enough vacancies to accommodate them after their tenure of office, or due to their inability or unwillingness to obtain the quaestorship or the tribunate of the plebs, the offices that normally brought entry into the *curia*) they were nevertheless enrolled into the senate of 216 so long as they met the objective criteria put forth by Fabius Buteo in his extraordinary *lectio senatus* of that year. On the mid-Republican *lectio* only allowing for the adlection of ex-magistrates – even in the emergency adlection of M. Fabius Buteo in 216 – see Barber (2020). Here, too, a demographic approach to the senate is useful. Based on the three model life tables employed above, it would have been necessary to enroll 55–60 new candidates following a *quinquennium* to sustain a senate of 300. The normal contingent of tribunes of the plebs and quaestors would have sufficed to fill these vacancies, though not without anywhere from 5 to 15 "leftovers" from those ranks who were not adlected – not to mention the many other *tribuni militum I–IV, Xviri stlitibus iudicandis* who would not make the cut and who were later unable to attain higher office.
46 Livy 23.23.5: *inde primos in demortuorum locum legit qui post L. Aemilium C. Flaminium censores curulem magistratum cepissent necdum in senatum lecti essent, ut quisque eorum primus creatus erat.*
47 Livy 23.23.6.
48 Presumably, no older ex-lesser magistrates would have entered the senate during a normal *lectio.*

senatores bore the dominance of this clique gladly at all times. Indeed, there is evidence that some of the surviving members of the "Lost Generation" did in fact attempt to rebel against Fabius Maximus almost immediately after Cannae, most conspicuously during the elections of 215. The response to this pushback nevertheless illustrates well the extraordinary power the senatorial *seniores* now held in the wake of the "Lost Generation's" passing. In these elections, T. Otacilius Crassus and M. Aemilius Regillus – praetors of 217 and survivors of Hannibal's decimation of the senate – put themselves forward for the consulship, presumably among a field of other candidates who were standing in that year. By Livy's account, the consul Q. Fabius Maximus oversaw the elections, though he contravened custom by doing so as a commander with *imperium*.[49] Under these circumstances, the *centuria praerogativa* cast its votes and announced for Crassus and Regillus, at which point Livy has Fabius halt the elections and berate the electorate in an attempt to change their votes. In reaction to this, Fabius' son-in-law, the propraetor and candidate Crassus, protested angrily, and Fabius answered with the threat of violence: Fabius is said to have asserted that his lictors still had axes, with the implication that they were authorized to use them. Crassus subsequently backed down, and immediately thereafter the centuries were called upon to re-cast their votes. The results are not surprising: Fabius was elected to his fourth consulship and Marcellus to his third. Fabius effected, in other words, nothing short of a coup under the threat of official violence against a propraetor – all without significant pushback from the ranks of the senatorial aristocracy.

In this way, those few surviving members of the "Lost Generation" were halted in their progress up the *cursus,* while the dominance of the leading troika had displayed itself in full. A generational perspective, backed up by the demographic data provided by Tables 9.1 and 9.2, can help to explain how this came to be. In normal years, those among the younger generation who hoped to one day run for the consulship might be expected to protest, alongside Crassus and Regillus, against the likes of Q. Fabius. While as individuals they did not possess nearly so much concentrated *auctoritas* when compared with a decorated ex-consul and *triumphator*, these younger *senatores* – who collectively made up the majority of the *patres* – could at least be expected to band their social and political capital together to ensure that they would compete only with one another instead of with the older *principes* of the senate. They would not be alone in their resistance. In a society as agonistic as Rome's, older *consulares* with further ambitions to the consulship should only expect to receive pushback from the wider group of ex-consuls and ex-praetors in the senate, who would be reluctant to see their careers outstripped and their children's chances circumscribed.

---

49 Livy 24.7–9. Fabius had not crossed the *pomerium* as had been customary in the past, and I take this as realistic, *contra* Drogula (2007, 438ff.) who claims that this is a late-Republican reading of this event.

The possibility of an embarrassing defeat against a united front of younger elites and their older allies was therefore likely to have been enough to keep most *principes* from competing for the top offices again and again, which helps to explain, for instance, the lack of consular iteration over the century preceding the Hannibalic War, particularly when compared with the consular *fasti* of the fourth century as a whole.[50]

Crassus, then, was confident enough to challenge a consul who still possessed *imperium* because he was likely laboring under the illusion that he could depend upon his own generational comrades in the customary way. Unfortunately for Crassus, however, they simply no longer existed, and the new divergence from the demographic norm now actively worked against him, Regillus, and the other members of the "Lost Generation." Indeed, because the "Replacements" of 216 ranged so dramatically in terms of their age and rank, including many who likely had held only lesser offices (such as the tribunate of the soldiers), they lacked the collective will, institutional experience, general *auctoritas*, and social and political capital needed to balance out the *potissimi seniores*, who could exert a great deal of these powerful forces in their own pursuit of office. Moreover, unlike those *patres* adlected under normal circumstances, the "Replacements" were on entirely different career trajectories, all but ensuring a lack of group cohesion which was itself necessary if they wished to oppose a smaller but more authoritative group of elder elites attempting to monopolize military commands.[51] The *principes*, on the other hand, had years of experience working together, and under the extraordinary circumstances of the Second Punic War, they had far more to gain than to lose by cooperating, particularly when faced with a divided field of competitors. Many of the older members of the "Replacements," it should be added, were likely to be grateful simply for their adlection into the senate and were in any case unencumbered by further ambitions to higher office. They were perhaps also more likely to defer to their age-mates among the *principes*, with whom they had never been able to compete in the first place. Thus, they could not offer the younger "Replacements" the degree of predictable and powerful aid that older *patres* had provided *iuniores* in the past. Moreover, these low-ranking *seniores* who entered in 216 were likely to have perished at irregular rates over the course of the next decade, and thus the normal age-distribution map in Table 9.1 was unlikely to be re-attained until the full cohort of 177 "Replacements" had themselves been

50 Indeed, the pattern of office-holding after 216, with a small number of powerful individuals holding alternating commands, appears more like Roman politics in the fourth century – before the relatively stable system of competition and power-sharing took hold in the first decades of the third century. For the earlier pattern: Cornell (1995, 370–73); see also Beck (2005, 96–98).

51 That those elder elites acted in this way during the Second Punic War is only natural given the ethos of the Roman aristocracy for primacy in all things. For mid-Republican aristocratic values, see Rosenstein (2006).

replaced. This, however, was likely to take decades, ensuring the long absence from the political arena of the inter- and intra-generational safeguards that helped younger senators resist the overawing strength of older *principes* in preceding decades.

Conclusions

While there are certainly a range of factors in play during this chaotic period, a lack of generational support goes some way in explaining the dominance of these elder elites in the years after 216 – particularly as they reappeared among the various *fasti* of the *imperium*-granting magistracies. Additionally, this demographic, generational approach to Rome's unusual patterns of office-holding in this period can also help to reveal a critical – yet almost entirely overlooked – mechanism that helped to sustain the long-term hegemony and stability of the senatorial elite under normal circumstances. Key here is the natural cycle of death and renewal that derived in large part from a combination of the standardized quinquennial interval between enrollments and the institutionalized and regularized features of the mid-Republican *lectio senatus*. Most importantly, this cycle simultaneously thinned the powerful elder ranks of the senate as it readjusted the relative distribution of *auctoritas* among the various age- and rank-cohorts of the *curia* – particularly as it reallocated a significant degree of influence, prestige, and standing to the new groups of ex-quaestors and ex-tribunes who replaced in the *lectio* those *seniores* who had perished. Because this cycle performed these tasks with considerable regularity, it allowed for the development of predictable patterns in the relative distribution of authority based on age- and rank-groupings. Over time, this cycle became a stabilizing force that ensured that inter- and intra-generational cooperation and competition remained within acceptable limits. It also organically reduced certain tendencies toward the creation of an unstable "oligarchy within an oligarchy" by acting as a deterrent to the domination of individuals, factions, and families, and thus obviated against the need for strict rules for aristocratic competition.[52]

But demographic crises that struck the senate would invariably upset this cycle, as was the case during the opening years of the Second Punic War. This, in turn, would portend disastrous results for the delicate balance of power within the Republican aristocracy, as offices and commands became increasingly concentrated in the hands of the few. As argued above, the processes at work are generational: when one "generation" of *patres* grew too strong due to the culling of another, the complex networks of power within the latter generation that allowed for collective resistance failed

---

52 Rosenstein (1990, 166–67) discusses the perils of this Aristotelian "oligarchy within an oligarchy," as well as various means by which the Republican elite were able to avoid this dangerous outcome.

while the competing generation – which had worked together to varying degrees as young officers, in elections to minor magistracies, in the senate, and elsewhere – was incentivized to pool their resources together once more to capitalize on the moment, a natural impulse for an aristocracy primed to strive for primacy. Only with the passage of time, and with the restoration of generational equilibrium, would normal competition resume, though consequences were sure to follow. These could take the form of new laws regulating competition, an intensification in the scale of electoral contests,[53] or general mistrust and antagonism within an elite that could thrive only so long as it presented a united front.

The Second Punic War would hardly be the last time that the senate's cycles of death and renewal would be violently interrupted, and the results would grow graver the more these cycles were unable to reassert themselves. The great civil wars of the first century, for instance, would decimate again and again the ranks of the *curia*, to the point where – as Tacitus famously laments – there were too few left who had seen the Republic to challenge the new regime of Augustus.[54] In order to explain the Republic's end, it might therefore be helpful to look back to its ascension, and to demography as a field that can make visible certain features of the ancient world that are otherwise relegated to the shadows.

---

53 The years following the Second Punic War saw the passing of several laws regulating aristocratic competition (for example, *lex Villia Annalis, lex Orchia*, etc.). There also appears to have been increased scrutiny over the awarding of triumphs. This is often seen in relation to the influx of wealth resulting from unprecedented imperial expansion (see Fronda this volume), but it may also be understood in light of the long-term effects of the disruption of aristocratic competition caused by the war.
54 Tac. *Ann.* 1.3.

10 Titus Quinctius Flamininus' "Italian triumph"*

Michael P. Fronda

Introduction

In 194, the proconsul Titus Quinctius Flamininus concluded his long command in Greece. After making a review and settling affairs in the province,
he consolidated his land and sea forces at Oricum, and then transferred
the entire army to Brundisium. From there, he returned to Rome, where he
celebrated an unprecedented three-day triumph. Our main sources for these
events –Livy (34.52.1–12) and Plutarch (*Flam.* 13.5–14.2) – provide lengthy
descriptions of the celebration, including a detailed inventory of the spectacular booty displayed in the procession. Briefer notices can be found in
the *Triumphal Fasti* (Degrassi 1954, p. 97), Cicero (*Verr.* 2.4.129, *Mur.* 31, *Pis.*
61), Plutarch (*Mor.* 197B), Valerius Maximus (5.2.6), Orosius (4.20.2), and
Eutropius (4.2.2).[1]

Flamininus' triumph is mentioned often in modern scholarship.[2] This is
not surprising given the lavishness of the spectacle, Flamininus' importance
in the political and military history of the Roman Republic, and the relatively
detailed description of the celebration itself preserved in the sources – which
yields rich source material for general discussions of the Roman triumph.
Yet, while the accounts of Flamininus' triumph have been carefully and
closely analyzed, one very peculiar and intriguing detail in Livy's version of

---

* It is my great fortune to have had Nate Rosenstein as both professor and dissertation su-
  pervisor, and later as mentor and friend. He is, in my estimation, one of the best historians
  of the Roman Republic in recent years: a careful, critical, and perceptive scholar, unafraid
  to offer an unorthodox thesis, and fully capable of defending it. I hope that this paper in
  some small way reflects what I have learned from him. Additional thanks to my co-editor
  Jeremy Armstrong. Any errors or omissions are, of course, my own. All dates are BC unless otherwise noted.
1 See also Just. 31.3.2; Euseb. *Chron.* 243; *MRR* 1.344; Itgenshorst (2005, 173, 177–80).
2 For example, Briscoe (1981, 128–29); Eckstein (1987, 312); Champion (2004, 52–53);
  Hickson-Hahn (2004, 48); Beck (2005, 386–88); Itgenshorst (2005, 173); Pfeilschifter (2005,
  116–17, 300); Bastien (2007, 279–80); Beard (2007, 150–51); Pittenger (2008, 122); Östenberg
  (2009, 22–23, 59–60, 134, 163–66, 268–69); Rosenstein (2012b, 246); Gauthier (2016a, 52,
  62–63).

DOI: 10.4324/9781351063500-10

This chapter has been made available under a CC-BY-NC-ND license.

events in the lead-up to his triumph has been almost completely overlooked by scholars. It is worth quoting the relevant section in full:

> And so when he had made a review of Thessaly, he came through Epirus to Oricum, from where he was about to carry across. From Oricum all forces (*copiae omnes*) were conveyed across to Brundisium. From there they came through all of Italy to the city nearly marching in a triumphal procession (*prope triumphantes*), with the train of things captured driven before him (*prae se*) no smaller than his own (*suo*) [marching army].[3]

According to Livy, then, Flamininus altered the Roman army's typical marching order so that the captured booty was driven (*acto*) in front of the general, rather than interspersed with the legions or placed at the rear of the column, while the soldiers, marching *prope triumphantes*, apparently followed behind their general.[4] In other words, he conducted his *reditus* as an enormous triumphal procession, beginning in Brundisium and continuing all the way to Rome – a march of around 600 km. This is a remarkable image, and (if historical) it must have been a stunning, even provocative, performance. And yet, as I just mentioned, this particular element of Flamininus' triumphal activities (broadly construed) is almost completely passed over in modern scholarship.[5]

This chapter will examine what I refer to as Flamininus' "Italian triumph" – that is, his triumph-like procession through Italy before, and in anticipation of, the formal triumph that he petitioned for, and was granted, in Rome – by placing it in the context of the political and military developments that unfolded in the generation following the Second Punic War. The great war with Hannibal had, of course, a profound impact on Rome and the rest of Italy. The Romans emerged from the war, bloodied but victorious, and

---

3 Livy 34.52.1–2 (trans. Fronda): *Ita cum percensuisset Thessaliam, per Epirum Oricum, unde erat traiecturus, venit. Ab Orico copiae omnes Brundisium transportatae. Inde per totam Italiam ad urbem prope triumphantes non minore agmine rerum captarum quam suo prae se acto uenerunt.*

4 Livy's language is difficult to disentangle here. Typically, we would expect the reflexives (*se, suo*) to refer back to the grammatical subject of the sentence, Flamininus' soldiers. If so, then the booty would have been driven in front of the soldiers, with the train of plunder as long as *their* own line. However, the larger narrative is focalized through Flamininus, and he is also the grammatical subject of the surrounding passage, which clearly marks him out as the natural antecedent to the reflexives, as I have done here: see Allen and Greenough 300.2 note; Woodcock 36 n. i. For similar translations, see Yardley's (Oxford World Classics) and Sage's (Loeb), who both take *se* as referring to Flamininus. However, even if the reflexive pronouns are understood as referring to the soldiers, the train of captured plunder (*agmine rerum captarum*) was then driven in front of *them* (see Briscoe 1981, 128) and the usual marching order again would have been altered. In either case, Livy clearly stresses that the arrangement of Flamininus, his troops, and their plunder had a distinctly triumphal appearance. For the typical marching order of a Roman army, see Polyb. 6.40; see also Goldsworthy (1996, 105–11).

5 Sumi (2005, 36) briefly notes Flamininus' long procession through Italy; Beard (2007, 266) mentions in passing that Flamininus and his troops marched through Italy *prope triumphantes*, emphasizing only Livy's coin of phrase.

they embarked almost immediately on a remarkable period of conquest and imperial expansion. The nature and motives behind these conquests have been discussed elsewhere.[6] I will focus instead on some of the consequences of Roman militarism and imperialism in the early second century on Roman political culture, and on Rome's relationship with its Italian allies.

Flamininus' procession *per totam Italiam* both reflects and sheds light on the deeply embedded relationship between warfare and politics in these pivotal years. Both his "Italian triumph" and his notably ostentatious formal triumph illustrate the profound importance for the Roman ruling elite to display military achievements in a grander and more innovative fashion to an ever-widening audience that included Latins and Italian *socii*. Indeed, Latins and allies also came to feature more prominently in the triumphal ritual itself. The motives behind including Italians as both participants and audience in a range of triumphal activities varied, as will be discussed below.

The sources

Livy's account is the only one to mention Flamininus' "Italian triumph." As a result, we are forced to ask ourselves, is the information authentic, or are we dealing with Livian elaboration or fabrication? Livy's larger narrative of Flamininus' activities in Greece and his return to Rome (34.48–52) expertly weaves together two source traditions – Polybius and the Roman annalistic tradition – though it is difficult to tell where one source ends and the other begins.[7] According to Briscoe, Livy follows Polybius at least up to 35.1, and indeed Livy cites Polybius directly as the authority for his background narrative (34.50.3–7) of 1,200 Roman prisoners-of-war who had been sold into slavery in Greece, were freed by Flamininus, and subsequently walked in his triumphal procession in Rome. Plutarch also mentions these freed Romans in his parallel account of the triumph (*Flam.* 13.3–6),[8] the information presumably also deriving from Polybius.[9] Plutarch cites Tuditanus as his source for the details of the booty displayed in the triumphal procession, quite possibly C. Sempronius Tuditanus (cos. 129).[10] Plutarch's and Livy's catalogs of

---

6 See now Terrenato (2019, 1–30), for an excellent diachronic survey of historiography on Roman imperialism from antiquity to the present; see also Armstrong and Fronda in this volume for a brief discussion.

7 Briscoe (1981, 124).

8 See also: Plut. *Mor.* 197B; Diod. Sic. 28.13; Val. Max. 5.2.6.

9 Smith (1944).

10 For the identity of Tuditanus: Cornell (2013, I.240–2). The manuscripts read περὶ τὸν ἰτανὸν or τουιτανὸν, which is usually emended to περὶ Τουδιτανὸν. See Cornell *FRHist* 10 F9 and Beck/Walter *FRH* 8 F6 *contra* Peter (τὸν Τίτον) and Cichorius (τὸν Ἀντίαν). The emendation to Τουδιτανὸν makes more sense paleographically, though it is admitted that Tuditanus would be a surprising source for a Livian triumphal notice, as he tends to derive triumphal notices from Valerias Antias. While Antias is often maligned, Rich (2005) has demonstrated that Antias' triumphal reports probably belong to the more reliable parts of his work that were derived from archival material. As such this information is likely

booty are very similar, with only minor discrepancies, suggesting that both authors are drawing on a common source.[11] If so, then Livy's description of Flamininus' *reditus* likely derives ultimately from second-century sources, either Polybius (presumably consulted directly) or Tuditanus (probably consulted indirectly). This, in turn, gives us some confidence in its historicity. It is my assumption for the remainder of this chapter that Flamininus paraded with his army and captured plunder from Brundisium to Rome in a manner that mimicked a triumphal procession in its organization and appearance.[12]

War, wealth, and aristocratic competition after Hannibal

Flamininus' "Italian triumph" fits the context of scaled-up Roman aristocratic competition which we see in the generation following the Second Punic War. As Rome fought more "big wars" against distant and wealthy opponents – especially in the East – Roman generals returned with greater wealth and more exotic spoils to display. The ancient sources agree that this period marked a turning point in Rome's history because of unprecedented riches pouring into the city. For example, M. Claudius Marcellus was reportedly the first general to bring back exquisite artwork and display it in his ovation, after sacking Syracuse in 211. Both L. Cornelius Scipio Asiaticus (triumph in 189) and Cn. Manlius Vulso (triumph in 187) are criticized for being the "first" to introduce luxury from Asia, which included such allegedly novel spoils as bronze beds, pedestal tables, and sideboards.[13] The discourse is moralizing, and we may rightly question whether the Romans were exposed to plundered luxury goods only as a direct result of these individuals. Nevertheless, no one doubts that the material rewards of empire had an enormous impact on Roman society.[14]

In terms of political competition, this vast influx of wealth allowed successful generals to put on more elaborate triumphs and furnished them with

---

to be more trustworthy. See also Rich's general introduction to Antias in Cornell (2013, I.298–304). It is, ultimately, impossible to determine the precise relationship between Livy, Plutarch, and their source(s), common or otherwise. Nevertheless, their correspondence in describing Flamininus' triumph is noteworthy. More importantly, Livy's telling of Flamininus' *reditus* is located among material drawn from Polybius, Tuditanus, and/or Valerius' more sound archival information.

11 Livy's report is fuller, containing items not mentioned by Plutarch. Figures given in both accounts are strikingly close. Livy reports 3,714 pounds of gold, Plutarch reports 3,713. Plutarch reports 43,270 pounds of silver were paraded in the triumph. The manuscript of Livy has 18,270 pounds of silver. Madvig assumed that 18 (XVIII) was a copyist's corruption of 43 (XLIII) and so amends Livy's figure to 43,270. Most accept the emendation: see Briscoe (1981, 128–29); Beard (2007, 171–72).

12 Even if the details of Flamininus' "Italian triumph" have been exaggerated through historiographical elaboration, there can be little doubt that Flamininus marched his troops and plunder across the peninsula – a performance that clearly struck observers as triumph-like.

13 Marcellus: Livy 25.40.3; Plut. *Marc.* 21. Scipio Asiaticus: Plin. *HN* 33.148; see Livy 37.59.1–6. Vulso: Livy 39.6.3–9 (specifying the introduction of *lecti aerati*, *monopodia*, and *abaci*).

14 On the influx of wealth, Hopkins (1981, 1–98) is still useful. Perhaps most famously, when L. Aemilius Paullus returned to Rome in 167, he dedicated so much plunder to the treasury that the Romans were able to cancel the *tributum*: Cic. *Off.* 2.76; Plin. *HN* 33.56; Plut. *Aem.* 38.

the resources to fund a variety of increasingly lavish public works, spectacles, games, and other *beneficia*.[15] The early second century saw a boom in temple construction, with approximately 20 new temples built between 200 and 167.[16] Besides temples, other forms of "triumphal architecture" began to transform Rome's urban landscape dramatically in the early second century, such as *fornices*, basilicas, and monumental porticoes. To these, we can add a slate of public works projects, including the building or improvement of bridges, roads, and sewers.[17] This competitive, triumphal building spilled into the private sphere, as Roman aristocrats aggrandized their houses and adorned them with booty.[18] Games (*ludi*) and other spectacles became more lavish – not only annual games dedicated to the gods but also occasional "one-off" festivities vowed by generals. In theory, such spectacles were publicly funded, yet the connection between victory games, the general who promised them, and the plunder that paid for them could not have been lost on the Roman audience.

Generals sometimes sponsored activities typically associated with triumphs – buildings, games, etc. – even when no formal triumph was awarded. For example, L. Stertinius famously sponsored the construction of the three monumental *fornices* in 196, each surmounted by gilded statues, paid for from 50,000 pounds of silver deposited in the treasury upon his return from Spain. According to Livy (33.27.3–4), Stertinius did not even bother trying to obtain a triumph. Similarly, in 191, P. Cornelius Scipio Nasica used spoils from his praetorship in Spain a few years prior to fund ten days of games, though he had not been awarded a triumph for these campaigns (Livy 36.36.1–2).[19] Both the construction of the *fornices* and the sponsoring of games must have been understood as victory commemorations that stressed the link between the general's military achievement and public benefaction.

---

15 Rosenstein (2012b, 245–43) provides an excellent, recent summary of the relationship between empire, money and aristocratic competition; increased frequency of triumphs following the Second Punic War: Bastien (2007, 277); Rich (2014).

16 Orlin (1997, 116–61, 199–202). Few temples built in this period were "manubial" constructions strictly speaking. Nevertheless, temples vowed in war were strongly linked to conquest regardless of who officially dedicated and paid for them, and virtually all such construction was made possible directly or indirectly by the influx of wealth from warfare.

17 Rosenstein (2012b, 246–47). On "triumphal architecture," see Roy (2017, 1–58, esp. 20–22, 226–50). While structures such as porticoes and basilicas were technically censorial buildings, Roy demonstrates that even these constructions could have strong martial connections, especially if the censors had won victories and brought in spoils when they held previous magistracies. See also Popkin (2016, 49–58, 187–95).

18 Welch (2006), arguing for a distinctly Roman "booty aesthetic."

19 Nasica may have opted not to seek a triumph because he anticipated receiving only an ovation, a lesser celebration granted to several Spanish commanders in the early second century. He also may have styled himself on his cousin Scipio Africanus, who also held games in 206 after his brilliant Spanish command. According to Livy (28.38.1–5), Scipio hinted to the senate that he desired a triumph but did not press his case since he suspected a triumphal petition would be denied on technical grounds, that is, he had not held a regular magistracy; see Rich (2014, 223–24).

Overall, increased wealth and resources heightened political competition, as successful generals were under pressure to sponsor more extravagant construction projects and other triumphal activities, and to find new and creative ways to commemorate their achievements.[20] This included putting on more opulent and striking triumphs, at least when circumstances allowed. The display of booty, especially plundered artwork, was a key aspect of this development. Not every triumph in the early second century featured jaw-dropping displays of booty, of course.[21] Indeed, a few examples were famous for their relative modesty and simplicity.[22] Still, there appears to be a general trend toward more occurrences of bigger, grander, and more creative triumphs, and associated triumphal activities, following the Second Punic War.[23] It is surely no coincidence that this same period saw several efforts to regulate the impact of wealth on aristocratic competition, including sumptuary laws restricting ostentatious displays of wealth and closer senatorial scrutiny over the proper distribution and use of spoils.[24] The increasing scale of triumphs and related triumphal benefaction prompted such responses.

Flamininus' "Italian triumph" reflects this drive to devise more memorable and impressive victory commemorations. By organizing his *reditus* in a triumphal (or triumph-like) form, he effectively extended his triumphal activities both temporally (as the march must have taken days or even weeks) and spatially (with his parade and the display of booty stretching all the way from Brundisium to Rome). In so doing, Flamininus appears to have been expanding on a precedent set a few years earlier, in 201, when Scipio Africanus conducted a similar *reditus* on his return from Africa. According to Livy,

> When peace had been secured on land and sea, and when the army had been boarded onto ships, he crossed to Lilybaeum in Sicily. For there, when a great part of the soldiers had been sent by ship, he himself arrived at Rome through Italy, which was rejoicing because of peace no less than because of victory, as not only were the cities emptied out in order to hold honors for him, but also crowds of rustic folk were blocking the roads. And he was carried into the city in the most brilliant triumph of all.[25]

20 Innovation in victory architecture: Popkin (2016, 46–91 esp. 58–75); see also Klar (2006).
21 See Rich (2014, 230–31).
22 For example, Livy 31.49.2–3 (L. Furius Purpureo), 45.43.1–8 (L. Anicius), 45.42.2–3 (Cn. Octavius), though Livy's emphasis on their modesty suggests grander triumphs were expected.
23 cf. Beard (2007, 163–73).
24 Rosenstein (2011); Rosenstein (2012b, 249–53). Distinction between *praeda* and *manubiae*, and the degree of the general's control over booty: Churchill (1999); Coudry (2009) *contra* Shatzman (1972); Shatzman (1975, 348–58, 391–92). Sumptuary laws: Clemente (1981).
25 Livy 30.45.1–2 (trans. Fronda): *Pace terra marique parta, exercitu in naves imposito in Siciliam Lilybaeum traiecit. inde magna parte militum navibus missa ipse per laetam pace non minus quam victoria Italiam effusis non urbibus modo ad habendos honores sed agrestium etiam turba obsidente uias Romam pervenit triumphoque omnium clarissimo urbem est invectus.*

Such spectacular *reditus* did not end with Flamininus. Indeed, L. Aemilius Paullus' return from the Third Macedonian War in 167 represents a further elaboration on the practice: he sailed up the Tiber on Perseus' flagship, which was decorated with captured Macedonian arms and royal tapestries. The sight amazed onlooking crowds, who had lined the river bank in order to catch a glimpse of the passing spectacle.[26] These flamboyant *reditus* represent yet another manifestation of this aristocratic drive for self-promotion and display of military success.

Moreover, such elaborately choreographed homecomings potentially headed off the sorts of political opposition that a general's success might generate. Triumphs were not awarded automatically; the sources record several examples of triumphal requests that were denied or granted only after considerable senatorial debate.[27] Triumphal debates were not simply a matter of politics, although they were a mechanism for maintaining aristocratic competition. Rather, the granting of triumphal awards was serious business. This is especially true in the generation following the Second Punic War, when the senate exerted itself to construct credible narratives of Rome's wars as always concluding in unambiguous, decisive victory.[28] Therefore, a general aspiring to obtain triumphal honors had to convince a skeptical senate to accept his petition. It benefitted a petitioning general to build his case in advance. News of victory arriving – by rumor, or through reports submitted to the senate, or by heralds sent in advance – would have laid the groundwork for the petition. It is likely that generals regularly sent letters to friends and potential allies to test the mood of the senate and to politick its members in order to ensure a favorable vote over a future triumphal petition. Indeed, even the most successful generals, like Paullus or Flamininus would have canvassed individual senators in advance by letter.[29] Sumi suggests that the general's return itself, full of pageantry and fanfare, generated excitement and built a certain momentum toward the formal awarding and celebration of a triumph, which could undercut political opposition.[30] In other words, we can interpret Flamininus' "Italian triumph" as an elaborate political maneuver, aimed at a senatorial audience, in order to overwhelm any possible opposition to his triumphal petition. Indeed, according to Livy (34.52.3), once Flamininus reached Rome, the senate met him outside the city, listened to his report, and willingly (*lubentibus*) decreed his triumph, with no indication of debate or controversy.

---

26 Livy 45.35.3; Plut. *Aem.* 30.1–3.
27 Pittenger (2008, esp. 303–7): approximately half of the 49 triumphal petitions between 218 and 167 were debated in the senate, and about 16% were denied. Another six commanders did not request a triumph despite military success, presumably for fear of rejection. See also Sumi (2005, 29–35).
28 Debates over triumphs: Pittenger (2008, esp. 33–53); Clark (2014, 94–134, esp. 95–100).
29 Beard (2007, 196–99), with appeal to Cicero's letter-writing campaign as a comparison.
30 Sumi (2005, 35–41).

Flamininus versus Scipio

As hinted at above, Livy's description of Flamininus marching "through all of Italy" (*per totam Italiam*) before entering Rome to celebrate a triumph finds distinct verbal resonance with his account of Scipio Africanus' *reditus* through Italy (*per Italiam*) in anticipation of his triumph. The connection between the triumphal activities of these two men can be pushed further. Livy (30.45.3–5), Polybius (16.23.5–7), and Appian (*Pun.* 65–66) provide accounts of Scipio's triumph. According to Polybius, the triumphal celebrations culminated with games and festivals that lasted for days, all funded by plunder; this is somewhat echoed by Appian, who concludes by mentioning the customary celebration of the triumphal banquet. Livy (30.45.5) concludes his version, however, by highlighting a unique feature of the triumphal procession: a young Roman senator, Q. Terentius Culleo, walked behind Scipio's chariot, wearing a *pileum* – the soft cap traditionally worn by freedmen. Culleo had been captured during the war with Hannibal and was among 4,000 Roman prisoners-of-war in Africa, whose liberation was stipulated in the treaty negotiated by Scipio. Livy reports that, for the rest of his life, Culleo honored Scipio as the author of his freedom. The story was apparently widely known, as Valerius Maximus (5.2.5) repeated it in his discussion of examples of acts of gratitude and ingratitude, noting that Culleo properly acknowledged his gratitude to Scipio by walking behind the triumphal chariot while wearing a *pileum*.[31] The sight of a young member of the senatorial aristocracy following Scipio's triumphal chariot while donning a freedman's cap must have been extraordinary. Livy's decision to close his account of the triumph with this detail, as well as its inclusion in Valerius Maximus' catalog, underscores its novelty and the strong impression it made on ancient viewers.

In fact, Scipio's decision to include Culleo "in costume" in the procession apparently set a precedent. In 197, Gaius Cornelius Cethegus celebrated a triumph over the Insubres and Cenomani (Livy 33.23.1). Cethegus' triumphal petition was supported by representatives from Cremona and Placentia, who testified that Cethegus had relieved these Latin colonies from the threat of a siege and freed many colonists who had been captured and enslaved by the Gauls (Livy 33.23.2). Cethegus' triumph was impressive enough, but what especially drew the crowd's attention were a large number of colonists who followed Cethegus' chariot wearing the *pileum* (Livy 33.23.4–9). Three years later, Flamininus utilized the very same motif, only on an even grander scale. According to Livy and Plutarch, 1,200 men followed behind Flamininus' chariot dressed in the appropriate costume, with their heads shaved and wearing the *pileum*.[32] These were Roman citizens

---

31 On Culleo, see also Plut. *Mor.* 196e; Cass. Dio fr. 57.86. For the *pileum* in triumphal performances and its association with *libertas*, see Weinstock (1971, 135–36).

32 Livy 34.50.1–7, 34.52.12; Plut. *Flam.* 13.4–6; cf. Val. Max. 5.2.6 (reporting, less plausibly, 2000 freed Romans); Plut. *Mor.* 197b; Diod. Sic. 28.13 (refers to the liberated as Italians).

who had been captured during the Second Punic War and sold into slavery in Greece. Flamininus had obtained their release before he departed the province for Italy, and their appearance reportedly added particular luster to the already stunning spectacle.[33]

Circumstances had provided Flamininus the opportunity to exploit this staging variation, and to play off the memories of both Scipio's and Cethegus' triumphs. Given the similarity of their dramatic pre-triumph processions through Italy, it is likely that Flamininus was gesturing particularly to Scipio's *exemplum*. Yet, Flamininus was not simply following in Scipio's triumphal footsteps. Rather, he drew attention to the great Africanus in order to show that he had surpassed him. For Scipio marched through Italy with a contingent of troops; Flamininus marched through Italy with his entire army, and with the captured plunder on display the entire route. Scipio's triumph was *clarissimus*, but lasted one day, while Flamininus' extended over an unprecedented three days. And instead of one freed Roman wearing a *pileum* walking behind the general's chariot, Flamininus showcased 1,200, highlighting Flamininus' overwhelming achievement as both liberator and conqueror.

The comparison to, and competition with, Scipio implicit in Flamininus' triumphal performance did not go unnoticed in antiquity. Indeed, Valerius Maximus explicitly juxtaposes the two. Immediately after concluding his account of Culleo walking in Scipio's triumph, he begins the very next section:

> But not one man followed the chariot of Flamininus when he triumphed over king Philip, but rather two thousand Roman citizens wearing the *pileum*. These men, who had been captured in the Punic Wars and were enslaved in Greece, after they were gathered together under his own administration, he restored to their former status. The honor of the general was thus double: at the same time enemies defeated by him and citizens preserved by him offered spectacle to the fatherland.[34]

Observers at the time would have made the same comparison that Valerius later highlighted, including veterans who had served under Scipio's command. At least 2,000 Scipionic veterans were recruited by Publius Sulpicius Galba in 200 for the war against Phillip, and another 3,000 were recruited

---

33 Livy 34.52.12; Plut. *Flam*. 13.6; Ehlers (1939); Bastien (2007, 279–80).
34 Val. Max. 5.2.6 (trans. Fronda): *At Flaminini de Philippo rege triumphantis currum non unus, sed duo milia civium Romanorum pilleata comitata sunt, quae is Punicis bellis intercepta et in Graecia servientia cura sua collecta in pristinum gradum restituerat. geminatum ita decus imperatoris, a quo simul et devicti hostes et conservati cives spectaculum patriae praebuerunt.*

by Flamininus in 198.[35] Some of these soldiers presumably remained in the legions until the army's return in 194, and thus would have marched in Flamininus' triumph. They surely did not miss the specific ways that Flamininus invoked and attempted to overshadow their former commander. Indeed, Flamininus may have taken the presence of Scipionic veterans into account when designing his triumphal performance. We can speculate plausibly that Flamininus' and Scipio's aristocratic peers, as well as members of the crowds who gathered to watch the procession, also picked up on the similarities.

A clue to the contemporary reception of Flamininus' triumph is found in Livy's (35.10.1–10) account of the consular elections in 193, when Lucius Quinctius Flamininus (Titus' younger brother), Scipio Nasica (Scipio Africanus' cousin), and Cn. Manlius Vulso competed for the sole patrician consulship. Lucius Flamininus had served under his brother in the east as legate, commanding the Roman fleet, and he too had returned to Italy in 194, presumably with his eyes on the elections the following summer. Meanwhile, Scipio Nasica had been praetor in 194 in Further Spain, where he achieved considerable military success.[36] That he would announce his consular candidacy in the next year's elections must have been expected. Livy reports that the election was hotly contested, as the voters turned their attention to the race between Nasica and L. Flamininus. Both Scipio Africanus and Titus Quinctius Flamininus campaigned on behalf of their respective relatives. Scipio was still Rome's preeminent war hero and statesman, having just held his second consulship in 194. Yet Flamininus' endorsement carried more weight, according to Livy (35.10.5), in part because his achievements were more current, and his triumph had been celebrated recently and was still fresh in the voters' minds.[37] Thus, Lucius Flamininus won the election. Read against the consular election of 193, Flamininus' carefully choreographed triumph may be understood not only as a means for Flamininus to enhance his own and his family's glory and fame more broadly but also as a performance specifically designed to support his brother in anticipation of a difficult political contest against the combined efforts of the Cornelii Scipiones. In any case, Livy's account of the election indicates that the contemporary audience, including voters in the consular election, recognized and responded to the comparison invoked by the specific staging of Flamininus' triumph, which included his elaborate pre-triumph procession through Italy.

A triumph was a sort of narrative text: its performance staged to tell a story about the triumphator and his victory. Yet individual triumphs, just like any other text, did not exist in isolation. They were linked to previous triumphs

---

35 Galba: Livy 31.8.6, 31.14.1–2, 32.3.2–7. The soldiers were still in the legions the following year when they mutinied because of long service. Flamininus: Livy 32.9.1; Plut. *Flam.* 3.3.
36 L. Flamininus: Livy 33.16.1–33.17.15, 34.26.11, 34.29.1–34.30.7, 34.40.7, 34.50.11; Zon. 9.16, 9.18; Scipio Nasica: Livy 35.1.3–12.
37 *[gloria] Quincti recentio, ut qui eo triumphasset.*

and other war commemorations. A knowledgeable audience could read the allusions between various triumphal texts: they would have picked up on the "intertextuality" – the dialogue and competition – between Flamininus' and other recent triumphs, especially that of Scipio Africanus.[38]

An Italian audience

When Flamininus led his army from Brundisium to Rome *prope triumphantes*, he probably followed the *Via Appia* past Tarentum and then north to Venusia and Beneventum. It is also possible that he took the path of the future *Via Traiana* from Brundisium toward Luceria and Beneventum. Between Beneventum and Rome, the procession would have traversed only Roman and Latin territory. Between Brundisium and Beneventum – no matter which route was followed – he would have passed through long stretches of allied territory, interspersed with Latin colonies, and perhaps also some lands confiscated by the Romans after the Second Punic War (see Maps 4 and 5). Flamininus presumably received a similar response as Scipio, with crowds of enthusiastic onlookers lining the route.[39] Indeed, it is hard to imagine that Flamininus would have orchestrated this triumph-like parade unless he expected a big crowd. But who comprised the audience?

Some Roman citizens, who had been settled along the route, perhaps came out for the procession.[40] Here, I come back to the Scipionic veterans. After the Second Punic War, those who had served in Africa or Spain were given land in Samnium and Apulia, viritane assignments of two *iugera* for each year of service (Livy 31.4.1–3). In 200, the consul P. Sulpicius took his army to Brundisium, where the Scipionic veterans joined his ranks, before departure to Macedon (Livy 31.14.1). Flamininus likewise departed from Brundisium in 198 after recruiting 3,000 reinforcements from among Scipio's veterans (Livy 32.9.1, 6). The men whom Sulpicius and Flamininus enlisted may have been settled nearby. If so, then perhaps Flamininus designed his procession from Brundisium to capitalize on the proximity of veteran settlers and their families. Indeed, both Flamininus and Sulpicius had been on the decemviral board that assigned land to the Scipionic veterans,

---

38 See Clark (2014), arguing for the centrality of the triumph to the formation of Roman victory narratives; Hölscher (2006), discussing the role of monuments in transforming moments of victory into lasting power; Popkin (2016) on the interplay of monuments, the triumphal route and the formation of Roman identity (esp. 46–91 on the period 264–146); Östenberg (2009) on the triumph as "ritualized play" (p. 7).

39 Compare to Cicero's *reditus* from exile, when (the orator claims) all of Italy came out to meet him on the road from Brundisium to Rome: *Att.* 4.1.4–6, *Sest.* 131–32, *Red. Sen.* 39, *Pis.* 22; cf. Livy, *Per.* 104.3; Vell. Pat. 2.45.3; Plut. *Cic.* 33.7–8; App. *B Civ.* 2.16; Cass. Dio 39.9.1.

40 Sipontum, settled as a Roman colony in 194 (Livy 34.45.3; see also 39.23.3–4), did not lie particularly close to Flamininus' route. Tarentum was not refounded as a colony until 122 (Vell. Pat. 1.15.4).

so they may have anticipated political support from the settlers. Flamininus was also a member of the triumviral board in 200 that enlisted settlers to reinforce the Latin colony Venusia, whose population had been depleted in the war with Hannibal.[41] These colonists, too, may have felt a sense of obligation or gratitude, and so could be expected to show up in large numbers to cheer the parade.

We can plausibly speculate that at least some inhabitants of allied cities, even those without a specific connection to Flamininus, also came to see the spectacle. Their motives for doing so would have varied. Here again, Scipio's *exemplum* is suggestive. Livy says that the crowds who honored Scipio rejoiced because peace had been restored, and some Italians no doubt legitimately saw Scipio as a hero: the general who finally drove Hannibal from southern Italy after so many years of fighting and devastation. Given the uneven loyalty of southern Italian communities during the war, some may have turned out as a public sign of fidelity to Scipio and Rome.[42] Similar reasons presumably motivated the onlookers for Flamininus' triumphal march. One publicly stated justification for declaring war against Macedon in 200 was that Philip posed a threat to the peace and security of Italy (Livy 31.3.4–6, 31.7.2–15). This is often dismissed as pro-Roman propaganda. However, given the highly competitive and conflict-ridden interstate environment – as well as the still-fresh memories of Hannibal's invasion and occupation of southern Italy – we should not easily disregard that some contemporary observers harbored this concern.[43] If some Italians feared that Philip was planning to invade Italy, they may have viewed Flamininus as their protector. Others may have wanted to curry Roman favor, while still others were simply drawn by the awesome spectacle itself, out of curiosity or for entertainment. Lastly, Livy's testimony that Flamininus landed in Brundisium with his whole army and then processed to Rome implies that both Roman and allied troops were included in the procession. If the allied soldiers were levied from towns and colonies along the route – and surely some were – then the onlookers would have included their family, friends, and townsmen.

To repeat, Flamininus must have expected big crowds for his "Italian triumph," and so presumably anticipated that the audience would have included a significant constituency of non-Romans. This performance, like Scipio's in 201, indicates the willingness of some generals in this period to present and perform displays of military achievement to a wider audience, including one that extended well beyond Rome. Along similar lines, the four known examples of triumphs *in monte Albano* (excluding Caesar's later reinvention of the ritual) also occurred in roughly the same period: in 231,

---

41 Livy 31.49.6; see also Plut. *Flam.* 1.4; see *MRR* 1.325–26.
42 The leading families in Locri probably had personal ties to Scipio after his recapture and administration of the city in 205: Livy 27.16.1–9; Fronda (2010, 313–14).
43 Roman security concerns: Eckstein (2008, 230–70, esp. 230–41) for thorough analysis and historiography; Eckstein (2006, 269–87).

211, 197, and 172.[44] The contexts were, of course, not identical, as triumphs on the Alban Mount were (at least presented as) acts of protest by generals who had been unjustly (in their eyes) deprived of a "proper triumph."[45] Nevertheless, these men were willing to celebrate a triumph outside of Rome before an audience presumably comprising all, or nearly all, *Latini*, suggesting that they perceived some benefit from the public recognition of their achievements by non-Roman Italians. This was, presumably, the same drive behind L. Mummius' decision a half century later to set up dedications of spoils around Italy.[46] That Flamininus orchestrated his "Italian triumph" to include *Latini* and *socii* indicates the similar calculus: display before an Italian audience, at least for some generals, was worth *something*.

The inclusion of allied soldiers in Flamininus' procession through Italy points to another aspect of his "Italian triumph": it allowed his non-Roman troops to participate in a triumph-like celebration. It is sometimes assumed (if the matter is considered at all) that *Latini* and Italian *socii* regularly took part in Roman triumphs.[47] Yet, the evidence for this practice is in fact quite limited. Livy reports five instances of generals who celebrated triumphs distributing donatives to both Roman and allied troops: Q. Fulvius Flaccus in 180 (40.43.4–7), Ti. Sempronius Gracchus and L. Postumius Albinus in 178 (41.7.1–3), C. Claudius Pulcher in 177 (41.13.6–8), and L. Anicius in 167 (45.43.1–8). In four of the instances, the allies were reportedly given shares equal to the Roman soldiers. Pulcher, however, infamously gave his allied troops only a half-share each. To show their displeasure, they followed the triumphal chariot in silence and thus put a damper on the celebration. Their reaction to receiving a smaller donative than their Roman counterparts implies that the allied soldiers expected equal rewards. More importantly, the passage indicates clearly that these *socii* marched in the triumphal procession in Rome. This is the only explicit reference to allied soldiers walking in a triumphal procession. In the other four cases, we hear only about the donatives (which generally occurred outside the *pomerium* and before the

---

44 C. Papirius Maso in 231: Val. Max. 3.6.5; Plin. *HN* 15.126. M. Claudius Marcellus in 211: Livy 26.21.1–10. Q. Minucius Rufus in 197: Livy 33.23.3, 8–9. C. Cicereius in 172: Livy 42.21.6–7.

45 See Brennan (1996); Bastien (2007, 265–68); Lange (2014).

46 Mummius: *CIL* 1².626, 627a, 627b, 628–630; Bizzarri (1973); *Imag. It.* Campanis/Pompeii 1; cf. Cic. *Verr.* 2.1.55; *Off.* 2.76; Livy, *Per.* 52; Frontin. *Str.* 4.3.15; Plin. *HN* 34.36; [Aur. Vic.] *De vir. ill.* 63.3. Mummius' dedications in Greece: *IG* 4.1183, 5².77, 7.433, 2478a, 1808, 2478; *SEG* 25.541; Philipp and Koenigs (1979); see Yarrow (2006).

47 For example, see Ostenberg (2009, 264–65): "A central function of the triumph was to perform and confirm Roman mastery, and the role-playing was fixed. Non-Romans played the defeated, Romans the victors. For the same reason, only Latins and Italian *socii* seem to have been allowed to walk with the Roman soldiers behind the triumphator." Interestingly, many standard works on the triumph do not mention the role of allies, even when discussing the army's part in the triumphal celebration: for example, Ehlers (1939, 509–10); Bastien (2007, 262–63); Beard (2007, 244–49).

procession began), though we may plausibly infer that those allies who received donatives also marched in the respective triumph.

To these we can add another example. In 187, M. Fulvius Nobilior triumphed over the Aetolians. According to Livy (39.5.13–17), on the day before his triumph, Nobilior assembled his army in the Circus Flaminius, where he lavished his troops with military gifts (donis militaribus), handing out awards to tribunes, prefects, centurions, and cavalrymen, both to Romans and allies (donavit...Romanos sociosque). Livy reports this example, but it is also referenced by Gellius (5.6.24–26), who quotes a speech of Cato the Elder criticizing Nobilior's generosity as a shameless ploy to win popularity. This Catonian fragment represents a contemporary response to Nobilior's actions and lends support to the authenticity of Livy's report. Again, we are not told explicitly that these allies actually marched in Nobilior's triumph, though it is probably safe to assume that they participated, if they had gathered in the Circus Flaminius.[48] That gives a total of six, more or less secure, instances of allies participating in a triumph between 187 and 167.[49]

Over that same 20 year span, Roman commanders celebrated at least 23 triumphal celebrations – 19 triumphs, one triumph on the Alban mount, and three ovations. All 23 are mentioned in the literary sources, and of these 23 triumphal reports, 17 make no reference whatsoever to allied participation. This includes several reports of donatives paid "to the soldiers" (militibus or in pedites) with no explicit mention of the allies receiving a share.[50] Admittedly, the surviving triumphal record can in no way be taken as a complete, accurate, and comprehensive catalog. Undoubtedly, some occasions of allies receiving donatives or walking in the triumphal procession have been omitted by Livy or escaped the notice of his sources. We also cannot exclude the possibility that a given generic reference to soldiers receiving a donative actually included both Roman and allied troops. Indeed, this is the case with Nobilior's triumph in 187: Livy (39.5.17) explicitly refers to allies receiving awards in the Circus Flaminius, but his report on the donatives does not mention the allies.[51]

---

48 Distribution of donatives and awards typically preceded the actual triumph: Zon. 7.21; cf. App. Mith. 116; Cass. Dio 51.21.

49 Livy refers to the recipients of donatives variously as "allies" (socii) and "allies of the Latin name" (socii Latini nominis), suggesting perhaps that sometimes only Latin allies joined in the triumph. Briscoe (1973, 77–78, with bibliography) argues, however, that Livy uses the terms socii ac nomen Latinum and socii nominis Latini interchangeably to refer to both Latins and Italians. I will not press the point here.

50 For example, in 187 (Livy 39.7.2), 181 (Livy 40.34.8), 179 (Livy 40.59.2), 167 (L. Aemilius Paullus' triumph: Livy 45.40.5; cf. Diod. Sic. 31.8.11–12; Plut. Aem. 32.4–34.7), and 167 (Cn. Octavius' naval triumph, with donatives paid to naval personnel: Livy 45.42.2–3). Typically, though not always, Livy reports double shares to centurions and triple to the cavalry.

51 militibus ex praeda vicenos quinos denarios divisit, duplex centurioni, triplex equiti ("from the booty he distributed 25 denarii to the soldiers, double to each centurion, triple to each cavalryman").

Nevertheless, the infrequency of references to allies in the triumphal record suggests that their participation may have been irregular. This should not be surprising given the wide variation in the celebration of the ritual,[52] including how many soldiers were present. Consider the following two examples. In 190, the proconsul M.' Acilius Glabrio's triumph over the Aetolians and king Antiochus was noteworthy because no soldiers marched behind his triumphal chariot. This is because he had turned his army over to the consul (Scipio Asiaticus) before returning to Italy and thus no troops were available for the triumph.[53] In 180, Q. Fulvius Flaccus discharged some of his troops, but handed over the rest of his army to the newly arriving praetor in Spain. Flaccus then sailed to Rome and marched into the city in a triumph accompanied only by those selected soldiers whom he had brought back with him (Livy 40.40.14–15). These instances do not pertain strictly to allied soldiers, but they do indicate how practical constraints could dictate what troops a general had available for a triumphal celebration. Put simply, a general could not always expect to march with his entire army in a triumph. Perhaps allied units peeled off and returned to their hometowns as the army marched back to Rome, especially if the donative promised was not large enough to be worth accompanying the general and waiting out the triumphal debate. Perhaps allied soldiers were sometimes not invited to join a triumphal procession. For whatever reasons, the evidence suggests that allied soldiers did not always feature in a triumphal celebration.

I would like to push the point further. All references to allies in the triumph are clustered between 187 and 167. The lack of examples after 167 can be explained by the loss of Livy's narrative and his relatively detailed triumphal reports. One suspects the practice continued, however frequently, and indeed a single ambiguous attestation points in that direction. According to Appian (*B Civ.* 1.46), the Marsi had such a reputation as fierce warriors that a well-known saying made the rounds in Rome: "no triumph over the Marsi, no triumph without the Marsi."[54] Appian reports the saying in the context of the Social War, indicating that it was already in circulation by that time. The dictum might indicate that the Marsi had previously taken part in Roman triumphs, though the precise meaning is not entirely clear.[55] More significantly, there are no references to allies in the triumph before 187, despite several notices of triumphal donatives distributed to soldiers as early as 295.[56] It is true that the loss of Livy's narrative covering the bulk of the period between 293 and 218, and with it the loss of as many as 64

---

52 Beard (2007) demonstrates efforts to reconstruct the "typical" triumph obscure the reality of a much more dynamic and variable ritual.

53 Livy 37.7.7, 37.46.1–6; Polyb. 21.5.13.

54 Οὔτε κατὰ Μάρσων οὔτε ἄνευ Μάρσων γενέσθαι θρίαμβον. Cf. Strab. 5.4.2; Livy 9.45.18.

55 See Rosenstein (2012b, 78). The phrase may have been coined only at the time of the Social War rather than being an age-old saying: Salmon (1967, 355).

56 Explicit references to donatives before 187 by year, as reported by Livy: 295 (10.30.10), 293 (10.46.14–15), 207 (28.9.16–17), 201, (30.45.3), 200 (31.20.7), 197 (33.23.7), 196 (33.37.11), 194

triumphal notices, makes it difficult to draw firm conclusions about the frequency of allied participation in the ritual in the third century.[57] Nevertheless, the clustering of recorded instances beginning in 187 implies either that allied participation in the triumph only began around this time, or that allies began to participate more regularly, in larger numbers, and/or with greater visibility in triumphal activities in these decades.[58] In other words, this small flurry of references to allies in the triumph reflects, I think, a distinct development in the ritual celebration that crystallized in the generation following the war with Hannibal.[59] If I am correct, then Flamininus' Italian triumph can be understood as an early experiment in incorporating allies in Roman triumphal (or triumph-like) celebrations.

Roman-Italian interactions

This development did not occur in isolation, but rather can be viewed in the wider context of Roman-allied affairs in the generation after the Second Punic War. Contact between Romans and Italians intensified in the second century, through a variety of interactions, including personal and family links, ties of patronage and clientage, political alliances, and business relationships.[60] This is highlighted by several events in the first third of the second century. For example, in 187 – the same year Nobilior doled out military awards to allies in the Circus Flaminius – the Roman senate responded positively to a request by delegates from Latin communities to address the growing number of *Latini* who had moved to Rome and obtained Roman citizenship *per migrationem et censuum*, which apparently made it harder for the Latin communities to meet their military obligations (Livy 39.3.4–6). The senate also responded positively to a similar request ten years later, in 177 (Livy 41.9.9–12). Also in 187, the senate decreed that both

---

(34.46.2), 194 (34.52.4–8), 191 (36.40.11–12), 189 (37.59.2–6; cf. Plin. *HN* 33.148). See Coudry (2009, 28–33, 71–79).

57 Itgenshorst (2005, 432–33) counts 64 triumphs from 291 through 219. Triumphs rarely appear explicitly in the *Periochae*. A total of five are mentioned, and never with details about the participants; see Livy, *Per.* 11, 17, 19.

58 Focusing just on the period after Scipio Africanus' triumph in 201 and prior to Nobilior's in 187: 17 triumphs are recorded, all mentioned by Livy. Of these notices, seven (41%) mention donatives explicitly; none refer to allies. It is worth noting, too, that the nature of our evidence changes for the period after the Second Punic War, as authors such as Livy have contemporary sources to consult (directly or indirectly). Given this, the inconsistent appearance of allies in triumphal notices is all the more striking. Livy may indeed have missed a couple of instances, but it goes too far, I think, to ignore the complete lack of references to allies in triumphs and thus assume they participated regularly before 187.

59 Recent scholarship tends to stress evolution, development and variation of the ritual over continuity: for example, Bastien (2007, 265–76); Beard (2007, esp. 80–105); Lange (2014); Lundgreen (2014), no fixed rules for obtaining a triumph; cf. Bastien (2007, 303–11); Rich (2014), varying frequency in triumphs awarded.

60 Fronda (2010, 316–20); Fronda (2011); see also Patterson (2006a, 2012).

Romans and *socii nominis Latini* should be immune from taxes and duties in the city of Ambracia (Livy 38.44.4). In 186, the Roman senate famously restricted the worship of Bacchus throughout Italy, their regulations binding on both Roman and allied communities.[61] Already in 193, the year after Flamininus' triumph, the senate undertook to address abuses of Roman credit laws, namely Roman creditors employing Italian middlemen to issue loans above the maximum legal interest rate. The generation after the Second Punic War, therefore, was one of accelerated contact between Romans and Italians.

This was true also of shared military service. In the 250-year period between 340 and the Social War, the proportion of allies serving in the Roman army was likely at its highest between 200 and 168, with allies comprising around 60% of Roman forces on average. The ratio may have been even higher in the first half of this period, between 200 and 180.[62] Livy mentions several instances in years immediately following the Hannibalic War of "Roman" armies comprised entirely of allies. For example, in 200, the provinces of Gaul and Bruttium were entrusted to two praetors, each commanding 5,000 allied and Latin soldiers, while a propraetor was sent to Sardinian with an army of 5,000 *socii et Latini* (31.8.8–10). In 199, the consul L. Cornelius Lentulus took over an army stationed near Ariminum, dismissed the Roman legions and then held the area with 5,000 allied soldiers (32.1.2–5). In 198, each praetor in Spain received reinforcements: 8,000 infantry and 400 cavalry levied exclusively from the allies (32.28.11). Conforming to the broader pattern, the reinforcements that Flamininus was given upon taking up his province in 198 had a ratio of 62.5% allied soldiers.[63] This suggests that his army in Greece – and thus the army that returned to march in his "Italian triumph" – comprised a similarly high proportion of allies.

There was, I suggest, a link between increased intensity of Roman-allied interaction, especially in the realm of military service, and the sudden appearance of notices of allies receiving donatives and participating in Roman triumphs (itself indicative of a development in the ritual, as I have just argued). Greater numbers of allies fighting in Roman armies meant additional opportunities for personal links and bonds to develop, most importantly among the elites serving in the cavalry and among the officers. In turn, connections between Roman and allied elites surely raised Roman awareness of, and attentiveness to, allied affairs. Sometimes, this resulted in greater Roman scrutiny: for example, cracking down on the cult of Bacchus throughout Italy. Other times, Roman elites responded positively to Italian appeals,

---

61 *CIL* 1^2.581; Livy 39.8.1–39.19.7; Cic. *Leg.* 2.37; Val. Max. 1.3.1, 6.3.7. The historiography is vast; see Fronda (2010, 321–22, with discussion and bibliography), *contra* Mouritsen (1998, 53–54).

62 Ilari (1974, 171–72).

63 Livy 32.8.2: in 198, Flamininus was given 3,000 Roman infantry, 300 Roman cavalry, 5,000 allied infantry, and 500 allied cavalry.

whether out of friendship, mutual interest, or a sense of duty or obligation. We may speculate that bonds forged in the context of military service were particularly powerful, fostering mutual devotion between the general and those who fought under and alongside him. If so, then allowing allies access to the triumph may be understood as an outgrowth of this dynamic.

At the same time, this and other (admittedly occasional) conciliatory Roman gestures may also represent shrewd political calculation, especially in the years immediately following the Second Punic War. This difficult and painful struggle was not a distant memory. Rather, the wounds of allied disaffection and Roman retribution were still fresh, and surely anger, bitterness, and mutual resentment remained. The Romans ruling class must have been wary of exacerbating residual tensions especially as they embarked on new and expansive imperial projects.

Indeed, Livy (31.8.11) comments in his discussion of the levy in 200, that the consuls were commissioned to enroll two legions to be sent anywhere they might be needed, "since many peoples in Italy had taken part in alliances in the Punic War and from that time were swelling with anger,"[64] which indicates not only lingering allied anger but also Roman awareness of it. Roman armies continued to be stationed in various regions of Italy for a couple of years after the war,[65] and subsequently Roman magistrates and promagistrates were assigned with some regularity to provinces within Italy (south of the Po River) through the 180s. The majority involved commands and/or investigations in southern Italy, in Bruttium, Apulia, or the vicinity of Tarentum.[66] Thus, Rome concurrently relied more heavily on allied military contributions, both to conduct distant wars and to police the peninsula during a time when the Romans continued to be concerned over the security (and perhaps loyalty) of allied territory, and especially of southern Italy. The willingness of some Roman commanders to accede to the apparent Italian desire to share in their triumphs may fit a larger tendency of the Roman elite not only to pay closer attention to the allies but also, to some degree, to keep them happy.

This brings us back once more to Flamininus, whose procession through Italy in 194 links to the inherent tension Rome's relations with their Italian allies. His "Italian triumph" was surely an awesome display of Roman power staged in allied territory that had seen fierce fighting during the war with Hannibal. Indeed, the first leg of his most likely route (as discussed above) would have gone from Brundisium to Tarentum,

---

64 Livy 31.8.11 (trans. Fronda): *multis in Italia contactis gentibus Punici belli societate iraque inde tumentibus.*

65 Bruttium in 200 and 199: Livy 31.6.3, 31.6.8; 32.1.7–11. Campania in 200: Livy 31.8.9. Etruria in 196, response to a slave uprising: Livy 33.36.2–3.

66 Provinces in southern Italy: 191 (Tarentum, Brundisium and the coast), 191 (Bruttium), 190 (Apulia and Bruttium), 189 (Apulia and Bruttium), 187 (Tarentum), 185 (Tarentum), 184 (Tarentum), 183 (Apulia), 181 (Apulia), 180 (Apulia). Most of the southern Italian provinces in the 180s pertained, at least in part, to investigating the Bacchanalia.

taking him past a city that had been brutally sacked in 209. The route also correlates with where the Romans subsequently investigated the Bacchanalia. More proximately, in 194 the senate initiated a major program of colonial foundations in southern Italy, including at least one colony on land confiscated in Apulia,[67] and in 190 the praetor M. Tuccius was given a very large force "to hold Apulia and Bruttium" (*ad Apuliam Bruttiosque obtinendos*).[68] In this context, it is possible that Flamininus' "Italian triumph" was intended, in part, to overawe both dissidents (actual or potential) and allies whose loyalty was uncertain. At the same time, however, his triumph-like performance was also a gesture to the allies, by including them as participants and spectators in a celebration imitating Rome's greatest commemorative ritual.

Conclusions

In 2007, Pfeilschifter argued powerfully against what was once the generally accepted position, that shared military service promoted blurred distinctions between Romans and Italians, promoting fellowship, shared identity, integration, and ultimately "Romanization": the army-as-melting pot model. More recently, Nate Rosenstein has proposed a very different way of approaching the problem. He suggested that the process of allied integration (at least with respect to shared military service) is better understood as a striving for equality in respect, dignity, and opportunity on the part of the allies, rather than a struggle to "become Roman." As Rosenstein argued, the allies, in fact, faced no significant disadvantage to their Roman counterparts in terms of distribution of plunder, decorations, and donatives, as well as in the opportunity to commend themselves on the battlefield.[69] To this, we can add increased opportunity to take part in Roman triumphs, the ultimate military honor, a process that began or at least accelerated in the 180s.

Yet, allowing non-Romans to march in the triumph meant extending the ritual community. If "[i]n the triumph, Rome defined herself by displaying others," and "if present, foreign soldiers would have played a Roman role" in the performance, as Östenberg asserts, how did the Roman audience respond to *socii et Latini* marching in the triumph?[70] Did they, in that ritual moment, distinguish between Roman and allied soldiers processing behind the triumphator's chariot? And did those Italians who took part, either the

---

67 Livy 35.45.3–5, 35.53.1–2. As mentioned above, Sipontum was founded in Apulia, and though it did not lie directly on Flamininus' route; nevertheless, Roman reorganization and assignment of lands to settlers may have exacerbated tensions in the region more generally.

68 Tucceius was given two legions plus more than 15,000 allied soldiers (Livy 37.2.6); his command was extended in 189 (Livy 37.50.13, 38.36.1).

69 Rosenstein (2012a).

70 Östenberg (2009, 262–66), quotations at 262 and 265.

soldiers in the procession, or (in Flamininus' elaborate celebration) the crowds stretching as far as Brundisium who made up the audience, feel, or come to feel, that they were part of the Roman community? If so, then the maybe the triumph of Titus Quinctius Flamininus *per totam Italiam* was a very small, first step on a very long and difficult march to *tota Italia*. But the Romans and Italians would first have to pass through the Social War before reaching that destination.

11 *Ager publicus*

Land as a spoil of war in the Roman Republic*

Saskia T. Roselaar

Introduction

The taking of spoils from a defeated enemy was a critical aspect of Roman warfare.[1] War spoils were expected to subsidize the cost of Rome's wars, and even to yield profits for the Roman people. Cato the Elder's famous saying, quoted by Livy (34.9.12) in reference to his ongoing campaign in Spain in 195, that "the war will feed itself" (*bellum se ipsum alet*) nicely captures this calculus. Indeed, the war profits that poured in, especially during the second century, had a transformative effect on Rome and Italy more widely – as is discussed elsewhere in this volume.[2] No doubt the Roman people grew to eagerly anticipate the spoils generated by successful warfare, which they benefitted from – both directly and indirectly.

However, the distribution of spoils was often not without problems. There were few written regulations about the way in which generals should dispose of the spoils that had been acquired under their command. The triumphs of 293 highlight the variation possible, even within a single year: the consul Papirius had won considerable amounts of silver and put this in the public treasury, which caused friction with both his soldiers and the people.[3]

---

* I would like to thank Nate Rosenstein for his continued support for my work, starting more than a decade ago, when I spent a semester in Columbus, Ohio, as part of my PhD studies which focused on *ager publicus*. I retain fond memories of our discussions in that period and in the years since then. All dates are BC unless otherwise noted.

1 See Rosenstein (2012b, 20–21, 106–12, 207–9, 245–56).
2 See, particularly, Fronda in this volume.
3 Livy 10.46.5–6 (Loeb translation, slightly modified): "Of bronze there were carried past 2,533,000 *ases graves*. This bronze had been collected, it was said, from the sale of captives. Of silver which had been taken from the cities there were 1,830 pounds. All the bronze and silver was taken to the treasury, none of the booty was given to the soldiers. The ill feeling this gave rise to in the plebs was increased by the gathering of the war-tax to pay the troops, since, if the consul had forgone the glory of depositing the captured money in the treasury, the booty would then have afforded the soldiers a donative, as well as providing for the pay."

DOI: 10.4324/9781351063500-11

This chapter has been made available under a CC-BY-NC-ND license.

His colleague Carvilius, on the other hand, distributed his spoils, which made him popular with the people and the soldiers. While one approach was clearly remembered as being more popular, both were evidently legal. Indeed, there is little evidence to suggest even the desire to create laws and fixed mechanisms to regulate the distribution of spoils during the middle Republic. While tensions around distributions were clearly possible, and indeed perhaps common, they were also seemingly accepted.

This is not to say, though, that the distribution of spoils was entirely unregulated. To the contrary, spoils were surrounded by quite strict (unwritten) social norms. It seems that the model to be followed by generals was to refuse to personally profit from spoils, even after a great victory, in order to adhere to the ideal of frugal living.[4] Additionally, Livy's lists of spoils carried in triumphs suggest that generals were expected to write down or otherwise record exactly how many spoils they collected, which shows a desire for transparency.[5] However, there was, as far as we know, no law that ordered generals to create such lists of spoils taken and deposited into the *aerarium*.

War spoils, therefore, played a key role not only in Roman state finances but also in the economic advantage of individual Romans. Spoils were also important to the interaction between the aristocracy and the people. The distribution of spoils, which led to both benefits and conflicts, was governed largely by an unsystematic combination of regulations, unwritten rules, and ad hoc measures, though in the second century there appears to have been some attempt to introduce (at least slightly) more senatorial scrutiny and oversight.[6] These statements pertain not only to moveable plunder, such as those seen in the example of the triumph of 293 – money, art objects, animals, and human captives carried off by the victors – which are typically stressed in discussions of spoils, but also to another category of war spoils that tends to be somewhat overlooked: the land that was taken from defeated enemies and became the property of the Roman state. This land is known as the *ager publicus populi Romani*, "public land of the Roman people."[7]

(Aeris gravis travecta viciens centum milia et quingenta triginta tria milia; id aes redactum ex captivis dicebatur; argenti quod captum ex urbibus erat pondo mille octingenta triginta. omne aes argentumque in aerarium conditum, militibus nihil datum ex praeda est; auctaque ea invidia est ad plebem quod tributum etiam in stipendium militum conlatum est, cum, si spreta gloria fuisset captivae pecuniae in aerarium inlatae, et militi tum donum dari ex praeda et stipendium militare praestari potuisset.)

4 This was tied in with a general ideology that presented an idealized "peasant" lifestyle as a model for aristocrats to follow. Elites supposedly spent their days in manual labor in the fields and were happy with little: for example, Cincinnatus (Livy 3.26.7; Val. Max. 4.4.6–7); Atilius Regulus (Val. Max. 4.4.6); Cato the Elder (Plut. *Cat. Mai.* 2.2; Festus, *Gloss. Lat.* 350.26L). On this ideal, see Adamo (2017).

5 Rich (2014). See Livy 39.7.1–2, 45.39 for examples of the treasure carried in triumphs.

6 See Fronda in this volume.

7 See Roselaar (2010) for extensive discussion of *ager publicus*.

This chapter will survey the Roman approach to *ager publicus* in the context of war spoils – from c. 400, when land became an important form of spoils, until the second century, when Roman conquest extended increasingly beyond the peninsula and other forms of war spoils (mainly moveable goods and war indemnities) became preferred – and it will consider some of the ways that this confiscated land impacted Roman politics and society. It will discuss how and when the idea of taking land as spoils originated, as this was by no means a natural choice, and will trace some major trends in the methods of land distribution employed over this period. In particular, it will concentrate on some of the structural problems regarding the acquisition and distribution of this land – for throughout the Republican period, the Romans never managed to find a method of dealing with *ager publicus* that satisfied everyone. As with other forms of spoils, the distribution of land regularly led to serious conflict between Roman politicians, and between the elite and the people. Over time, the Roman state increasingly regularized the process of distributing land, for strategic reasons but also perhaps in response to the political conflicts land distribution caused. However, the state never "solved" the problem, and allocation of land remained contentious.

The early Republic (c. 400–338): Why *ager publicus*?

Roman society underwent a major transition in the period c. 450–400, as the Roman state developed an innovative military regime, including new strategies to fund warfare.[8] It is this context that helps explain the shift from spoils comprising predominantly moveable goods to the increased importance of non-moveable spoils (most importantly land confiscated and turned into *ager publicus*). The Roman practice of mulcting defeated communities of land developed rapidly after the conquest of Veii, as did the systems of exploitation and distribution. In this period, land became a vital part of Rome's military equation.

Until the mid-fifth century,[9] small farmers in central Italy seem to have lived in an interconnected world of small microecological niches,[10] in which they could apply a variety of survival strategies. In order to protect the future quality of the land, cooperation and mutual dependency between small and large farmers were essential.[11] In a sense, the poor may have served as vassals of some sort for richer farmers; part of a system founded on social and economic bonds of mutual obligation.[12] In this type of society, warfare

8 Armstrong (2016c, 183–232). See also Drogula and Tan, in particular, in this volume.
9 For the following discussion, see also VanDerPuy in this volume. Fifth-century developments are hazy; see Roselaar (2010, 20–31) for a survey.
10 Horden and Purcell (2000, 61–62); Hughes (2014).
11 See, for example, Colum. 2.1.6–7, 2.2.13.
12 Cornell (1995, 291, 330); Horden and Purcell (2000, 84–86); Bernard (2016). See Smith (2006a, 190–92) on the *gens* as a system of hierarchical dependency, including the control of land.

consisted of short-term raiding for (typically) portable spoils, with only rare occurrences of large-scale participation in armies recruited by the state. It seems that most spoils consisted of items like cattle, clothing, and slaves, rather than the precious objects that we encounter in the later Republic.[13] Trade over longer distances occurred, especially in crisis situations (such as famines), and was mediated through the elites. In this system, elites took on the role of protectors of their dependents, which included the task of finding food through trade when necessary.[14] The power of the lower classes was vested mostly in the labor they could provide, especially during the harvest. Although somewhat unequal, this system created a stable and mutually beneficial relationship between rich and poor, ensuring that the "peasants" were not entirely subject to the whims of the elite.[15]

Between the mid-fifth and mid-fourth centuries, however, Rome experienced a transition from a more collective to a more individualistic society,[16] as many old clan structures disappeared, and individuals became bound to the state through their civic obligations.[17] Around 450–440, the Roman state more forcefully asserted its power, indicated by the codification of law in the Twelve Tables and the creation of the censorship. The latter allowed the state to calculate more precisely the demands that it could make of its citizens. Households now owned their own land. They were assessed individually, and their military and taxation obligations determined individually, rather than collectively. This had the effect of highlighting the distinctions between individual households.[18]

Since the state had inserted itself into the lives of its individual citizens, seemingly in place of the previously existing clan- and obligation-based structures, the citizens began to expect the state to deliver in terms of providing for their subsistence. This can be seen in a number of ways. As Helm argues,[19] the fourth century saw a shift from private warbands to more heterogeneous state-based forces, bolstered by increasing numbers of plebeians, as indicated by finds of cheap weapons belonging to light infantry. The state seems to have increasingly become the nexus for military power.[20] This, however, raised the problem of sharing spoils equally: light-armed soldiers assumed less of the burden of fighting, but, compared to wealthy heavy-armed soldiers, they received a larger share of the spoils in relation

---

13 Armstrong (2016c, 218–19).
14 Livy 2.23, 4.12, 4.25; Dion. Hal. *Ant. Rom.* 10.54.1–2, 12.1.2. See Crone (1989).
15 Crone (1989, 110). See Rosenstein (2004, 66–69) on family life cycles, where there is often a phase in which the typical family has more labour than necessary to work its own farm. At this point, young men could be sent away to work elsewhere, for example, as soldiers or wage labourers.
16 See VanDerPuy in this volume.
17 Armstrong (2016c, 272–80).
18 See VanDerPuy this volume; see also Cornell (1995, 188); Armstrong (2016c, 184, 231, and 241).
19 Helm (2017); see Armstrong (2016c, 183–232).
20 See Drogula in this volume.

to their economic situation. Small-scale warfare and raiding presumably played an important role for these troops, since it enabled them to improve their economic situation by gaining spoils. Thus, it seems likely that plebeians would have been more eager to participate in war, which may have contributed to more regular occurrences of warfare from the fourth century onwards. The state also began to reward soldiers with the *stipendium*, most likely introduced in or around 406.[21] This underscores that the army was increasingly viewed as an extension of the state, rather than under the control of individual warlords.[22]

In an interesting correlation during this period, where Rome's citizen soldiers were increasingly dependent on the state for their survival, they were also increasingly rewarded for military service in land which had been confiscated from defeated enemies, instead of the more traditional, more portable forms of spoils.[23] Whether this shift was driven by demand from the soldiers, or the new mechanics of state-based warfare is unclear – but the association is marked. This dynamic took off with the defeat of Veii (c. 396) and the subsequent confiscation of its territory (perhaps doubling the extent of the *ager Romanus*), which set an important precedent for fourth-century policies.[24]

In the course of the fourth century, the outlines appeared of a more regulated system for dealing with defeated peoples and the land that was confiscated from them. The Roman state was primarily interested in widening its "tax base," and therefore devised various methods to gain as much as possible from its citizens and allies, in the form of money or men.[25] In the fourth century, Rome employed different ways of dealing with defeated peoples. The first method was to grant defeated peoples a form of Roman citizenship, for example Tusculum in 381 and Velitrae around the same time.[26] The inhabitants of these towns were then obliged to pay *tributum* to Rome and contribute soldiers in the levy.[27] Thus, while Roman citizenship has often been interpreted as a reward,[28] in the fourth century it was arguably much more a mechanism of subjugation to Rome.

Another option which the Romans utilized was to take part of the land from the defeated people and either hold it as *ager publicus* or distribute it to Roman citizen settlers. The granting of confiscated land to poorer Roman citizens allowed a potentially wide cross-section of Roman society to share in this increasingly important form of non-moveable war spoils. In turn,

21 See Tan in this volume.
22 Crawford (1985, 22–23).
23 Bernard (2016).
24 Importance of Veii: VanDerPuy in this volume.
25 See Tan in this volume.
26 Tusculum: Livy 6.4.4, 6.26.8: Velitrae: Livy 6.17.7, 6.21.3, 8.14.5.
27 On *tributum*, see also Nicolet (1976); see also Tan in this volume.
28 For example, Galsterer (1976, 65).

these land grants helped the Roman state to maintain a high number of *assidui*, who were liable both for military service and to pay the *tributum*, and thus contribute financially to Rome's wars. As mentioned, the first case of large-scale distribution of confiscated land occurred in Veii in 396. The distribution of land in the fourth century usually occurred in viritane settlements. In this case, the land was distributed to existing Roman citizens, who were assigned individual plots. The new settlements were organized in *tribus*, the voting tribes that made up the *comitia tributa*. New *tribus* were created periodically in the fourth century (see Map 3): in 386 (*Stellatina, Tromentina, Sabatina,* and *Arnensis*), 358 (*Pomptina* and *Publilia*), 332 (*Maecia* and *Scaptia*), and 318 (*Falerna* and *Oufentina*). In the third century, fewer were created: the *Aniensis* and *Teretina* in 299 and the *Velina* and *Quirina* in 241.[29]

Despite what appears to be the emergence of a rather loose "system" for dealing with defeated people and their land, decisions to share out *ager publicus* with the Roman people were taken in a largely ad hoc manner. Given the apparent increasing importance of this land to the Roman people, its irregular distribution led to repeated conflicts between the state and its citizens. Indeed, as I have shown elsewhere, ancient sources record continuous conflicts between the state and its citizens regarding the land that was taken in wars.[30] Indeed, the confiscation of land was not very regulated at this time: the Roman state did not always confiscate land from the defeated, and when it did, there was no fixed amount of percentage of land that was taken.[31]

Some peoples, such as the Hernici, were not made Roman citizens, but remained allies (*socii*).[32] These communities remained independent from Rome, and so their citizens were not obliged to pay *tributum* – although they did supply soldiers to the Roman army.[33] Moreover, the soldiers of allied states were not paid *stipendium* by the Roman state; rather they were paid by their own states, if they received any pay at all.[34] Thus, allied status was another means for the Roman state to spread the financial burden of warfare. The allied states did not always suffer land confiscations, although many did; indeed, the allies appear to have been better off than those communities that received Roman citizenship, since the latter had to supply soldiers

---

29 Livy 7.15.12., 8.17.11, 9.20.6, 10.9.14; Livy, *Per.* 19.15; Diod. Sic. 19.10.1. According to the sources, 13 colonies were also founded between 510 and 383, but it remains unclear what a colony at this time represented, see Bispham (2006b).

30 See Roselaar (2010, 25–31) for more detail.

31 One-third or one-half of the land of the defeated party is often considered the standard amount taken by the Romans (for example, Hopkins 1978, 60); but the case of Frusino in 303 is actually the only one in which one-third is specified as the amount seized (Livy 10.1.3).

32 Livy 9.43.23–4.

33 Brunt (1971a, 545–49); Cornell (1995, 361); Erdkamp (2007a). They were sometimes given land as well, for example, Livy 4.11.3–4, 4.51.3. See Roselaar (2010, 75–76).

34 Pfeilschifter (2007, 31).

and pay taxes. This may have contributed further to tensions and discontent among the Roman people, who may have perceived that vanquished foes got a better deal.

How the Romans decided what status to give to (or impose upon) their allies and defeated peoples is discussed elsewhere in the volume.[35] Whatever factored into their calculations in each specific case, by the second half of the fourth century, the Romans felt compelled to adapt their practices with respect to conquered peoples in Italy. These new policies impacted how the Roman people gained access to what had become a very desirable albeit controversial form of spoils: land. The critical turning point was the Latin War.

Developments after the Latin War (c. 338–200)

During the Latin War (341–38), the Romans finally established their authority over the Latins, who were punished for their "rebellion" in a number of ways, including through loss of land.[36] After this war, the Romans gradually embarked on the conquest of Italy as a whole, leading to the creation of a large-scale system of allies subject to Roman hegemony. The Latin War seems to have been a watershed in this regard, as a new system of dealing with the defeated peoples of Italy was devised – or at least cemented – that made use of methods already in place, while adding new strategies for maximizing both state income and the supply of soldiers. Patterns in the distribution and use of *ager publicus* also changed. On the one hand, in contrast to how the Romans dealt with moveable spoils, it seems that from the Latin War onwards the process of distributing land became somewhat more structured. On the other hand, there was a move away from assigning *ager publicus* to Roman citizens and creating new tribes, and by the Second Punic War, the Roman state appears to have retained communal ownership of large amounts of public land.

As we have seen, by the middle of the fourth century, the Romans had created a system of taxation and recruitment among the Italian peoples which they had conquered, which was designed – at least in part – to attract both maximum tax revenues and the maximum number of soldiers for the army. This system included distributing land to Roman citizens, making some defeated peoples into Roman citizens, and sometimes leaving land with trusted allies. From around the time of the Latin War, the Romans began to experiment with this system. Most notably, they created the status of *civitas sine suffragio*. The rationale behind the creation of this status is much debated. Tan argues in this volume that citizenship without the vote was devised to exploit, to the maximum extent, those allies who were wealthy

35 See Tan in this volume.
36 Livy 8.11.14, 8.14.8.

and able to pay taxes, but could not serve in the legions.[37] Rich cities in Campania are good examples, and their inhabitants were indeed made *cives sine suffragio*.[38] Yet exceptions existed: some *civitates sine suffragio*, such as Privernum, Fundi, and Formiae,[39] were not exactly wealthy, or at least we do not possess data to prove that they were. Other towns, such as Praeneste and Tibur, were perhaps too powerful and too close to Rome to alienate; therefore, they were not made *civitates sine suffragio*. However, some land was taken from these towns and became *ager publicus*. A further anomaly were the colonies with Roman citizen rights: their inhabitants were too poor to pay *tributum*, since they only received two *iugera* of land, so they could not be subject to the *dilectus*. However, they were given a special role as defenders of the coast, through the grant of *vacatio militiae*. This meant that they fulfilled their assigned military role, without receiving *stipendium*.[40] Whatever the reasons behind the creation of the status of *civitas sine suffragio*, it is clear that the Roman state was experimenting with, and making modifications to, its policies and practices regarding conquered peoples that it had devised in the early fourth century.

We also see a distinct development in the handling of *ager publicus*: an apparent major decline in viritane distribution occurred, as indicated by the low number of new tribes created after the turn of the third century. As we have seen, only the *tribus Aniensis* and *Teretina* (in 299) and the *Velina* and *Quirina* (in 241) were created in the third century. It is remarkable, too, that many *tribus* were created long after the actual conquest and settlement of the land: the *Falerna* and *Oufentina* were created 22 years after the distributions had taken place in 340, while the *Velina* and *Quirina* were located on land distributed in 290. The reasons for this development are not entirely clear. It may have been in part based on political calculations, as the distribution of land to Roman citizens and the (eventual) creation of new tribes may have caused political tensions. For example, in the case of the Velina and Quirina, it may be that the senate opposed the creation of these *tribus*, because it would have given Dentatus, the conqueror of the area in which they were settled, too much personal influence in the voting assemblies. He had evidently settled his own soldiers in the territory and could therefore influence the way these *tribus* voted.[41]

The proposal in 232 to distribute the *ager Gallicus* likewise provoked much opposition (See Map 4). Cicero is fond of emphasizing this, and uses Flaminius as an example of a demagogue bent on wooing the people with

---

37 See Mouritsen (2007, 156–57), who, however, struggles to find an explanation for the creation of this status; Stewart (2017) points to the importance of language in the decision whether to grant *civitas optimo iure* or *sine suffragio*.
38 On the wealth of the Italians, see Roselaar (2019).
39 Livy 8.14.10, 8.21.9–10; Vell. Pat. 1.14.3.
40 Roselaar (2009).
41 [Aur. Vict.] *De vir. ill.* 33.

land distributions: "Gaius Flaminius ... when tribune of the people sedi-
tiously proposed an agrarian law against the wishes of the senate and in general contrary to the desires of all the upper classes."[42] It has been suggested
that the Senators themselves had occupied this land; it had been conquered
by the Romans 50 years before and it is hardly likely that *ager publicus* would
have remained untouched all this time.[43] On the other hand, it is difficult to
imagine that the involvement of senators in this area was widespread, since
northern Picenum was far away from profitable markets, and large estates
were not common in Picenum in the third century. It is more likely that
the opposition against Flaminius was motivated in part by fear of the personal influence he would gain by binding to him, as clients, the people who
received land. Especially after the restructuring of the *comitia centuriata*,
around 240, a *homo novus* with a large clientele in one *tribus* would have a
significant amount of power. The distributions in the *ager Gallicus* therefore
took place without the establishment of new *tribus*, even if this had still been
a (reluctantly) acceptable option in 241, less than a decade earlier.[44]

Instead, the Roman state increasingly moved to create more colonies with
Latin status for settlers, rather than new *tribus* where the settlers would be
Roman citizens. The reason behind the shift to Latin colonies may have
been largely strategic. Latin colonies were likely intended as military outposts in enemy territory, often far from Rome. Therefore, they needed to
be able to act independently in case of attack. It made little sense to subject
them to the *dilectus* in Rome, since that would presumably require their
inhabitants to travel to Rome first, before going back to the colony to fight.
It is usually assumed that settlers in Latin colonies could include both Romans and Latins – that is to say, people who already had Latin rights, for
example because they were from communities in Latium which possessed
this right before 338, or had earlier become settlers in a Latin colony.[45]

Some scholars assume that Italian allies, like Latins, could also become official colonists in Latin colonies throughout the Republican period,
as a result of treaties which were concluded with them and allowed them a
share in collective spoils.[46] The literary sources give some evidence for
the admission of allies into colonies after the Second Punic War,[47] but it is
highly unlikely that the practice occurred on a large scale before this time.[48]

---

42 Cic. *Inv. rhet.* 2.52: *C. Flaminius,...cum tribunus plebis esset, invito senatu et omnino contra
   voluntatem omnium optimatium per seditionem ad populum legem agrariam ferebat.*
43 Humbert (1978, 237).
44 Roselaar (2010, 56–57).
45 For example, Sherwin-White (1973, 27). Around 70,000 colonists were sent out between
   334 and 263: Cornell (1995, 367). Many of them may have been Latins, but the majority
   were most likely Roman citizens in origin.
46 Cornell (1995, 367–68).
47 Livy 33.24.8–9; 34.42.5–6.
48 If an Italian settled in a Latin colony, he would receive Latin rights, but in the middle
   Republic the Romans were reluctant to share such privileges with others. Furthermore,

Indeed, most of these allies had only recently lost land, which had been taken from them as *ager publicus* at the time of the conquest. It would make little sense to take part of their land and then reward these same allies with land elsewhere in Italy in Latin colonies – although, given both the mobility of Italy's population and the largely geographic/community-based nature of citizenship categories during the Republic, this situation was likely possible in real terms.[49]

In order to maintain connections between Rome and the Roman citizens who moved to Latin colonies and thus lost their Roman citizenship, certain privileges in their contacts with Rome were created. These were the *ius commercii* (the right to acquire property in Roman territory and conduct trade with Romans), the *ius conubii* (the right to marry Roman citizens), and the *ius migrationis* (the right to move to Rome and receive Roman citizenship there).[50] These rights presumably made it more attractive to join a colony: colonists received land in exchange for giving up Roman citizenship, but were not completely cut off from their families and could return to Rome if they wished.[51] An additional benefit of the creation of Latin colonies was that those colonists who had originally been Roman citizens no longer qualified for *stipendium*, which lessened the burden on the Roman state. Yet *tributum* could still be collected from the land which the colonist had previously owned, as this was paid by the new owner.[52]

Besides using conquered land to establish Latin colonies, there was another possibility for in the use of land taken from defeated enemies. Most of the areas in Italy which were subjected to Roman hegemony in this period remained allied.[53] In many cases, their defeat was accompanied by confiscations of land by the Roman state. As mentioned, some of this was used to settle colonies. However, other confiscated land remained *ager publicus*

---

only from the late third century were the obligations of the allies laid down in the *formula togatorum*; only this gave them some right to a share of the spoils, including participation in colonies: Erdkamp (2011b, 121–22).

49 Italians did, of course, receive shares of moveable spoils. This neat system on the one hand maximized the contributions of the defeated Italians in money and men, and on the other hand rewarded them with a part of the spoils: Latins could receive land in Latin colonies, Italian allies usually received moveable wealth. On spoils in the Roman Republic in general, see Roselaar and Helm (forthcoming).

50 Roselaar (2013a, 2013b).

51 Armstrong (2016c, 249) considers the establishment of colonies to have been "not in the interest of the developing Roman state," as it broke the ties between colonists and their mother community. However, as Tan argues in this volume, from a fiscal point of view this did make some sense. Furthermore, the creation of *commercium*, *conubium* and *ius migrationis* ensured that ties remained between Rome and the colonies, see Roselaar, 2019.

52 On the decision to establish Latin colonies, see Tan in this volume.

53 The Italian allies did not pay taxes to Rome, but they did furnish soldiers. The reasons why they were not subjected to taxation varied: some were not very wealthy; for others their wealth was difficult to measure in terms of individual landed property, as the census did, since they did not use the same systems of land ownership as the Romans. Rome therefore did not bother to impose *tributum* on them, and simply availed itself of their most valuable resource, their men. See Tan in this volume.

and was not distributed to Roman or Latin settlers, but instead was owned and held by the Roman state. As I have argued elsewhere, this land often remained in the hands of the previous Italian owners despite the change in ownership.[54] It is clear, however, that the Romans did view the land as *their* resource, even if they did not exploit it immediately. During the Second Punic War, the Roman state alienated several tracts of *ager publicus* in new ways in order to raise money for the war. These methods differed from colonization and viritane distribution, which had been aimed mostly at subsistence-level agriculture. The new methods of distribution were aimed at richer farmers, allowing them to acquire land with a secure title in return for their support for this financially exacting war. The first of these methods was the sale of land by the quaestors; this occurred in 205, when a relatively small part of the *ager Campanus* was sold: "Since the war was facing a financial deficit, the quaestors were instructed to sell off an area of Capuan farmland between the Fossa Graeca and the coast."[55] That the state did not often resort to the sale of land may be explained by the fact that it preferred to maintain control over its land; after the sale, it would effectively become the private ownership of the buyer. Indeed, the sale of land in the Second Punic War seems to have been an emergency measure to raise money, and it occurred only rarely.

A further way in which *ager publicus* was privatized in the Second Punic War was the through so-called *ager in trientabulis*. In 210, many citizens gave their gold, silver, and jewelry to the state to finance the Second Punic War:

> The senate then adjourned, and each man brought his own gold and silver and bronze into the treasury, while such rivalry was aroused to have their names the first or among the first men on the public records, that neither were the commissioners equal to the task of receiving nor the clerks to that of making the entries. The equestrian class followed this unanimity of the senate, the plebs that of the equestrian class.[56]

It was decided that this money would be paid back in three instalments,[57] but in 200, when the time came for repayment of the second, there was no money available. Therefore, the senate decreed that

> they should be granted the opportunity of using public lands within a fifty-mile radius of Rome. ... The private citizens were happy to accept

---

54 Roselaar (2010, 113–19).
55 Livy 28.46.4. See Roselaar (2010, 121–17).
56 Livy 26.36.11–12 (Loeb translation, slightly modified): *Senatu inde misso pro se quisque aurum et argentum et aes in publicum conferunt, tanto certamine iniecto ut prima aut inter primos nomina sua vellent in publicis tabulis esse, ut nec triumviri accipiundo nec scribae referundosufficerent. Hunc consensum senatus equester ordo est secutus, equestris ordinis plebs.*
57 Livy 29.16.1–3.

this compromise, and the land involved was given the name *trientabu-lum* because its granting accounted for one-third of the public debt.[58]

Unfortunately, we do not know how much land belonged to this category. It is likely that most of the land in the environs of Rome had already been privatized before 200, and as a result the amount involved cannot have been large. Its proximity to Rome, however, likely meant that it was valuable, particularly for those wishing to produce for the market in Rome. There is no indication that anyone ever exchanged his land for money, and the *trientabula* are mentioned as an existing category of land in *the lex agraria* of 111.[59] It had never been taken away from its possessors, who had now held it for almost 90 years. Thus, the creation of this land had been another easy way for the rich to gain control of *ager publicus* and acquire on it a title which secured their possession.

Lastly, *ager publicus* could be assigned to individuals on lease while preserving its public status. As this was usually arranged by the censors, this land is known as *ager censorius*. Again, this did not occur very often; the only land known to have been rented out was the *ager Campanus*. In 210, "a motion was brought to the plebs, which gave its consent, that the two censors should lease out the farmland of Capua."[60] However, since the state did not in fact collect this rent, in 173 officials needed to be sent into Campania to sort out which land was public and which was private.[61]

All this indicates that not all of the *ager publicus* owned by the state was put to immediate use; even in the late third-century public land was still available in central Italy, the most attractive region in Italy for commercial agriculture. Clearly, the Roman state did not dispose of all its land. In fact, since most of the colonies were located at some distance from Rome, and viritane distribution no longer took place, it was especially *ager publicus* near Rome that created something of a problem for Rome. In the end, it proved convenient for the Roman state to hold onto this land, as in the Second Punic War its sale or lease was an effective way of raising money. Whether it had also been rented out previously, unfortunately, cannot be determined. In any case, the remaining public land came to present more and more of a problem, as we will explore in the next section.

Public land after the Hannibalic War (200–133)

The enormous strain caused by the Second Punic War, followed immediately by a period of unprecedented overseas imperial expansion and the

58 Livy 31.13.6–9: *agri publici qui intra quinquagesimum lapidem esset copia iis fieret.... laeti eam condicionem privati accepere; trientabulumque is ager, quia pro tertia parte pecuniae datus erat, appellatus.*
59 *CIL* 1².585.31–33; see Roselaar (2010, 127–28).
60 Livy 27.11.8; see Livy 27.3.1.
61 Livy 42.19.2; see Roselaar (2010, 128–33).

corresponding influx of wealth, caused the Romans to rethink several policies and practices – including those related to the distribution and exploitation of land as spoils of war. In the years immediately following the Second Punic War, the Roman system of alliances, in which Italian allies fought in Rome's wars and received part of the spoils in return, was under great stress.[62] The war had shown that the loyalty of many allies was not unlimited, as indeed many had joined Hannibal's cause. Rome had finally wrestled them under control, but it needed to quickly and decisively restore order in Italy.

First, the Romans confiscated significant stretches of land, especially in southern Italy, from various communities who had joined Hannibal.[63] Indeed, the Second Punic War presented Rome with the first serious opportunity in decades to confiscate land in Italy and create more *ager publicus*. This indicates that the Romans still considered the taking of land from defeated enemies a normal way of both symbolizing their defeat and also increasing Roman power and wealth. Land was, apparently, later confiscated in Africa and Greece as well. Nevertheless, as will be discussed further below, defeat and confiscation did not always go hand-in-hand. In Italy, some of these recent confiscations in Samnium and Apulia were settled by veterans of Scipio's army, with the land distributed in viritane allocations of two *iugera* for each year of service.[64] The Roman state also implemented a new viritane distribution in Cisalpine Gaul in 173.[65] The return of viritane distribution is somewhat surprising, given that the Roman state had mostly avoided this type of land allocation since the early third century.

And yet, it appears that a great deal of *ager publicus* remained undistributed, available by lease to anyone who wished to occupy it, as Appian explains:

> The Romans, as they subdued the Italian peoples successively in war, used to seize a part of their lands and build towns there, or enrol colonists of their own to occupy those already existing, and their idea was to use these as outposts; but of the land acquired by war they assigned the cultivated part forthwith to the colonists, or sold or leased it. Since they had no leisure as yet to allot the part which then lay desolated by war (this was generally the greater part), they made proclamation that in the meantime those who were willing to work it might do so for a toll of the yearly crops, a tenth of the grain and a fifth of the fruit. From those who kept flocks was required a toll of the animals, both oxen and small cattle.[66]

---

62 See Fronda in this volume.
63 Most notably in Campania (Livy 26.16.6–8; App. *Hann.* 43), Lucania (Livy 32.7.3, 34.45.2), Bruttium (App. *Hann.* 61), Apulia (Livy 31.4.1–2, 34.45.3) and Samnium (Livy 31.4.1–3). See Roselaar (2010, 320–24); Fronda (2010, 307–11).
64 Livy 31.4.1–3.
65 Livy 42.4.3.
66 App. *B Civ* 1.7: Ῥωμαῖοι τὴν Ἰταλίαν πολέμῳ κατὰ μέρη χειρούμενοι γῆς μέρος ἐλάμβανον καὶ πόλεις ἐνῴκιζον ἢ ἐς τὰς πρότερον οὔσας κληρούχους ἀπὸ σφῶν κατέλεγον. καὶ τάδε μὲν

There are several intriguing points in this. First of all, Appian mentions that a rent (*vectigal*) was due on *ager publicus* which could be occupied by individuals to work, so-called *ager occupatorius*. Many modern scholars have accepted this statement without question,[67] but in fact it is quite puzzling. If a rent was asked for occupied land, this implies that an administration was required to keep track of who had occupied land, or at least of the amounts of produce harvested from this land. However, one of the characteristics of this land was that it was not measured in any way; Appian describes how the state "did not have time" to allot it. Usually, land was not measured until it was used by the state for colonies or viritane distributions or as *ager quaestorius* or *censorius* (discussed above). Measurement simply for the sake of collecting a rent is unlikely, since if the state had no time to distribute the land, it would have had no time to measure it either.[68] It is possible that rent was applied based on production and not the amount of land used, but this is not attested anywhere in the sources.

After the Second Punic War, the Romans also strengthened the existing colonies and established new ones to help secure the peninsula. At this time, the Roman state even allowed loyal Italian allies to join the colonies, something which it had not explicitly allowed before.[69] This may have been viewed as a reward for loyal allies. However, within a few decades, the Roman state seems to have faced a different problem: some settlers evidently tried to leave their colonies and return to Rome, a practice which would clearly endanger the strategic position of the colonies. As discussed before, the instruments of *commercium, conubium* and *ius migrationis* had been granted to the Latin colonists as mechanisms to keep distant colonists connected to Rome. At the same time, if a Latin colonist decided to move (back) to Rome, and he had children, he was supposed to leave behind an adult son in the colony. This served to keep the population of Latin colonies up to strength in the long term. However, in the early second century, people managed to get around this obligation. As Livy explains, in 177,

> Individuals had engaged in two kinds of fraud to change citizenship. The law entitled the Latin allies to become Roman citizens as long as they left a son of their own at home. In abusing this law some men committed an injustice against the allies and some against the Roman

---

ἀντὶ φρουρίων ἐπενόουν, τῆς δὲ γῆς τῆς δορικτήτου σφίσιν ἑκάστοτε γιγνομένης τὴν μὲν ἐξειργασμένην αὐτίκα τοῖς οἰκιζομένοις ἐπιδιήρουν ἢ ἐπίπρασκον ἢ ἐξεμίσθουν, τὴν δ᾽ ἀργὸν ἐκ τοῦ πολέμου τότε οὖσαν, ἣ δὴ καὶ μάλιστα ἐπλήθυεν, οὐκ ἄγοντές πω σχολὴν διαλαχεῖν ἐπεκήρυττον ἐν τοσῷδε τοῖς ἐθέλουσιν ἐκπονεῖν ἐπὶ τέλει τῶν ἐτησίων καρπῶν, δεκάτῃ μὲν τῶν σπειρομένων, πέμπτῃ δὲ τῶν φυτευομένων. ὥριστο δὲ καὶ τοῖς προβατεύουσι τέλη μειζόνων τε καὶ ἐλαττόνων ζῴων.

67   For example, Lintott (1994, 54).
68   On *ager occupatorius*, see Roselaar (2010, 113–19) in detail.
69   For example, at Cosa in 197, see Livy 33.24.8–9.

people. To avoid the necessity of leaving a son at home, men would hand their sons over as slaves to anyone with Roman citizenship, on the condition that the sons would be manumitted; as freedmen they would become citizens. Men with no offspring to leave behind adopted sons to become Roman citizens.[70]

It appears that some inhabitants of Latin colonies moved to Rome, as was their right by the *ius migrationis*,[71] but that they then devised ingenuous legal constructions in order to move their sons to Rome as well. This obviously negated the efforts of the Roman state to maintain the strength of the Latin colonies. It appears that the Roman citizenship had become more popular in the early second century – how and why this development took place cannot be discussed here, but it presented a serious problem for the state.[72]

The Romans decided, therefore, to create a new type of colony from 184 onward, once again modifying the distribution of confiscated land. These new colonies were Roman colonies, in the sense that the settlers retained their Roman citizenship, but instead of the small outposts utilized previously (which held only 300 men), these were much larger establishments. It is assumed that 2,000 colonists received land in each Roman colony from then on.[73] The allotments distributed here were much smaller than in contemporary Latin colonies, varying from five to ten *iugera*,[74] but this was most likely a sufficient amount of land to maintain the status of an *assiduus*.[75] In this period, the importance of Roman citizenship, or at least the desire to live in Rome, seems to have increased, and therefore to have become more exclusive; this is also shown by the fact that those Latins and allies who had moved to Rome were expelled in 187.[76] Yet, this new type of colony was only implemented a few times; the last one was either Luca or Luna, established in 177.[77]

---

70 Livy 41.8.8–10: *Genera autem fraudis duo mutandae viritim ciuitatis inducta erant. lex sociis nominis Latini, qui stirpem ex sese domi relinquerent, dabat, ut cives Romani fierent. Ea lege male utendo alii sociis, alii populo Romano iniuriam faciebant. Nam et ne stirpem domi relinquerent, liberos suos quibusquibus Romanis in eam condicionem, ut manu mitterentur, mancipio dabant, libertinique cives essent; et quibus stirps deesset, quam relinquerent, ut . . . cives Romani fiebant.* See Roselaar (2013a, 2019) for detailed discussion of this passage.

71 Broadhead (2008).

72 See Roselaar (2019).

73 The number 2,000 is reported only for Mutina and Parma: Livy 39.55.6.

74 Livy 39.44.10, 39.55.6, 40.29.1.

75 See Galsterer (1976, 59); Rosenstein (2004, 68–69); Roselaar (2010, 204–8).

76 Livy 39.3.4–5.

77 There has been much confusion between the colonies at Luna and Luca, both located on land taken from the Ligurians and both reportedly founded in 177: Roselaar (2010, 325 n. 141). Auximum was established at some date in the second century; the year 157 is given in Vell.

And then Rome stopped distributing land altogether, until the Gracchan land reform in 133 more or less forced the senate's hand. There are several possible reasons why land distribution and colonization stopped after 173. One likely explanation is the fact that there was no longer any military reason to found new colonies because the whole of Italy had been pacified. No further colonization was necessary to make sure the defeated peoples remained loyal. It has also been suggested that the nobility feared the power of the men who founded new colonies because the inhabitants of these colonies would become their clients – as was indeed the case for viritane distributions. However, there is no reason why such a position would suddenly have been viewed as a danger from the 170s onwards, when this had not been the case in the previous decades. It may be, nevertheless, that increasing competition within the elite in the second century was a more important factor, judging from the number of *leges sumptuariae* which were passed around the same time.[78] This may have contributed to the cessation of land distributions.

Another possible explanation for the sudden end of colonization may be deduced from the census figures preserved for the second century. The census figures show a quick recovery from the decline during the Second Punic War: the census of 169/168 recorded an impressive rise to 312,805 citizens, more than 40,000 above the last pre-war figure. If the Roman elite believed that land distributions were an incentive to stimulate population growth, it would have assumed that it was now no longer necessary to distribute land for this purpose. Therefore, rapid population growth may have contributed to the decision to end distributions of land.[79]

The fact that *ager publicus* often remained undistributed for a long time suggests that the Roman state did not usually feel the need to dispose of the land that it had taken as spoils. As we have already seen, even in the third century, some of the *ager publicus* remained in the state's possession and was used only periodically to raise money, such as during the Second Punic War. This was not considered a problem; there are no indications in the sources that Roman politicians felt it necessary to always distribute all available state-owned land.

Lastly, the changing nature of Roman imperialism may have contributed to a shift in Roman attitudes toward the confiscation and distribution of land as spoils. As mentioned above, confiscation did not always follow conquest. This was especially the case in the second century, when Rome expanded its conquests outside of Italy. In the cases of overseas conquests, Rome usually did not directly confiscate land. In Sicily, for example, the Romans left intact the preexisting arrangements with regard to tribute. The

---

Pat. 1.15.3, but this is not certain: Roselaar (2010, 318–19 n. 100). Heba may also have been a colony, but this is even less certain: Roselaar (2010, 316 n. 87).

78 Patterson (2006b, 202).

79 See discussion in Roselaar (2010, 149–53).

land was not confiscated; the inhabitants remained in place and paid taxes to the Roman state.[80] The same arrangement was created in Asia in the later second century. The conquests in Greece were accompanied by the taking of great amounts of moveable spoils, rather than the confiscation of land. Similarly, Roman exactions in Spain comprised movable booty and various taxes. Carthage paid an annual indemnity for 50 years after the Second Punic War. I would suggest that, in these cases, the Romans chose the most expedient, most practical, and most profitable approach. It was not necessary to confiscate land in the period when these areas were conquered. In the third century, when Sicily was conquered, there was still land available in Italy for the settlement of colonies, while in the second century the settlement of colonies had been halted and therefore overseas land was not needed. Taking overseas land as *ager publicus* without distributing it would serve no real purpose, apart from humiliating the defeated population, as it was impractical to collect rents from it.[81] Therefore, the Romans moved to more convenient methods of exploitation, such as the imposition of indemnities and permanent taxation collected through *publicani* rather than directly by the state, which brought in large profits and benefits at low cost.

Conclusions

Whatever the causes for the temporary halt of land distributions in the second century, it did create a problem for many Roman citizens – and potentially for the Roman state. It was especially important for the *cives optimo iure* to have enough wealth to qualify to serve in the army, provide at least some of their own weaponry, and to pay the *tributum* – that is, to remain *assidui*. Landownership is assumed to have been the citizen's most important asset. Thus, in order to remain *assidui*, citizens generally needed to own land.[82] This was not only good for the citizen but also for the state, since it guaranteed a supply of soldiers and tax revenue. Similarly, in order to maintain the position of their colonies as defensive and offensive bulwarks, Latin

---

80 Prag (2013, 59–63).
81 The exceptional confiscations of land in Africa and from Corinth, mentioned in the *lex agraria* (*CIL* 1^2.585) are consistent with this suggestion. The utter destruction of these recalcitrant enemies in 146 and corresponding confiscations were clearly meant to send a signal.
82 See Rosenstein (2004) for an analysis of strategies used by Roman households, in which young men first served in the army, before settling down on a plot of their own, often (but not always) in colonies of veterans. This idea has recently been challenged by Adamo (2017), who argues that the traditional "citizen-soldier-peasant" was rare. He argues that land distribution schemes were not intended to make the distribution of property more egalitarian, and that, due to their poverty, peasants were unable to keep hold of them. Also, starting a new farm in unexploited territory was expensive. However, I think this reconstruction is overly negative, and that at least until the mid-second century, most citizens aspired to own a plot of land.

colonists needed land as well. In this context, it made sense for the Roman state to distribute land to those who fell below the census threshold for *assidui*. The *ager publicus* was the best source of land which could have served to maintain the numbers of *assidui* available for the Roman state. Thus, the use of land that was taken as spoils in return could have served to maintain the army's strength.

In the early second century, the Roman state may have thought that there were sufficient *assidui*, as indeed most people had been able to find land immediately after the Second Punic War through one of the recent land distributions, and the network of colonies was fully up to strength. However, throughout the second century, the population continued to increase while the amount of commercial agriculture also grew. This increased competition for land, especially in central Italy, so that not enough land was available for everyone in Italy. The *assidui* were the first to suffer the effects.[83] And so, after a period of quiet in the mid-second century, the debate surrounding *ager publicus* exploded spectacularly in 133, when Tiberius Gracchus proposed to distribute the remaining *ager publicus* to the landless poor. From this point until the collapse of the Republic, disputes over the distribution of *ager publicus* continued to rage.

Similarly, the idea of settling colonies was not completely abandoned and, for the first time, proposals were made to establish colonies outside of Italy. Gaius Gracchus first proposed settling a colony in Carthage. Although this may have failed,[84] further settlers were established in Africa by Marius.[85] Later, during and after the civil wars of the first century, colonies were settled in the provinces as a rule. Colonies were also settled in Italy in this period, often on land confiscated from communities which had supported the opposing general.[86] This brings us back to the earlier confiscations of land from defeated enemies. Perhaps the land taken by the generals of the first century did not have the legal status of *ager publicus populi Romani*, as it was simply taken by the victorious general and distributed to his men, without the approval of the popular assembly. Furthermore, the urgency for the generals to find land was greater than in the earlier periods, as without immediate reward their veterans would revolt.[87] Nevertheless, this shows that the confiscation of land was still a prime mechanism of subjugation, which suggests that, in this respect, little had changed in Roman thinking about conquest and spoils in comparison to the fourth and third centuries.

---

83 See Roselaar (2010, 180–220).

84 App. *B Civ* 1.24; Solinus 27.11; Plut. *C. Gracch.* 10.2–11.2; Vell. Pat 2.7.7–8. The colony is described as a failure in the sources, but some land seems to have been distributed here, see App. *Pun.* 136; Obseq. 33; Fronto *Ad Marc.* 2.1. A possible Gracchan boundary stone was found here (*CIL* 1^2.696 = 8.12535 = *ILS* 28 = *ILLRP* 475).

85 *BAfr.* 56.3; [Aur. Vict.] *De vir. ill.* 73.1. Other Italians settled here in the wake of the Marian-Sullan civil war; three boundary stones in Etruscan, found near Carthage, may have been set up by these people; see Heurgon (1969).

86 See Keppie (1984b) for an overview.

87 See on land distributions in the first century Brunt (1962); Schneider (1977).

Appian sees the events surrounding Tiberius and Gaius Gracchus as the beginning of the fall of the Roman Republic,[88] and not without reason. We may conclude that, since the value of land was much greater than that of moveable spoils, the debates surrounding it were necessarily greater as well. Even the enormous portable wealth taken from Greece in the early second century, when distributed to a large group of soldiers, amounted to a few hundred *denarii* each at most.[89] Land, on the other hand, would yield a stable income for many years, provided it was well cared for. Thus, people receiving land would be more loyal to the man who had distributed it than those who had merely received money, and arguably for longer. Indeed, one of the reasons why the Gracchi were opposed was the fear that their influence over the people would become too large.[90]

However, as in the early Republic, personal ties between elites and others remained strong, in the form of patronage or more incidental assistance, as in the case of land distribution. Thus, even though VanDerPuy (in this volume) is likely correct that relations between the Roman state and its citizens became more individualized from the fifth century onwards, we cannot forget that these relations were still part of a much wider network of personal relations between elites and their dependents. This limited the power of the state to act in the absolute best interests of the state and its citizens. Personal power remained important for the elites and therefore interfered with the aims of the state as an independent actor.

The treatment of *ager publicus* is a good example of how the state was often constrained by the personal interests of individual politicians. This led to uncertainty and conflict within the Roman aristocracy about the way in which *ager publicus* should be used. Thus, even though *ager publicus* was perhaps the most valuable spoil of war taken by the Roman state in the Republican period, internal social and political configurations of power made it impossible to use it in the way that served the citizens and the state the best. These tensions were not fully resolved until more land in the provinces became available in the first century, though this advance came only with the breakdown of the Republican political system.

88 App. *B Civ.* 1.2: "The sword was never carried into the assembly, and there was no civil butchery until Tiberius Gracchus, while serving as a tribune and bringing forward new laws, was the first to fall a victim to internal commotion" (ξίφος δὲ οὐδέν πω παρενεχθὲν ἐς ἐκκλησίαν οὐδὲ φόνον ἔμφυλον, πρίν γε Τιβέριος Γράκχος δημαρχῶν καὶ νόμους ἐσφέρων πρῶτος ὅδε ἐν στάσει ἀπώλετο).
89 Gauthier (forthcoming).
90 Plut. *T. Gracch.* 14.2; Sall. *Iug.* 31.4; Cic. *Amic.* 12.41, *Mil.* 27.72, *Rep.* 6.8, *Off.* 2.12.43, 2.23.80, *Phil.* 8.4.13; Val. Max. 6.3.1b–1d.

12 The manipular army system and command decisions in the second century*

Jeremiah McCall

Introduction

The armies of the Roman Republic did not win every pitched battle they fought, but they won most in the period 218–100.[1] Yet, the yearly commanders of most of those armies, typically consuls and sometimes praetors, frequently won elections without being able to claim significant command experience, if any at all. While they could make claims of superior service and lineage, they often could not claim any superior command ability in order to win the electorate's votes.[2] Nor did defeat necessarily mean political disaster: defeated consuls were just as likely to be elected to a second consulship – rare as that was for any Roman aristocrat – as those who had not suffered defeats.[3] It is counterintuitive, but in this period of the Republic, a commander's actual and perceived ability to make sound command decisions were quite distinct from, and relatively insignificant to, his political status as a Roman aristocrat.

These points raise the question: since the elected commander of a Roman army typically had little command experience, and was not often held accountable for command decisions, what role did such a commander play in the ultimate success or failure of his army while on campaign? Or, to put this another way, to what extent could the Roman army in this period operate effectively without the need for skilled command decisions from the general? The argument this chapter will pursue is that the Roman army of the middle Republic consisted of interconnected systems – soldiers, supplies,

---

* All dates are BC unless otherwise noted.

1 This period is the focus of the chapter for two reasons. First, the historiographical tradition, where the strength of our main sources – Polybius and Livy – increases. Second, though the maniple and manipular army may have been a gradual and less formal transition from earlier styles of combat initially (see Armstrong's chapter in this volume), the period from 218 to 100 probably best represents the period of the fully formed manipular army, before it began to be phased out in the Late Republic. However, see Gauthier in this volume for questions around these later developments.

2 Rosenstein (1990, 114–52, 2011, 132–36).

3 Rosenstein (1990).

DOI: 10.4324/9781351063500-12

This chapter has been made available under a CC-BY-NC-ND license.

weapons, positioning, terrain, morale, etc. – that, through tradition, habit, and training, tended to operate in certain ways, often without much need for the typical general's direct intervention at all.

This idea of systems based on conventions of behavior benefits from some unpacking. The assertion that the Roman army in this period functioned as a "system" does not mean that the army functioned like a machine – consistent, regular, and precise in its operations. Rather, the term "system" is used here in its basic meaning: a set of interconnected elements that operate together to carry out tasks. The Roman army of our period certainly fits that definition. Soldiers, officers, logistics, and many other elements interconnected and interacted in ways that ultimately led to victory or defeat on the battlefield and to the success or failure of campaigns. Those human parts of the Roman army system had established ways of doing things: habits, traditions, conventions, and practices. These could and did change over time, but still they existed; the Romans did not reinvent anew travel and camping procedures, command structures, and battlefield practices for each campaign.[4] Rather, they relied upon the conventional practices and, at times, wisdom, stored in the collective memory of the veteran soldiers and those many officers who had seen service before. This is what is meant when this chapter speaks of systems in the army – built-up customs and practices for how the parts of the army were to function, not a precisely established set of rules or procedures – though, of course, some of these too may have existed. And, as I suggested, many of these conventions and rules – these systems – operated typically with very little input from the commander of the army.

In addition to the systems of army operation that developed over time via the soldiers and officers who did the fighting, the practices and constraints of Roman battles, which focused on central lines of heavy infantry, further reinforced some conventional practices and systems. Therefore, we should ask systemic questions about the army's operation in this period: what were the required steps needed to get an army to a battlefield, how did the army tend to operate on the march and on the battlefield, and where within these typical army systems were the fundamental command-decision points? At the same time, when considering a particular commander's importance to battlefield success, the exercise of decision-making and agency by those outside of the commander – senators, military tribunes and legates, as well as centurions and common soldiers – must be considered. When were important command decisions typically made by those other than the commander? To what extent could a general rely on others' experience and insight when making decisions? A Scipio Africanus or Gaius Marius might, perhaps, involve themselves in all matters of command and

---

4 Though, as Milne notes in this volume, this does not mean there was anything like a fixed or "standing" army in the middle Republic.

generally have excellent results to show for it. But what about the "average general," the amateur elected with some experience of battle but not necessarily any experience or skill at commanding an army? Did that "average general" have to make many skilled command decisions to have a successful military campaign?[5] The evidence suggests, and this chapter will argue, that in most situations a general could rely a great deal on conventional military systems and practices of the armies in the middle Republic and did not have to make much in the way of skillful command decisions to achieve military success.

Before the Battle 1: province and army assignments

At the start of a campaign, the senate designated a commander's province, the forces allocated to him, and, often, his intended foes. These instructions could be limited, conditions in a province of operation could change, and commanders might stray from their initial assignments. Still, the senate provided important parameters for where and how a commander was to operate. As the second century progressed, the senate increasingly came to expect magistrates to limit their operations to their assigned provinces. Not all commanders complied, but the point is that the senate provided considerable direction in its yearly assignments of forces, provinces, and commanders.[6] These highest-level decisions that ultimately led to a battlefield, in short, were usually not determined by the commander at all.

Furthermore, the elected commander may not have been typically involved in selecting his direct subordinates, the military tribunes.[7] These were the direct commanders of the legions and the cavalry, tasked with levying troops, executing orders, organizing, commanding, and inspiring their soldiers, and, optimally, keeping them effective in battle.[8] Every year, Polybius asserts, the assemblies elected the 24 tribunes required for the four legions that made up the two standard consular armies; army commanders appointed others.[9] A perusal of Livy's levy notices reveals that, typically, consuls levied and then commanded new legions, though sometimes they did not and simply took over command of existing legions.[10] Commonly,

---

5 Sieges unfortunately cannot be considered in this small space, nor cavalry, except for their role in infantry battle decisions.

6 Eckstein (1987, xx–xxii); Rich (1993, 55–64); Roth (1999, 246–47); Brennan (2014, 32). Livy 31.3.2–3, 31.19.2–4 gives an excellent example of senatorial planning in the Second Macedonian War.

7 Also see Helm's discussion in this volume, on the important social role of tribunes in the army of this period.

8 Keppie (1984a, 39–40).

9 Polyb. 6.19.1. Livy 43.11–12 reinforces the normal practice of electing tribunes.

10 Some examples where the consuls did not both command new legions: Livy 33.25.10, 35.20.4–5, 37.50.4, 39.20.1–3, 39.38.10.

then, the consuls would have had little or no choice of the tribunes in their army. Even in cases where the consuls assumed command of existing legions, it is not at all clear that they would have been able to pick their own tribunes. The 24 elected tribunes of the year had to go somewhere. Equally important, many tribunes already serving in an existing legion would have experience and ties to their soldiers that might best be preserved. When full replacement might require selecting as many as 48 new tribunes, commanders likely relied on existing officers – perhaps appointing only a few, as needed, when elected tribunes did not fill those posts. The same probably applied also to those praetors holding military commands.

While the senate determined the number of soldiers to levy or retire for the year, the military tribunes actually levied the soldiers who fought the battles.[11] Polybius explicitly states this.[12] Livy agrees. Though he often says the consuls conducted levies, this is likely just shorthand for the real work of the tribunes, as his detailed account of the levies of 171 demonstrates. Livy reports (42.32.6), "The consuls were conducting the levy with by far more painstaking care than usual."[13] A reader might suppose the consuls personally selected the recruits. However, in the next section (42.32.7), he specifies, "When the military tribunes who were appointing centurions were assigning men as they came, 23 veterans who had held the rank of chief centurion on being named appealed to the tribunes of the people."[14] Later still, he indicates that military tribunes were actually selecting the troops.[15] The process, in other words, did not regularly involve commanders handling the selection of individual soldiers.[16] Nor did commanders typically choose those critical unit officers who fought alongside the soldiers, the centurions and *optiones*. According to Polybius, at least in the second century soldiers selected their own unit officers personally.[17] While that may have been normal when insufficient veteran officers were available, Livy's account of the levy for 171, as quoted above, suggests the tribunes were expected to enroll former unit officers to a position at least comparable to their prior rank.

---

11 Veterans spared the levy: Livy 31.8.6, 32.8.3.

12 Polyb. 6.19–21.

13 Livy 42.32.6: *Dilectum consules multo intentiore quam alias cura habebant.*

14 *Cum tribuni militum, qui centuriones sed primum quemque citarent, tres et viginti centuriones, qui primos pilos duxerant, citati tribunos plebis appellarunt.*

15 Livy 42.33.5: *deprecatus est deinde, ne in nouo bello, tam propinquo Italiae, adversus regem potentissimum, aut tribunos militum dilectum habentis inpedirent...*("then he [the consul] made a request that, in a new war, at so little distance from Italy, against a very powerful king, the people should not hinder the military tribunes who were holding the levy..."). See also Livy 42.34.14, 42.35.2.

16 Polybius' account of the levy is streamlined and problematic (see Armstrong and Helm in this volume), but the citizens who must have been levied by other agents away from Rome must also have typically been selected without the commander's input.

17 Polyb 6.24.1–2.

Up to this point in war preparations, the typical commander provided very little command input. Whether neophyte or veteran officer, he simply relied upon the competence – or lack thereof – of many other individuals to set the army properly in motion. The senate, with its collective years of experience in military affairs – as soldiers, officers, commanders, or all three – determined the size of the army and where it would fight. Tribunes levied the actual soldiers and enrolled centurions. If additional centurions were needed, the soldiers voted for them. All these decisions, large and small, loomed large in battlefield successes or failures and were normally outside the general's purview.

One exception, where the typical commander might exercise significant input, was the decision to train inexperienced troops. A classic example of this comes from Fabius Maximus' dictatorship in 217. Polybius explicitly notes Fabius' plan to avoid pitched battles for a time was partly "with the view of gradually strengthening and restoring by partial successes the spirits of his own troops, broken as they were by the general reverses."[18] Other occasions, when commanders trained their troops, demonstrate that this practice was not unique to Fabius.[19] Clearly, when such training was effective, it helped soldiers keep their formations and stand their ground in the stresses of the killing zone.

Before the Battle 2: getting the army to the battle

On the march, the general had more command tasks, but could still rely on the army's systems and support from officers in carrying these out. Leaving aside sieges, and assuming successful overseas transport, the main task at this stage was to march the army safely to engage the enemy at a suitable time and place. Several points of command input were important in this process. The first was ensuring a sound route of march. Critical to this were logistical operations. Much of a Roman army's supply system, however, fell outside the commander's control and under senatorial authority. Field commanders had some control over their supply lines, but often this amounted to haggling with merchants or recalcitrant praetors in command of naval forces. Generals, not infrequently, delegated operational command over important parts of the supply process to subordinate officers, and elected quaestors often played important roles managing supplies.[20] Roman armies did not regularly fail in this period due to faulty supply systems – the result of successful procedures developed over the centuries. The typical commander could delegate and generally count on the supply system to function without making particularly sophisticated command decisions. When not

---

18 Polyb. 3.90.4: ἅμα δὲ τὰς τῶν ἰδίων δυνάμεων ψυχὰς προηττημένας τοῖς ὅλοις διὰ τῶν κατὰ μέρος προτερημάτων κατὰ βραχὺ σωματοποιεῖν καὶ προσαναλαμβάνειν.

19 Polyb. 10.20.1–8; App. *Hisp.* 65, 86; Livy 34.13.1–3, 44.1.4; Sall. *Iug.* 44.1–4.

20 Roth (1999, 246–60).

relying on supply lines, Roman soldiers foraged and raided to supply the army. There were a number of factors involved in executing foraging work properly, but it is far from clear that typical generals needed to manage this work personally to ensure success.[21]

In addition to well-developed supply procedures, the Romans had an organized protocol for camping, which was directed by the tribunes.[22] Here, it is worth taking a moment to consider the reliability of Polybius, our main source both for the Roman marching order and camping procedures in this period. There is good reason to suppose Polybius idealized the organization and function of the Roman army in his day, just as he idealized the functioning of the Republic itself, and there are some areas in his account that scholars suggest are problematic, undermining his seemingly rational and highly organized approach.[23] It is perfectly legitimate to read Polybius with caution and recognize that the army likely did not function so mechanistically, so cleanly and orderly, as he suggests.[24] Nevertheless, there is no warrant for rejecting his eyewitness account altogether. For example, to suspect that Roman camps were not always laid out exactly as Polybius said, does not justify the conclusion that the Romans lacked camping patterns and procedures altogether. Indeed, archaeological evidence shows a high degree of structure and consistency in camp layouts at Numantia, which testifies to Roman organization in these matters.[25] The camp layouts do not always perfectly match Polybius' well-ordered description, but they do seem to confirm his account overall. The evidence suggests that the Romans had procedures for laying out camps, albeit procedures that could be adapted to specific landscapes and circumstances. To suspect that the Roman marching order was not always organized the way that Polybius describes likewise does not justify the conclusion that Romans had no orderly procedures for marching. In the absence of developed arguments against it, we can and must suppose that Polybius provided a reasonably reliable description – not prescription – of Roman army practices, and that that description included a fair degree of organization and habitual procedures.

Within the maniples of a legion, each unit/class type was numbered from one to ten (for example, first maniple of *hastati*), and (according to Polybius) each occupied a set place within the camp relative to their comrades in other maniples and legions every night.[26] Each soldier slept, ate, and mustered

---

21 Erdkamp (1998, 122–40).
22 Polyb. 6.27–42; Dobson (2008, 50–51, 54, 68–70).
23 See Champion (2004) and Scanlon (2015, 202–36) on Polybius' schematizing. See Miltsios (2013) on his narrative devices. See also Armstrong in this volume for discussion with relevance to the army.
24 See Armstrong in this volume.
25 Dobson (2008).
26 General layout of the camp: Polyb. 6.29–31, 6.40–41. Numbered maniples: Polyb. 6.24.1–5, 6.29.9, 6.40.11; Livy 26.5.15, 27.14.8.

next to those soldiers who would hold the line alongside him. The soldiers of every unit customarily knew their camp positions beforehand.[27] Clearly, these camping practices, beyond eliminating the need to make many major decisions on a daily basis, also did much to reinforce the spatial organization of the legion and the connections between and within units.[28] Equally clearly, the tribunes were in charge of setting up the camp, not the general.

The organized procedures, directed by tribunes in the camp, apparently extended to marching. Each legion and allied wing reinforced its organizational integrity by marching as a unit, an integrity that they would need to maintain on the battlefield. A two-legion consular army on the march, says Polybius, followed this order: *extraordinarii*, allied right wing, the two Roman legions, and allied left wing. These positions rotated so that each could lead in turn and access the cleanest water and best forage.[29] The preservation of grand tactical units in the daily marching order reinforced unit identity and cohesion, and the process needed little commander oversight. Furthermore, though here direct evidence is scarce, since protocols governed the transition from march to camp, it is reasonable to suppose protocols dictated how marching columns deployed for battle to avoid an ad hoc scramble of units from column to line. The Romans also had a special marching order in dangerous country. In these cases, the *hastati*, *principes*, and *triarii* of the army tended to march in three parallel columns so that the army could swiftly deploy to the left or the right of the marching route, with at most the *hastati* having to shift positions to face the enemy.[30] The tribunes, as the ones who supervised the camp and sometimes – if not regularly – carried out deployments, likely managed these changes in order and organization, presumably with help from centurions and *optiones*.[31]

Though the ancient sources generally do not specify reasons for Roman defeats, they identify two major causes during the march: ambushes and camp attacks. Ambushes presumably resulted from ineffective reconnaissance, poorly chosen routes of march, or both. Deploying scouts was certainly an important command decision, though it is quite conceivable that the actual deployment of scouts, and the specific circuits they took, were decisions also relegated to the tribunes. Ineffective scouting on the march could, of course, prove disastrous. The most infamous example of this is the battle of Lake Trasimene. The source tradition about the consul Flaminius is generally hostile, though over time an account developed of his honorable deportment in the face of impending death.[32] Whether that hostility included inaccurately attributing the defeat to his poor command decisions

---

27 Polyb. 6.41.10.
28 Culham (1989, 193). See also Rosenstein (2012a) and Helm in this volume.
29 Polyb. 6.40.9.
30 Polyb. 6.40.10–14.
31 See below on tribunes carrying out deployment.
32 Rosenstein (1990, 77–78, 116–17).

is far from clear. In both Polybius' and Livy's accounts (the latter heroizing the consul), Flaminius willfully ignored his advisors' pleas for caution when approaching a mighty enemy with superior cavalry forces. They asked him to wait for the other consul to arrive, but Flaminius would not. He failed to ensure the route was properly scouted and led the Roman army into a lakeside ambush from which they could not recover.[33] The Carthaginian army sprung the trap, assaulting the Roman marching columns in the mist by the lake, and Polybius asserts:

> ...the Roman centurions and tribunes were not only unable to take any effectual measures to set things right but could not even understand what was happening. They were charged at one and the same instant from the front, from the rear, and from the flanks, so that most of them were cut to pieces in marching order as they were quite unable to protect themselves, and, as it were, betrayed by their commander's lack of judgement. For while they were still occupied in considering what was best to do, they were being slaughtered without realizing how.[34]

Two crucial points surface in the narrative. First, the general allegedly ignored his advisors, pursued a formidable enemy with insufficient reconnaissance, and stumbled into an ambush. One may be concerned that these accusations are false, but there is no evidence to seriously support discarding them. If the hostile tradition of his disastrous decisions had some truth, Flaminius was more of a derelict general than a mediocre one, overriding the safeguards and practices that ordinarily allowed the Romans to fight a pitched battle. Second, the ambush prevented the officers from properly organizing the men into functioning units at all, and this disruption of deployment practices increased the catastrophe. The sources note other instances of ambushes in this period, and these too must have resulted from poor reconnaissance or route choices.[35] Similarly, those instances of Roman armies attacked while either in camp or pitching camp can be attributed to the command system's failure (and perhaps the commander's) to set effective

---

33 Polyb. 3.82–84; Livy 22.3–8.
34 Polyb. 3.84.2–5: οὐχ οἷον παραβοηθεῖν ἐδύναντο πρός τι τῶν δεομένων οἱ ταξίαρχοι καὶ χιλίαρχοι τῶν Ῥωμαίων, ἀλλ᾽ οὐδὲ συννοῆσαι τὸ γινόμενον. ἅμα γὰρ οἱ μὲν κατὰ πρόσωπον, οἱ δ᾽ ἀπ᾽ οὐρᾶς, οἱ δ᾽ ἐκ τῶν πλαγίων αὐτοῖς προσέπιπτον. διὸ καὶ συνέβη τοὺς πλείστους ἐν αὐτῷ τῷ τῆς πορείας σχήματι κατακοπῆναι, μὴ δυναμένους αὐτοῖς βοηθεῖν, ἀλλ᾽ ὡς ἂν εἰ προδεδομένους ὑπὸ τῆς τοῦ προεστῶτος ἀκρισίας. ἔτι γὰρ διαβουλευόμενοι τί δεῖ πράττειν ἀπώλλυντο παραδόξως.
35 L. Manlius Vulso in 218 : Polyb. 3.40.11–4; Livy 21.25.8–14); C. Flaminius in 217 : Polyb. 3.82.1–84.5; Livy 22.3.1–7.5; L. Postumius Albinus in 216/15 : Livy 23.24.6–13; L. Cincius Alimentus in 208: Livy 29.36; Cn. Baebius Tamphilus in 199: Livy 32.7.5–7; Q. Marcius Philippus in 186: Livy 39.20.5–10; Q. Fulvius Nobilior in 153: App. *Hisp.* 45–47; C. Vetilius in 147: Livy, *Per.* 52; App. *Hisp* 63; L. Cassius Longinus in 107: Livy, *Per.* 65; Caes. *BGall* 1.7.4, 1.14.

sentries.[36] Unlike at Trasimene, however, it is not generally clear in these other ambushes and the associated failure to take adequate reconnaissance was primarily due to the general, the officers, or even the scouts themselves. Still, reconnaissance and the choice of marching routes were important command tasks where human errors at any point could destroy an army.

Before the Battle 3: choosing the battleground

If the army successfully made contact with the enemy with the intention of engaging them, then the choices of time and place to engage in a pitched battle were the two most important command decisions open to a general. Terrain, weather, and the size and position of the enemy forces could have a significant impact on the success of a Roman army in battle. However, our sources rarely suggest the legions lost due to unfavorable deployments, although it did happen. Ti. Sempronius Longus' decision to send his hungry troops across the frigid Trebia to fight Hannibal was clearly a poor deployment choice.[37] At Cannae, Varro chose ground that Aemilius had dismissed as unsuitable.[38] In 185, C. Calpurnius Piso sent his troops to support Roman foragers who had begun to skirmish with Spanish foragers. Perhaps a reasonable order given the importance of foraging, but the decision sparked a full-scale battle on unfavorable terrain.[39] In 104, M. Titinius engaged a slave army with an inferior Roman force on poor terrain, and his small army was routed.[40]

The sources, perhaps, do not mention this more frequently because of the manipular army's flexibility. Polybius suggests that, compared to the Hellenistic phalanxes, the division of the Roman manipular army into small, independent units made it readily adaptable to different terrains, and Roman armies certainly did operate effectively on varied terrains.[41] Still, when the Roman strategy was to engage the enemy army in a decisive battle, as was often the case, a commander generally had to, if possible, select ground that did not patently favor the enemy. This was certainly not an insignificant task, but it was often a reasonably straightforward assessment of level ground and obstacles.

Before the Battle 4: commanders, military councils, and legates

The prior discussion clarifies that the typical commander had two command tasks: (1) determining a sound route of march, including attendant supply lines, effective reconnaissance, and suitable water and forage opportunities; and (2) choosing a suitable time and place for battle. When making

---

36 M. Claudius Marcellus in 196: Livy 33.36.4–15; L. Aemilius Regillus in 190: Livy 37.2.11, 37.46.7–8; A. Manlius Vulso in 178: Livy 41.2; C. Marcius Figulus in 156: App. *Ill.* 11.
37 Polyb. 3.72.3–5; Livy 21.54–55.
38 Polyb. 3.112.2.
39 Livy 39.30–31.
40 Diod. Sic. 36.3.5.
41 Polyb. 18.32.10–12; Polyb. 18.22–26; see Livy 33.9–10, 41.18.

decisions to execute each of these important tasks, however, the general was not simply left to his own devices. Ever present was his military council, and often present was an experienced legate, who was able and expected to offer competent advice.

Judging by its frequent mentions in the sources, the commander's military council played a very important role in Roman command decisions.[42] The first centurions of each maniple apparently held a place on the military council, along with the military tribunes and, presumably, any legates who happened to accompany the commander.[43] These were usually all veteran campaigners. The councils were a critical link between the commander and the army: at these meetings, the commander relayed instructions to the officers in the council so that they could pass these on to the troops in the units.[44] More than just a command link though, the council provided a sounding-board and a source of advice for all kinds of critical military issues: routes of march, when and where to engage the enemy, what towns to attack, changes in strategy, responses to emissaries, truces; the list goes on. Essentially, military councils offered counsel on exactly those most important decisions the commander faced prior to the actual clash of soldiers. For example, when P. Cornelius' army attempted to intercept Hannibal near the Rhône, he discussed the most suitable locations for a battle with his tribunes.[45] After the Syracusans repulsed a Roman army, Ap. Claudius' council unanimously decided to forego any future attempts to take the city by assault.[46] In the Second Macedonian war, the consul P. Villius consulted his council to determine whether the army should march through a gorge – risky but direct – or take a less direct, but safer route.[47] When L. Scipio failed to goad Antiochus into engaging in a pitched battle, he consulted his military council, which, in turn, decided he should launch an attack.[48] In the Third Macedonian War, the consul P. Licinius Crassus summoned his council to determine where the army should operate in Thessaly.[49] Livy and Polybius sometimes note differences of opinion in councils, times when the commander was persuaded, and times when he was not. Their testimony indicates that the military council was an important, regular part of decision-making, providing the commander valuable input. Though responsibility for success or failure in a campaign would not be laid at the council's feet, this group provided important advice for the general.

---

42 References to military councils: Polyb. 3.82.4–5, 3.89.3, 8.7.5, 14.2.11, 14.9.1, 21.14–15, 21.16–17, 27.8.6; Livy 22.3.8, 24.45.2, 26.15.1–6, 27.20.1, 27.46.5, 30.5.1, 30.36.10, 37.14.4–15.9, 42.57.1, 44.35.4, 45.7.5–8.7.
43 Polyb. 6.24, 8.9, 8.7.5.
44 Livy 30.5.2–3, 37.5.2.
45 Polyb. 3.41.8.
46 Polyb. 8.7.5.
47 Livy 32.6.3.
48 Livy 37.39.1.
49 Livy 42.57.1.

In addition, the use of legates increased steadily during this period. Legates were experienced subordinate commanders, chosen by the general, and placed in commands of higher authority than that wielded by military tribunes. They provided additional command experience and ability that the commander could draw upon when on campaign.[50] They often performed important command tasks, ranging from commanding detachments, to commanding large segments of the battle line, and to standing back with the general to monitor a battle. Judging from Livy's references for the second century, they were generally of high rank, praetorians or, frequently, consulars.[51] When consular, they would often have more command experience than their commander – surely an asset for decision-making.

The Battle 1: deployment and order of battle

With the place and time of battle decided, the army deployed. The Romans of the middle Republic, who had a set order to pitching camp, fixed places for bunking units, and set procedures for breaking camp and marching, also had a standard battle deployment – at least in its basic form. Polybius refers several times to a customary order of deployment. Sometimes he does this to indicate the Romans deployed that way, and occasionally to indicate a deviation from the norm, as at Cannae.[52] Here, again, one may question the degree to which Polybius has over-rationalized deployment and presented it as excessively orderly, though there is little need to worry that his basic picture is not sound. When Polybius described a customary deployment, he meant just that: customary, a conventional deployment pattern for the Romans. Livy's descriptions of battle deployments also support that a conventional deployment existed. Indeed, the nature of battle, based on heavy infantry battle lines clashing, dictated a general shape of deployment. This deployment consisted of cavalry on the wings, heavy infantry maniples in the center, and skirmishers in the front; indeed, that is regularly how the Romans deployed in the late third and second centuries. Tellingly, when Polybius described the mustering of allied troops for Roman field armies, he noted that they were divided into a left wing and right wing, surely denoting their common positions in the battle line.[53] Each legion and wing, and their constituent maniples, maintained their

---

50 Keppie (1984a, 39); Rosenstein (2011, 136–37). See *MRR* 1.237–573 for known legates in the period 218–101.

51 Command tasks: Livy 31.3, 31.21, 31.27, 31.44, 32.28, 34, 17, 34, 50, 36.17, 37.1, 40.27.3–6, 40.39. Consular and praetorian ranks: Livy 32.28, 36.1, 36.17, 37.1, 40.27.3–6.

52 Polyb. 1.33.8–9, 2.28.2, 2.30.1, 3.27, 3.72.10, 3.113.1, 14.8.5. See also Serrati's chapter in this volume.

53 Polyb. 6.26.

integrity in their camp positions and on the march, reinforcing their need for integrity in the battle line.[54]

The Roman legions commonly occupied the center of the battle line, with the allies on the flanks, but this was not always the case. The legions could occupy the flanks, or one legion and one wing might occupy the battle line while the others remained in reserve.[55] Clearly, the size of the battlefield must have played a role in deployments like these. Still, the deployment of heavy infantry in the center, cavalry on the wings, and skirmishers out front, did not vary greatly in the middle Republic.[56] The placement of this legion or that allied contingent in the main battle line certainly could contribute to the battle's outcome, but it is not at all clear that any general could accurately assess the unit cohesion of two comparable legions or allied infantry wings before battle. In short, decisions about the composition of the main battle line largely came down to two questions: which heavy infantry units should occupy the front and which units, if any, should be kept in reserve?[57] These were not necessarily taxing command decisions.

What did not seem to change from commander to commander was the structure of the Roman legionary part of the central battle line. The Roman heavy infantry was made up of maniples arranged in at least three lines – those of *hastati*, *principes*, and *triarii* – possibly providing a built-in mechanic for relieving ineffective Roman troops in the killing zone.[58] While a rare commander may have adjusted the normal manipular spacing or depth, there do not appear to be any references to commanders deploying maniples in anything shallower than the standard three lines.[59] Though it is hardly certain, there is a good reason to suppose that, in the second century, even the allied Italian heavy infantry was organized into maniples, or maniple-like units.[60] Polybius states that the allies used the same levy selection methods as the Romans and makes no distinction between allied Italian and Roman heavy infantry when discussing battles, suggesting that at least *he* thought that there was no tactical distinction. Indeed, he uses the manipular terms of *hastati*, *principes*, and *triarii* to refer to all the heavy infantry.[61] In addition, Livy and Plutarch suggest Italian and Roman infantry

---

54 Dobson (2008, 66–121); Polyb. 6.40. The hand-picked *extraordinarii* were the exception: Polyb. 6.40.4–6.
55 Some examples: Livy 27.13.15; 31.21.7; 34.15.3.
56 On skirmishers, see Anders (2015).
57 Since the Romans had an orderly rotation of the army's marching order, perhaps they also had an orderly rotation of the units occupying the front line.
58 Livy 8.8.9–14.
59 On tactical deployment of the Roman army, with varying frontage and line depth, see Taylor (2014).
60 See Armstrong's chapter, in this volume, for the possible flexibility of this term.
61 Erdkamp (2007a, 49–55). See also Helm's chapter in this volume for more discussion of the social and regional make-up of legionaries in this period.

had at least roughly comparable equipment.[62] Even supposing, however, that the allied heavy infantry was organized fundamentally differently than the maniple system, there is no reason to suppose this influenced the conventions of grand deployment.

The commander must have regularly delegated the actual details of deployment to the tribunes. Polybius gives an example at the battle of the Telamon (225) where both consuls explicitly instructed their tribunes to draw up the infantry battle lines while they proceeded with the cavalry.[63] Since the infantry battle line of even half of a consular army would have conservatively stretched over one mile – quite out of voice range and indeed quite likely out of effective sight – the tribunes had to manage the actual deployment.[64]

Though it happened rarely, if a general unwisely tinkered with the conventional deployment system, catastrophe could result. C. Terentius Varro's deployment plan at Cannae seems the clearest example. Varro accepted battle when Aemilius would have looked for more favorable ground, then doubled down on his error by deploying the maniples in deeper-than-normal attack columns, positioned closer-than-normal to adjacent maniples. The soldiers were packed too tightly to fight effectively.[65] Hannibal's tactical brilliance certainly contributed significantly to the Roman defeat, but Varro aided matters by overriding the maniples' normal deployment. A failure to deploy in normal battle order may also have caused Cn. Fulvius Flaccus' 212 defeat at Herdonea, though Flaccus may not have initiated the faulty deployment. His reportedly unruly soldiers were so eager to fight that they impulsively deployed with little regard for their assigned positions and refused to reform properly when the tribunes pointed this out. Granted, Livy's narrative of this battle is a notorious doublet suspect, and a skewed source might have blamed Fulvius' lack of control or absolved him and blamed the soldiers.[66] Either way, the testimony suggests the resulting battle line was not planned and poorly formed. The Romans could not withstand the Carthaginian charge, and some 16,000 soldiers died.[67] These exceptions, however, suggest the rule. The system functioned properly without micromanagement when left untampered with – though, of course, this by no means necessarily resulted in victory.

---

62 Polyb. 6.21.5; Livy 34.38–39; Plut. *Aem.* 20.
63 Polyb. 2.26.3, 2.27.4.
64 A rough estimate: 1 legion = 10 maniples each of 120 men in lines of *hastati* and *principes*. Positing a depth of three men in each maniple and 4.5 feet occupied by each soldier in the line, each maniple extended 180 feet (120 men / 3 ranks = 40 men × 4.5 spacing = 180 feet long). 10 maniples + 9 maniple-sized gaps in between = 3420 feet of infantry. Half of a consular army would extend well over a mile (± 6840 feet not including cavalry). When both legions and wings formed the main battle line, it would extend 13,680 feet without cavalry.
65 Polyb. 3.113.3, 115; Livy 22.47.8–10.
66 Erdkamp (2006a, 549–51).
67 Livy 25.21.1–10.

The Battle 2: The infantry clash and battlefield dynamics in the "killing zone"

Historians in recent decades have analyzed the mechanics of the manipular army in battle.[68] At the macro level, pitched battles between the Romans and their enemies from 218 to 100 consisted of clashes between battle lines of, more-or-less, close-ordered infantry. As units in the battle lines engaged in the limited "killing zones" of hand-to-hand and missile-weapons, each sought to disrupt their opponents so those enemy units would fail to hold position, become disordered, and, optimally, disintegrate – their constituent soldiers no longer resisting and instead fleeing or dying. At this scale, terrain and the positioning and maneuvering of units could play an important role. Ideally, a heavy unit faced one foe in one direction, as units along the battle line would, when not flanked or encircled. Attacks to the side and rear by infantry or cavalry tended to increase the disruption in a unit as soldiers felt compelled to respond not only to a single direct threat to their front – the default and anticipated vector of enemy attack – but also multiple attacks from multiple vectors. Under such stressors, units in a battle line could fail to withstand the enemy. When enough units failed, so did the battle line.[69]

Many questions remain concerning the behavior of the soldiers in the killing zone. To address some of these, Sabin has developed an informal model of combat accounting for four features of Roman battles: (1) their length of many hours; (2) the far greater casualties suffered by the defeated, suggesting both sides sustained relatively few casualties until one side broke; (3) the infantry's ability to backpedal for significant distances yet remain in the fight; and (4) the importance of multiple lines of soldiers in combat. He proposed that, in the killing zone, Roman battles did not consist of soldiers jammed into a shoving match of locked shields, such as with the usual image of the traditional Greek-phalanx style *othismos*. Nor did the infantry engage in a single, continuous, hours-long match of psychologically and physically exhausting hand-to-hand dueling – a physical impossibility. Instead, infantry combat consisted of a series of pauses with some space between opposing front lines – a default state of rest – punctuated by flurries of hand-to-hand combat when the lines clashed. During the pauses, the front-rank fighters would rest and, optimally, regain the strength and determination to clash again.[70]

At the level of the individual soldiers and smallest tactical units in the killing zone, morale (the willingness to stay in the fight) and unit cohesion (the capacity of a unit's soldiers to maintain their positions in formation and resist enemies) were the critical factors in the success or failure of an

---

68 Culham (1989); Sabin (1996); Sabin (2000); Goldsworthy (2000); Zhmodikov (2000); Goldsworth (2001); McCall (2002); Quesada Sanz (2006); Koons (2011); Rubio-Campillo, Valdés Matías, and Ble (2015); Anders (2015); Slavik (2017).
69 Culham (1989) is the landmark description of this system.
70 Sabin (2000). See also Quesada-Sanz (2006).

ancient army locked in battle – a point Culham noted decades ago.[71] Soldiers in units that maintained their space, kept formation, and were able to withstand clashes with enemy infantry, would succeed against those who lost their ability to resist attack, physically or, more often, psychologically. Units of soldiers that lost the capacity to resist, deformed, disintegrated, and fled.[72]

The success of any formation depended on its unit cohesion. Soldiers who stayed in their ranks, if not actively attacking, then at least defending themselves and their nearby comrades, collectively made stronger and more stable unit formations. These, in turn, enabled the main battle line to maintain its formations and hold its ground. Defeat in battles, a point well attested in ancient sources and emphasized by modern scholars, came with the disruption and turning of one army's units. The stressors of close combat were enormous, as soldiers fought and died in the noise, dust, and stink of the battlefield. The safety and effectiveness of those soldiers' formations depended on the individuals in it resisting fear and panic and staying in their place alongside their comrades. Once enough soldiers in a unit reached that turning point, where fear and the accompanying panic and hope for self-preservation overwhelmed any desire to stand firm with one's comrades, the unit lost its cohesion and disintegrated, either during one of the pauses in combat or during a melee with the enemy. After this threshold moment of disintegration, the soldiers of the defeated formations fled the battlefield, opening themselves to slaughter. On the larger scale, a critical breaking point was reached when enough soldiers abandoned the safety of their formations, and the battle line itself turned and broke. Then the losing infantry formations collapsed, losing their spatial integrity as individuals sought to save their own lives. The defeated fled, and the victors often pursued, killing those unfortunate enough to be caught in flight.[73]

These interactions between soldiers in and around the killing zone must be understood not as a chaotic system but rather as a complex system.[74] Rubio-Campillo, Valdes, and Ble make the critical distinction:

> Warfare is not a chaotic system; the situations studied by military historians and conflict archaeologists are robust enough to minimal variation on the initial conditions, as they will not produce major changes on the dynamics of the system. Even though some authors suggest the contrary, by its mathematical definition a chaotic system is not a good model of human interactions, because the sensitivity of the system to minimal changes on initial conditions is not as extreme as to be impossible to predict.[75]

---

71 Culham (1989); discussed as "horizontal cohesion" by Brice in this volume. See also Helm in this volume.
72 Culham (1989); Goldsworthy (1996, 206–27); Sabin (2000); McCall (2002, 13–20).
73 Culham (1989, 196–202); Sabin (2000, 14–15); McCall (2002, 13–20); Koon (2011, 91–93).
74 Culham (1989) employed the term "chaotic."
75 Rubio-Campillo, Valdés Matías, and Ble (2015, 246).

In a chaotic system, the authors note, changing the number of combatants in an army of thousands by one soldier would have extreme effects – an unlikely proposition. And so, complex systems theory provides a better framework for modeling ancient combat than chaos theory:

> [Complex systems] portray a situation where the interactions between the components of the model are non-linear. This means that some properties of complex system cannot be detected in any individual part but emerge from the relation of their components. These emergent properties are difficult to predict, but not chaotic.[76]

The battlefield systems, the clash of weapons, horrific sounds, sights, and smells, all affecting the bottom line of soldiers' morale and units' cohesion, developed in a complex and non-linear fashion. The condition of an individual soldier, his morale, his willingness to stick with the unit and stay in the fight, and the extent to which their comrades nearby perceived this, all affected those comrades. The affected comrades, in return, influenced the individual with their own projections of fear. The system was a complex set of feedback loops, increasing or decreasing unit entropy. If the entropy in a portion of a unit was too great, the morale of one or more of the soldiers there too low, those soldiers would lose their ability to defend their space and keep formation, crowding against their comrades and surrendering ground. Unit cohesion diminished. At the breaking point, soldiers fled. If this flight panicked enough other soldiers, the unit disintegrated. As the small units collapsed so too, ultimately, did the larger units of the army and the battle line itself.[77]

The field of Roman battle studies has to date not produced broadly persuasive, detailed, and formal – that is, mathematical – models of ancient combat dynamics.[78] Still, Rubio-Campillo, Matías, and Ble's effort to develop a simple one helps us visualize the systems at play in the killing zone that has been proposed in historians' various informal models. Several identifiable factors must have determined whether a formation remained combat-effective, and maintained its space and shape on the battlefield: (1) the physical condition of each soldier, including levels of fatigue, hunger, and wounding; (2) the psychological condition of each soldier, including resistance to battlefield stressors generated by friend and foe and the willingness, conscious and unconscious, to stay with comrades in formation, which is what we mean by morale; (3) the presence of nearby comrades and their own psychological condition; and (4) the presence of veterans, and unit officers, like centurions, to the extent they served to inspire and steady

---

76 Rubio-Campillo, Valdés Matías, and Ble (2015, 247).
77 Culham (1989).
78 Though for an investigation of how video games provide the features of formal models of combat, see McCall (forthcoming).

nearby soldiers through setting an example. The effectiveness of a battle line came down to the individual soldiers. Their individual ability to withstand pressures and harm, and maintain their space ultimately determined the cohesion of the maniples. The cohesion of the maniples determined, ultimately, the ability of the battle line to withstand the enemy. Generally, then, a successful battlefield army would consist of soldiers that (1) maintained their formations while interacting with terrain, enemy formations, and the stressors of battle; (2) put physical (wounding, killing, sometimes shoving) and psychological pressure on soldiers in enemy formations so that those soldiers lost morale and their formations disintegrated; and (3) capitalized on the disintegration of enemy formations by killing and capturing significant numbers of the enemy, optimally crushing further resistance in that engagement and campaign.

In a complex system, such as that which existed in the killing zone, a commander had little control over the performance of soldiers and the elemental units of the battle line.[79] Instead, the critical task of keeping the soldiers in the core units together and the men in the fight came, first, from the centurions and then from the tribunes. The Romans recognized this, acknowledged Polybius, in their criteria for selecting effective centurions:

> [The Romans] wish the centurions not so much to be venturesome and daredevil as to be natural leaders, of a steady and sedate spirit. They do not desire them so much to be men who will initiate attacks and open the battle, but men who will hold their ground when worsted and hard-pressed and be ready to die at their posts.[80]

Anecdotes confirm the potential of centurions, and even tribunes, to provide heroic, low-level leadership and keep their soldiers in the fight.[81] It was, perhaps, their most important function and, critically, they normally had to operate without direct oversight from the commander.

The Battle 3: the commander in battle

Still, Roman commanders had some limited decisions available to help soldiers and officers remain orderly and in the fight. They could (1) inspire a section of the battle line through their personal presence; (2) support a flagging section through the deployment of reserves; (3) add additional vectors

---

79 Culham (1989, 199–201).

80 Polyb. 6.24.8–9: βούλονται δ᾽ εἶναι τοὺς ταξιάρχους οὐχ οὕτως θρασεῖς καὶ φιλοκινδύνους ὡς ἡγεμονικοὺς καὶ στασίμους καὶ βαθεῖς μᾶλλον ταῖς ψυχαῖς, οὐδ᾽ ἐξ ἀκεραίου προσπίπτειν ἢ κατάρχεσθαι τῆς μάχης, ἐπικρατουμένους δὲ καὶ πιεζομένους ὑπομένειν καὶ ἀποθνῄσκειν ὑπὲρ τῆς χώρας.

81 Tribunes: Livy 27.14.8; 34.46.11–12, 41.2.9. Centurions: Livy 25.14.4–5; 26.5.12; 34.46.11–12; 39.31.9.

of attack against the enemy through flank and rear attacks by cavalry and unengaged infantry; and (4) help maintain order after the battle, especially when pursuing defeated enemies. These will be considered next.

1. Inspiring

Commanders in the middle Republic, as Rosenstein noted, were not expected to be particularly skilled at command decisions, but were expected to be outstanding models of *virtus*, martial manliness, in battle.[82] This ethos reflected the practical reality that offering moral support was often the only thing a general could do once the battle lines clashed, a task requiring character and empathy, not tactical skill. No doubt the presence of the general, facing danger, sharing risks, and urging his soldiers on, could provide a great boost to surrounding soldiers' morale. Certainly, multiple examples exist of generals providing moral support to a segment of the battle line.[83] The length of battle lines, the din of battle, and the grimly absorbing work of killing or being killed must have ensured, however, that such commander support was limited to a small section of the line.[84] Additionally, when the general committed to rallying soldiers at points along the front of the line, he sacrificed any ability to monitor the battle as a whole.[85]

2. Deploying reserves

Beyond the relief systems built into the three lines of maniples, Roman commanders sometimes kept additional troops in reserve to relieve units faltering in the main battle line.[86] This deployment of reserves at key moments in the action could be one of the general's most important command tasks in battle. Unsurprisingly, reserves that were effectively deployed could tip the balance, by bringing fresh troops into the killing zone and allowing comrades weakened by fatigue, wounds, and stress to retire. Assessing when and where to deploy reserves could be critical.[87] The command itself, however, was not enough to guarantee a successful reinforcement and an ineffective relief operation could lead to the collapse of a battle line. Livy suggests such a collapse occurred under M. Claudius Marcellus at Numistro against Hannibal. Marcellus had kept the 18th legion in reserve and deployed it to relieve

---

82 Rosenstein (1990, 114–52).

83 See Rosenstein (1990, 188–120 and n. 11). Goldsworthy (1996, 146–63); Livy (34.14) gives an excellent example of Consul M. Porcius Cato rallying men.

84 Livy (41.18.11–12) notes when the consul Petillius was killed in front of the standards, rallying his troops, only a few saw the disaster; the rest of the army was unaware. Before our period, P. Decius Mus' self-sacrifice at Sentinum was not perceived by his colleague Fabius and the troops on the right of the battle line (Livy 10.29.5).

85 Goldsworthy (1996, 149–70).

86 Some examples: Livy 31.21.7, 34.15.1, 35.5.1–2; App. *Hisp.* 40.

87 Cato the Elder in Spain is an excellent example: Livy 34.14.

the allied right wing and *extraordinarii* when they faltered. Something went terribly wrong. The press of allies falling back and the legion moving forward dissolved into disorder. The whole segment of the line collapsed, and the Romans lost the battle.[88]

3. Outflanking by tactical maneuver – the cavalry and infantry

Though tactical assessment and maneuver have been touted as the critical skills of a great commander, once an army was deployed, a general could often do little to execute such tactical maneuvers.[89] When they did occur, these maneuvers were intended to outflank the enemy formations and attack them from the flank or rear, additional vectors that further strained soldiers already fighting an enemy to the front.

Most often, however, flank and rear attacks in this period occurred, not through an infantry maneuver, but through the success of the Roman cavalry.[90] Beyond pursuing a defeated enemy, cavalry fulfilled two critical tasks on the battlefield. They defended the flanks of the Roman heavy infantry battle line and sought to harass the flanks and rear of the enemy battle line. This latter function often required engaging and driving off enemy cavalry who were similarly tasked.[91] The Roman cavalry of the Republic were generally effective at this. There is little reason to suppose, however, that their maneuvers were specially controlled by the commander, except in cases – increasingly rare in the second century – when the army commander rode with the cavalry. Their long range and high speed of operation prohibited this. Rather, the cavalry functioned according to the principles held for centuries, perhaps reinforced by the general at the start of a battle, but standard nonetheless: guard the flank and look for ways to attack the enemy's flank and rear.

Authentic Roman infantry flanking movements, where the commander maneuvered infantry to attack the sides or rear of the enemy battle line, are not common in the sources.[92] Some of the greatest Roman victories in this period suggest that even the most skilled Roman generals did not always engage, or need to engage, in such maneuvers to win.[93] Scipio's planned and executed double-flanking movement at Ilipa stands out as an exception of complex tactical outflanking maneuvers.[94] He did not repeat himself at Zama; there his only major maneuver, if it can be called that, was not to

88 Livy 27.12, though note that Plutarch (*Marc.* 24) makes no mention of the defeat.
89 Goldsworthy (1996, 169–75).
90 See McCall (2002, 53–62). Examples: Polyb. 2.30, 2.34, 15.9–14 (Livy 30.33–34); Livy 31.21, 33.36 (possibly), 35.5 (probably), 37.42, 39.31, 40.40 (probably).
91 McCall (2002, 13–25).
92 Some exceptions: Polyb. 10.39; Livy 34.14, 38.26, 40.32.
93 Taylor (2017b) has helpful overviews.
94 Polyb. 11.22–23.

flank at all but to recall his pursuing *hastati* – who, it should be noted, had succeeded against Hannibal's infantry with blood and steel and struggle, not through any tactical maneuver – and reform his battle line so that the *principes* and *triarii* occupied the wings, with the *hastati* still in the center.[95] Ultimately, Laelius and Massinissa knew the role of cavalry well as they brought their troopers home to strike the Carthaginian rear after driving off enemy cavalry. They sealed the victory.[96]

When infantry successfully outflanked the enemy, a sub-commander was often responsible, not the overall commander. At the Metaurus (207), Claudius Nero, commanding on the right wing and finding the right largely unengaged by the enemy, detached some inactive cohorts and marched them behind the Roman battle line so that they arrived to support Livius Salinator by attacking the enemy's right flank.[97] A sub-commander also exploited an opportunity at Cynoscephalae. Though the battle narrative is difficult to disentangle, Polybius insists (and Livy concurs) that the Roman victory sprang from a tribune who, on his own initiative, led 20 maniples in an attack on Macedonian right flank from the rear.[98] At Magnesia, Roman and auxiliary forces defeated the Syrian wings, including a stalwart defense by a subordinate officer at the Roman camp. No significant heavy infantry flanking maneuvers happened that day, and the commander, L. Scipio, seems not to have directed any tactical maneuvers at all in this victory.[99] Even at Pydna, the commander Aemilius Paullus reportedly did not order the outflanking of the Macedonian battle line, but noticed gaps in the Macedonian line as it drove the Romans back and ordered the Roman soldiers to work into those gaps and attack the less maneuverable phalanx in these weak spots.[100] These examples do not suggest that a commander initiated tactical flanking maneuvers often or that they were regularly a decisive part of battlefield victory. They do illustrate, however, that effective sub-commanders could often initiate such maneuvers and thus do a great deal to make a general shine.

4. Pursuing the defeated

Not uncommonly, the victorious army would pursue the defeated enemy, continuing to deal death and potentially shattering the defeated army beyond recovery. This pursuit, however, could be a hazardous affair. The victors could become disordered in pursuit. Under effective leadership, an enemy in flight might rally and take advantage of the Roman disorder. Appian attests

95 Polyb. 15.9–14; Livy 30.33–34.
96 Polyb. 15.12–14; Livy 30.33–35.
97 Livy 27.48.12–14.
98 Polyb. 18.26.1–3, Livy 33.9.7–9.
99 Livy 37.40–44.
100 Livy 44.41, Plut. *Aem.* 20.

to this several times in wars against the Spanish tribes. The Romans drove off the enemy, grew disorganized in pursuit, and were defeated when the enemy rallied.[101] This may have been a result of the often-rugged Spanish terrain, but the danger must have existed in theaters outside Spain. Keeping the troops orderly in pursuit was an important command task. Still, subordinate officers must have been critical in this. Scipio at Zama, for example, used a bugle call to make his *hastati* stop their drive on the enemy and reform ranks. The *hastati* listened. That they did must have reflected their training and the quality of their officers. What if they had not listened? Would Scipio have had any real control beyond that point?

Conclusions

Some years ago, as Nathan Rosenstein's graduate student, I read in *Imperatores Victi* that the aristocracy could not allow skill at command to be an important factor in the electability of a praetor or consul. It would make for an uneven playing field and privilege a small number of aristocrats, when the aristocracy collectively sought to maintain the flow of offices and honors to a larger number of the elite. And so, the Romans tended to explain military defeat in three fundamental ways that did not involve the skill or ineptitude of the commander: his poor display of *virtus* on the battlefield, the insufficient *virtus* of his soldiers in the battle, and his failure to secure the gods' blessings through proper sacrifice and observation of omens.

I underestimated the Romans at the time, thinking they avoided reality and expected far too little actual skill from their generals. Twenty years later, I suggest they understood all too well that any commander had quite limited control over the outcome of a battle, and that limited control was best in a system where usually it was elected amateurs who commanded armies. What this brief investigation suggests most of all is that the elaborate system of protocols combined with – often frequent – command inputs from subordinates, buffered the typical general from disaster. They made it so the typical commander needed very little command skill to expect a positive outcome from his year in command – or at least to avoid a disaster. A mediocre commander, who did not insist on micromanaging things his own way, could rely on the system. The province and enemy, as well as the army and officers, were set by the senate, electorate, tribunes, and soldiers. Logistics were overseen by the senate and quaestors, and could regularly be delegated to subordinates. The commander did have to make important decisions about routes of march and reconnaissance, and the place and time of battle, but a military council, and quite often legates – not to mention tribunes and centurions – were at hand providing experienced advice, and the general could go with the wisest counsel.

---

101 App. *Hisp.* 56, 58, 64 (a feigned flight), 67.

Once the actual battle commenced, a commander typically required little tactical skill. The tribunes deployed and managed legions and wings in battle, the manipular system had built-in reserves, and the battle line remained constant with one or more legions of heavy infantry in the front, the rest in reserve, cavalry on the wings, and skirmishers in front. Centurions were the point officers in the killing zone, assisting the soldiers in their formation, helping them hold firm, rallying for new clashes with the enemy close by. Tribunes were generally nearby to provide greater direction if needed, sometimes even winning battles by initiating tactical maneuvers. Cavalry had a clear task. The general was left to intervene personally to shore up morale and, perhaps, to issue the order for reinforcements if the triple-maniple line was insufficient. Very occasionally, he might order a unit to flank or exploit an enemy weakness, but this was not common. Above all, infantry holding the line and cavalry attacking flanks and rear accounted for most Roman victories. Finally, if the Romans soldiers proved victorious, the general might command an orderly pursuit of the defeated, though this too would fall to the sub-officers to execute.

At best, Roman battles – from the command perspective – were loosely controlled mayhem. Victory was never guaranteed in a complex system like a Roman battlefield. A well-trained army, talented officers, and a skilled commander could still lose a battle. Events at one spot in the killing zone could magnify into the collapse of a line. The typical general could do very little to bring about victory or defeat once the battle began.

A Roman commander could rely upon little – except for the system: skilled officers, tested conventions for units, camps, deployments, and procedures. The manipular system, developed over the decades, generally operated efficiently and effectively regardless of the commander. The average general could, if he chose, rely on the system and the experience of others at most steps in the process: the selection of army, officers, and province, the camping, marching, and supply of the army, and even the time place and deployment for battle. And so Roman aristocrats could happily compete for the consulship, knowing that if they did secure the office, they would not generally require any special qualifications, other than simply being an aristocrat, to avoid disaster and probably even secure some level of victory in their year of command.

13 Anecdotal history and the Social War*

Jessica H. Clark

Introduction

The Social War, fought between Rome and many of its Italian Allies (*socii*) in the early first century (91–88), has long been grounds for debate among ancient historians. The war's causes and consequences are of major import for understanding the dramatic final decades of the Roman Republic, but our evidence is remarkably elusive. Although at least three ancient writers recounted the war in some detail (Cornelius Sisenna, Diodorus Siculus, and Livy), and despite its manifest significance, modern scholars have only disarticulated fragments from which to craft a coherent narrative of the Social War. The reconstructions that result can give us a sense of the war's strategic progression – a matter of no small value to our understanding of relations between Rome and Italy.[1] What they cannot do, however, is restore access to the story of the war as Romans would have told it;[2] we lack the framing

* It is a great pleasure to acknowledge here my gratitude to Nathan Rosenstein, both for his formative role in shaping the current state of the study of the Roman Republic and for his generous support of students and colleagues alike. The present chapter owes much to the insights of the participants in a graduate seminar on "Civil War and Civic Violence" at Florida State University in 2016, including my colleague John Marincola, and the panel "New Directions in Roman Military Studies" at the 2017 meeting of the Celtic Conference in Classics; I would especially like to thank the panel organizers and volume editors, Michael Fronda and Jeremy Armstrong, as well as the anonymous readers for this volume. All dates are BC unless otherwise noted.

 1 The Social War is the subject of two recent monographs: Kendall (2013) and Dart (2014); see also Keaveney (2005); Mouritsen (1998), offering an important revisionist critique of traditional interpretations of the war's causes. Fronda (2010, 324–29) provides a clear summation of the war in the context of Rome's (and the Romans') larger relations in Italy. I agree that we are generally better served by following Appian than attempting to reconcile diverse and fragmented textual sources for individual campaigns, with the key caveat of Westall (2015, 146–47) (on the "dramatic lack of chronological framework"). Isayev (2017, 311–51) approaches the war from the perspective of human mobility and its reflection of contemporary conceptions of place and culture, with references to current scholarship.
 2 That Romans were thinking about the war's textual commemoration in the first century is well illustrated by a fascinating anecdote (Plut. *Luc.* 1.5), in which Lucullus, the orator Hortensius, and the historian Sisenna (who may all have served in the war as young men) discuss undertaking a history of the Social War in (alternatively) Latin or Greek, in prose or in verse.

DOI: 10.4324/9781351063500-13

This chapter has been made available under a CC-BY-NC-ND license.

devices, the careful juxtapositions, the speeches, and other tools which ancient historical writers used to shape their audiences' reception of their own past.[3] In short, we do not know what the war meant.

While there could never be a single, or simple, solution to that proposition, it may be that our existing evidence has more to tell us than we have yet appreciated. The aim of this chapter is thus to begin an assessment of the commemorative legacy of the Social War, and to suggest a method for approaching its particular historiographic difficulties. Much that follows is, admittedly, impressionistic, and makes no claims to comprehensiveness. In place of resolving contradictory evidence into a rational, military-historical narrative, I instead propose several ways of understanding the legacy of the war for some of its survivors, and the generations that followed. Because this is one area in which our historicizing reconstructions have proven less successful, the pursuit of the ways in which Romans understood the war justifies the consideration of other methods and other goals. Thus, I begin from the premise that the sum of causal and temporal connections among the war's events is a separate matter from, and potentially incompatible with, the narratives that influenced and informed the people living in post-Social War Italy.

Although the Social War is less well-studied than many of Rome's conflicts, its importance is not in question. Diodorus Siculus may have been objectively wrong to label it the greatest of all the wars ever fought, but thus he did:

> In the time that people's deeds have been handed down by recorded history to the memory of posterity, the greatest war known to us is the "Marsic," named after the Marsi. This surpassed all that preceded it both in the valorous exploits of its leaders and in the magnitude of its operations.[4]

---

3 This is not to say that scholars have neglected the war's legacy; thus especially Mouritsen (1998, 9–10); Dart (2014, 24–40). Brennan (2000, 371–84, 584–86) highlights the political innovations prompted or hastened by the war; Rosenstein (2011, 150) notes the considerable financial consequences of the war for the next generation. See also Farney (2007, 220–25) on the selective reinterpretation of the "rebel" Italian past for self-promotion by the next generations, and Marincola (2010) on the *Aeneid* as informed by the Social War. Gauthier (this volume) discusses the consequences of the war for the Roman army and its use of auxiliaries.

4 Diod. Sic. 37 fr. 1: Ἀφ' ὧν χρόνων αἱ τῶν ἀνθρώπων πράξεις διὰ τῆς ἱστορικῆς ἀναγραφῆς εἰς αἰώνιον μνήμην παρεδόθησαν, μέγιστον ἴσμεν πόλεμον τὸν Μαρσικὸν ὀνομασθέντα ἀπὸ Μαρσῶν. οὗτος γὰρ πάντας τοὺς προγεγονότας ὑπερεβάλετο ταῖς τῶν στρατηγῶν ἀνδραγαθίαις καὶ τῷ μεγέθει τῶν πράξεων. Here and following, the text and numbering is that of P. Goukowsky (2014), with translations adapted from the Loeb edition. This excerpt forms part of the tenth-century *Excerpta de sententiis* compiled, alongside numerous collections on other themes, under the auspices of the Byzantine emperor Constantine VII Porphyrogenitus. The same sentiment was also excerpted by Photius in the ninth century ("Diodorus declares that the so-called Marsic War, which fell in his lifetime, was greater than all that came before." Ὅτι τὸν Μαρσικὸν ὀνομασθέντα πόλεμον ἐπὶ τῆς αὐτοῦ ἡλικίας Διόδωρος

Diodorus cannot have thought that his contemporary readers would greet the claim with as much incredulity as we do now, and the size of the gap between this assessment and the state of our evidence invites extraordinary measures in our efforts to understand this strange, short, terrible war.

This chapter will discuss three sets of anecdotes. The first, a pair of stories set at the Italian town of Pinna, provides a salutary illustration of the perils of our sources. The second compares the literary record of two towns, Aesernia and Grumentum, which both experienced sieges, and, despite having other points in common, produced interestingly different historiographic legacies. The third pair of anecdotes highlights two of the individuals whose names we know from the war, and sets their stories beside those previously discussed. Together, this collection of fragments, anecdotes, and side comments allows us to approach the war as Diodorus knew it: a conflict of surprising, and surpassing, magnitude, the legacy of which far outstripped the chronological and geographic limits of its campaigns.

Pinna

The Social War does not lend itself well to maps.[5] Some towns and battle-sites occupied locations unknown to us now, and the boundaries between tribal or cultural groups – and their loyalties – were both inconsistently defined and subject to change over the course of the war.[6] Adding to the difficulties are towns like Pinna (modern Penne), a settlement of mid-dling size near the Adriatic. Pinna was a town of the Vestini – an Oscan people whose territory bordered that of the Sabines in the central Apennines, and who had entered into an alliance with Rome by the end of the fourth century.[7] We have little information about the 200 years in which the Vestini were at peace with Rome, and thus we are not well

---

μείζονα πάντων τῶν προγεγονότων ἀποφαίνεται). This makes a notable contrast with Appi-an's ambivalence about the Social War (*B Civ.* 1.34.151), on which see Bucher (2000, 437). Fronda (this volume) highlights the deep antecedents of the war – a crucial point to bear in mind when evaluating its resonance on a human scale.

5 Though most towns discussed in this chapter may be found on Maps 3–5.

6 This is as applicable to intangible boundaries, such as those of identity and affinity, as it is to physical demarcations of landscape; on the question of Roman and Italian identi-ties during this period, see Wallace-Hadrill (2008, 78–81); Dench (2005, 55–69); Hermon (2007); Isayev (2011); and Neil (2012). These caveats being noted, a broad sense of divisions within Italian territory, in contrast to the *ager Romanus*, can be gained from Map 5.

7 Pinna is now the subject of a magisterial two-volume collection of studies on its history and material remains: Franchi dell'Orto, Agostini, and Buoncore (2010); Buonocore, Staffa, and Franchi dell'Orto (2010).

positioned to understand their internal dynamics. Their lands included some of the highest peaks in the Apennines (topping 7,000 feet) but also river valleys and accessible coast, and both zones were home to a number of small cities as well as dispersed settlements. The main settlements were Pinna, between the mountains and the sea; Aternum, on the Adriatic; and Peltuinum and Aufinum, in the Apennines. Although at least two of the towns in the western area of Vestinian territory seem to have been granted Roman citizenship (*sine suffragio*) in the third century, local administrative systems, language, and material culture continued to unite settlements throughout the region.[8] According to Livy (*Per.* 72) and Appian (*B Civ.* 1.39.175), the Vestini allied against Rome in the Social War; we know nothing about the actions of those towns believed to have Roman citizenship before the war.[9]

Pinna has been taken as an example of the loyalties divided by the war: while the Vestini opposed Rome, Pinna opposed the Vestini.[10] The evidence is a grim tale, for which the main source is Diodorus Siculus. The tenth-century *Excerpta de sententiis* relates the story in three separate excerpts: in the first, as an example of premonitions, the women of Pinna mourn, in advance, the loss of their children; in the second, an Italian army besieging Pinna had somehow captured "all the children of the Pinnans" (τὰ τέκνα τῶν Πιννητῶν ἅπαντα) and threatened to kill them if the defenders did not surrender, whereupon their fathers replied that they would be able to have more children if they remained loyal to Rome; and in the third, we find a chilling

8 The inhabitants of Peltuinum and Aveia were probably admitted to Roman citizenship (in the tribe *Quirina*) at some point before the Social War. On their status as *praefecturae* and relations to other towns of the Vestini, see Humbert (1978, 226–33); cf. Taylor 2013 (1960, 65–66), on the tribal affiliation. A Latin inscription from the third or second century, found in Praeneste, records a dedication to Hercules by one L. Gemenius L. f. Pelt[——] (*CIL* 1².62 = *ILLRP* 132); the text is plausibly restored as *Peltuino domo*, establishing the domicile from which Gemenius traveled to make his dedication. A roughly contemporary inscription which seems to represent a local Oscan dialect, found in Navelli (in western Vestinian territory), records a dedication to Hercules, by T. Vettius: "Titus Vettius bestowed this as a gift on Hercules, Jupiter's son, for favours granted," *T Viitio duno didiit Hiirclo Iovio brat data* (*CIL* 1².394 = *ILS* 3431, trans. Warmington).

9 The Italians' brief capital, Corfinium (renamed Italica for the duration of the war), was scarcely ten miles south of the one-time border of Vestinian territory; see Isayev (2017, 320–23) on the selection of Corfinium. The elevation and isolation of (for example) Peltuinum may account for its lack of mention by our sources.

10 Although the Vestini do not loom large in our surviving narratives of the war, Obsequens, who excerpted Livy's text for its lists of prodigies, notes a rain of stones indoors in Vestinian territory in 94 ("Among the Vestini it rained stones within a country house," *in Vestinis in villa lapidibus pluit*, 51) and, explicitly connected with the prodigies at the outbreak of the Social War, includes a seven-day rain of stones and potsherds in 91 ("Among the Vestini there was a rain of stones and sherds for seven days," *in Vestinis per dies septem lapidibus testisque pluit*, 54).

description of the children's vain pleas before the city. A separate excerpt, in the *Excerpta de Virtutibus et Vitiis*, summarizes the event; thus,

> The people of Pinna were caught in a terrible dilemma. Having an unshakeable alliance with the Romans, they were compelled to detach themselves from the emotions of their soul and to watch while their children were deprived of life before the eyes of those who had given them it.[11]

The event comes to us devoid of context; we do not know whether it occurred closer to the beginning or to the end of the war, what precipitated it, who commanded either side and what soldiers committed the violence, or what happened next. The massacre at Pinna stands in brutal isolation, despite the assumed – but little documented – horrors of the war.[12] While many historians have deployed it as such with little commentary, the most recent editor of the fragments of Diodorus Siculus pointedly notes the absence of any confirming literary source and the silence of the earlier excerptor, Photius, on Pinna.[13]

A Latin writer contemporary with the Social War, the anonymous *Auctor ad Herennium*, provides evidence that the Vestini in Pinna performed in a praiseworthy manner (2.45):

> Again it is a fault in making a comparison to think it necessary to disparage one thing when you praise the other; for example, if the question should arise, who are to be held in greater honor for services to the Roman republic, the Albensians or the Pinnensian Vestini, and the speaker should attack one or the other.[14]

---

11 Ὅτι οἱ Πιννῆται δειναῖς συνείχοντο συμφοραῖς. Ἀμετάπειστον δ' ἔχοντες τὴν πρὸς Ῥωμαίους συμμαχίαν ἠναγκάζοντο κατεξανίστασθαι τῶν περὶ ψυχὴν παθῶν καὶ περιορᾶν τὰ τέκνα στερισκόμενα τοῦ ζῆν ἐν ὀφθαλμοῖς τῶν γεγεννηκότων. Goukowsky sets this following the three excerpts from the *de Sententiis* (37 fr. 28–31), and before another excerpt from the *de Sententiis* (37 fr. 32) which praises the "excess of virtue" (τῇ τῆς ἀρετῆς ὑπερβολῇ) of a group of the besieged in the face of more numerous besiegers, but gives no clear indication of the identities of either party; while it may represent Diodorus' praise of the Pinnans, the next excerpt opines on the Romans and Italians in general terms, and thus we should resist (with Goukowsky) grouping fr. 32 with frs. 28–31, on Pinna.

12 There are notable exceptions, chief among them the problematic excerpt from Cassius Dio, on the Picentines' treatment of those who did not join the rising (98.3); the lack of detail with regard to the massacre at Asculum, however, is interesting (thus Cic. *Font.* 41; App. *B Civ.* 1.38.173–174; Diod. Sic. 37.12; Vell. Pat. 2.15.1) and contrasts with the details provided for the fall of Praeneste and Norba in 81 (App. *B Civ.* 1.94.434–439, with Gabba [1967] ad loc.); cf. Thein (2016) on Sulla's treatment of Italian towns. The fragments of L. Cornelius Sisenna's historical account of the war suggest that he provided his readers with a vivid, and detailed, narrative of its violence.

13 Goukowsky (2014, 365 n. 121).

14 *Auctor ad Herennium* 2.45: *Item vitiosum est in rebus conparandis necesse putare* [mss.: *putari*] *alteram rem vituperare* [mss.: *vituperari*] *cum alteram laudes; quod genus, si quaeratur utris maior honor habendus sit, Albensibus an Vestinis Pennensibus, quod rei publicae populi Romani profuerint, et is qui dicat alteros laedat.*

The *Auctor* presents the Vestini of Pinna as loyal to Rome, and assumes that readers will understand the reference. Taking this to refer to the same events as related by Diodorus, we assume we understand as well – but we are thus led into difficulties. First, we know of no services rendered to Rome by the town of Alba Fucens during the Social War, beyond a brief notice in the *Epitome* of Livy Book 72 that it (and Aesernia, to which we will return below) was besieged by the Italians. Second, the shorthand reference to the benefits conveyed to the Roman people by the residents of the two towns implies some potential difference in the services rendered, but not one that would obviously elevate one over the other. Alba and Pinna are explicitly comparable, but the creativity of the rhetorician could yet distinguish between them in some meaningful way. I do not wish to belabor this one passage, but it is worth noting that if the service to which the *Auctor* refers is the Pinnans' willingness to sacrifice their children, then that was a decision which the *Auctor* could imagine speakers alternatively praising and censuring in a public context where honors or rewards were at stake. Thus, this brief remark invites us to consider that such debates indeed took place in the aftermath of the Social War, and that they were one means by which the urban population of Rome encountered stories from the war.[15]

We cannot participate directly in the lesson here because we do not know the referents nor what, though it must be something, was at stake.[16] Nonetheless, if our goal is to integrate the experience of the people of Pinna within a chronological narrative of the Social War, we might seem to have enough to do so: the Vestini rose against Rome, but at least one of their cities did not, and came to illustrate, in the starkest possible terms, the price that the Italian confederacy exacted from dissenters within. To use Pinna thus, however, we must discard another piece of textual evidence. Valerius Maximus gives Pinna as the setting for a somewhat different tale:

> The same affection armed with strength of mind and body a young man of Pinna, surnamed Pulto, in the Italian War. He was in charge of the defenses of his besieged town, and the Roman commander had placed his father, who was a prisoner, before his eyes surrounded by soldiers with drawn swords, threatening to kill him unless Pulto let his assault party through. Single-handed, Pulto snatched the old man out of their clutches. He is to be commemorated for a double piety, because he was both his father's preserver and no traitor to his fatherland.[17]

---

15 That Romans spoke in recognition of towns' and peoples' services in other contexts is well documented; for the ways in which such recognition might focus attention on select anecdotes, see Clark (2014, 76–78).

16 Hilder (2015) discusses the use of historical references in the *ad Herennium*, which – though generally short – are often more informative than this.

17 Val. Max. 5.4 ext. 7: *Eadem caritas Italico bello Pinnensem iuvenem, cui Pultoni erat cognomen, tanto animi corporisque robore armavit ut cum obsessae urbis suae claustris praesideret, et Romanus imperator patrem eius captivum in conspectu ipsius constitutum destrictis*

Here, Pinna is under siege by a Roman (*Romanus imperator*), who has at least one local prisoner whom he does not shy from exploiting. A native of the town was in charge of its defenses, which Valerius presents as vulnerable to a forward assault party seeking ingress.

The actual event is not easy to imagine, however. Is this a secret negotiation, attempting to suborn young Pulto? What did Pulto do after he grabbed his father from the soldiers' grasp? That is, did he go back inside Pinna and continue its defense, or was it enough, for the moral point of the tale, that he did not actively participate in the betrayal of his town? The excellent phrasing of Valerius' closing *sententia* cannot obscure the fact that we are missing most of the story here.[18] What little we do have is made suspect by its lack of detail, insofar as what Valerius has selected to omit includes everything that would give this event a place in the strategic trajectory of the war. Nevertheless, although one cannot deny that Valerius Maximus sometimes mistook his details, we should not be too quick to dismiss this story because it appears to contradict Diodorus.[19] Both tales, in fact, share a common valence of unreality, offering variations on the theme of parents, children, and choices.

Diodorus' story is, also, not necessarily more plausible. It seems to have included a mass premonition: the mothers of Pinna's children foresee what they cannot prevent. We might question, also, how the Italians besieging Pinna managed to acquire *all* the children (τὰ τέκνα τῶν Πιννητῶν ἅπαντα) and separate them from their parents? This calls to mind, as perhaps it was meant to do, the tale told in book five of Livy's history, in which a schoolmaster from the Italian town of Falerii attempted to use his young students to buy himself safe passage, when M. Furius Camillus was besieging the town. In that exemplary anecdote, the Romans rejected such unseemly hostages (and the concomitant treachery), and won Falerii as an ally through their own honorable conduct.[20] Diodorus' version – at least as presented in summary excerpts – gives us no hint as to who perpetrated the massacre at Pinna, or who was in charge; the focus is on the men defending the town and how their reactions should be perceived, and we can be sure Valerius' Pulto was not among them.

We can, of course, tell a story that reconciles this evidence: Pulto, perhaps, saved his father, but was unable to secure his city; perhaps Pinna itself was divided in its loyalties. Appian records that Italian polities exchanged

---

militum gladiis circumdedisset, occisurum se minitans nisi irruptioni suae iter praebuisset, solus e manibus senem rapuerit, duplici pietate memorandus, quod et patris servator nec patriae fuit proditor.

18 Rawson (1979, 339) suggested the lost history of L. Cornelius Sisenna was a plausible source for this and other dramatic anecdotes from the war.

19 Kendall (2013, 328 n. 74) is judicious.

20 Livy 5.27; for the importance of the story for Livy's contemporary audience, and with reference to further ancient sources for the famous tale, see Gaertner (2008, 36).

hostages at the beginning of the war, and that may explain the seeming availability of older parents and young children to be exploited by the different sides.[21] If the town changed hands or faced an internal battle over its choice of loyalties, perhaps Diodorus' besieging Italians did what they did as reprisal for whatever happened earlier. Perhaps the people outside Pinna and the people inside Pinna were both Pinnans.[22] Here, though, we must stop. The most valuable data point is probably what is tersely implied by the anonymous *Auctor*: different towns – all Roman, now – served in different ways, and we do not need to weigh the details along some grand balance of past suffering.

In historical terms, the anecdotes set at Pinna might seem to pose only a little problem, and one easily solved by declaring Valerius Maximus either to have misunderstood his information or to have followed a faulty source.[23] Neither would be particularly surprising. Taken together, however, Diodorus and Valerius, in fact, give us the same information about the legacy of the war: the residents of Italian towns were caught between the loyalties demanded by blood ties and those chosen by treaties, they responded honorably (by their definitions) in situations orchestrated to provoke the opposite, and they did so regardless of what side they were on. Historians can reconstruct the war with shifting allegiances and towns won, lost, and won again, but that big picture need not obscure the moral landscape of these texts. In the memory of the war, men's affinities were absolute.

Aesernia and Grumentum

Diodorus' story of Pinna has become metonymic for the assumption that atrocities were a tactical reality of the war, inasmuch as we often assume that similar horrors recurred in places where our record is silent.[24] Two other besieged towns, Aesernia and Grumentum, offer examples in which – whatever actually happened – the textual transmission of their fates offers complications and emphasizes "not-quite-horror," or "not-quite-so-simple" judgment. Aesernia, a Roman colony, appears in the *Epitome* of book 72 of Livy's history paired with Alba Fucens as sites besieged by the Italians (*Aesernia et Alba coloniae ab Italicis obsessae sunt*). We know nothing further

---

21 App. *B Civ.* 1.38.170; on the role of hostages in negotiations, see Álvarez Pérez-Sostoa (2015), with references.
22 Thus Fronda (2010, 328–29 n. 157).
23 See Morell (2015) for the benefits of working with the inconsistencies of the record on the Social War, rather than rejecting or "explaining away" the evidence (in this case, of Appian on events of 91).
24 Dart (2014, 3, 126) adduces Pinna as the site of such horrors. Interestingly, Catiline (if, indeed, he served during the war) is not reprehended for any blameworthy activities; the evidence for Catiline's presence is the attestation of an L. Sergius on a contemporary inscription (the "Asculum bronze"): *ILS* 8888 = *CIL* 1^2.709 = *ILLRP* 515, ed. and comm. Criniti (1970).

about Alba, beyond the brief reference in the *ad Herennium* (discussed above), which suggests that its inhabitants did something that the Romans might value, and the fact that Alba is not included in Florus' (2.6.11) list of towns "laid waste by fire and sword" (*ecce Ocriculum, ecce Grumentum, ecce Faesulae, ecce Carseoli, Aesernia, Nuceria, Picentia penitus ferro et igne vastantur*).[25]

We can say more about Aesernia, and it is interesting that in this case we have four sources for the city's siege. A fragment of the contemporary historian L. Cornelius Sisenna refers to the siege of Aesernia and the inhabitants' want of provisions.[26] More specifically, according to Diodorus, hunger drove them to eat dogs and other animals of the sort which it was not their custom to consume, and also to expel their slaves from the city (presumably in order to conserve resources for the free population). They did this through some sort of trick, we are told, and the pitiable slaves were taken in by the besiegers.[27] In the *Epitome* of book 73 of Livy, we read that Aesernia, and with it, M. Marcellus, fell to the Samnites.[28] Appian (*B Civ.* 1.41.182) has the city besieged by Vettius Scato (to whom we will return below), and two Roman commanders (named as L. Scipio and L. Acilius) escape through the Italian lines, disguised as slaves, before the city fell (through hunger, not assault).[29]

These excerpts come to us as examples of hardships, but it is worth noting two things. First, Roman readers would have been well aware that the Aesernians stopped short of the final measure of desperation; they are emphatically not cannibals in this tale, and the expulsion of the slaves is, in light of

25 Orosius noted briefly (5.22.17) that Alba held out valiantly when besieged by Sulla's forces: "The city of the Albans, besieged and suffering terribly from hunger, was saved by the surrender of its wretched survivors. Scipio, the son of Lepidus, was captured there and put to death," (*Albanorum civitas, obsidione oppugnata atque excruciata fame ultima, miserabilium reliquiarum deditione servata est; ubi tunc Scipio, Lepidi filius, captus atque occisus est*). An undated inscription from the town honors Sulla (*CIL* 1².724); the Social War is not the only first-century conflict for which we have problematic evidence. Dart (2014, 226–29) compiles the known sieges of the war.

26 Sisenna *FRHist* F13, as quoted by Nonius (trans. Briscoe): "at the same period, the inhabitants of Aesernia, surrounded by a double ditch and rampart, having eaten the corn which was brought into the town from the threshing floors," (*iisdem temporibus Aesernini, duplici fossa valloque circumdati, frumento adeso quod ex areis in oppidum portatum est*).

27 Diod. Sic. 37 fr. 26 (expulsion of slaves) and 27 (consumption of animals). Appian (*B Civ.* 1.42.186, 190) describes several instances of the slaves and prisoners of Roman towns being enrolled in Italian armies.

28 Livy, *Per.* 73: "That the fortunes of war might be fickle, the colony of Aesernia, along with Marcus Marcellus, fell into the hands of the Samnites," (*et ut varia belli fortuna esset, Aesernia colonia cum M. Marcello in potestatem Samnitium venit*).

29 See further Gabba (1967, 134); Kendall (2013, 249–50, 296–98) explores the range of possible interpretations of our scattered evidence for Aesernia's siege(s).

that alternative, not an unmitigated cruelty.[30] Second, it may not have been a cruel act at all. It is impossible to know whether the escaping Romans took advantage of the townspeople's expulsion of the slaves, or if the slaves were expelled to cover their flight. The Italians occupying Aesernia thereafter, according to another reference ten chapters later in Appian's narrative, provided a refuge for the Italian commander C. Papius Mutilus, defeated by Sulla in 89.[31] Within Photius' long summary of Diodorus' account (37.2.10), Aesernia was the Italians' capital in the final year of the war.

Unlike in the case of Pinna, our sources here do not contradict each other, but as with Pinna, they are not as illuminating on closer examination as they appear from a distance. The summary of Livy has Aesernia besieged and then coming under the control of the Samnites, along with its presumed commander, M. Claudius Marcellus. Diodorus' excerpts describe a long siege; Appian suggested that (at least some of) the Romans abandoned the town, which surrendered. Florus listed the city among those devastated during the war, but its continued occupation (and a dearth of archaeological evidence for destruction) makes it difficult to date, or even confirm, this. Rather than sort our handful of fragmented notices into a plausible order, we might be better served by contemplating what is absent from (the story of) Aesernia's experience of the war. Working only with the information we do have, it appears as if everyone, on all sides, behaves rather well (within the parameters of war). The Aesernians enact their loyalty with demonstrable endurance, but they break no taboos. They allow their slaves to leave the town alive, and the besieging Italians, for their part, give them refuge.[32] Their surrender seems to be accepted by the Samnites, who do not execute Marcellus. The anecdotal excerpts from Diodorus present the events at Aesernia in a negative light, but the comings and goings of Roman legates make clear that the town's situation was far from straightforward. Again, loosely grounded detail overshadows – or, is made to overshadow – a more complicated story that we cannot read.

30 Compare, for example, the exigencies to which the people of Calagurris were driven by the siege of their town in the mid-70s: Val. Max. 7.6. ext. 3; Juv. 15.93–109; Sall. *Hist.* fr. 3.60 R/3.86 M: "when, after committing and suffering many abominations respecting their food," (*ubi multa nefanda esca super ausi atqui passi*); further associated detail: fr. 3.61 R/3.87 M.

31 C. Papius Mutilus is a particularly interesting character; the coins minted in his name provide a rare material dimension to discussions of the Italian side in the war, on which see Briquel (2010); Dart (2014, 113–15). On Aesernia, note also a reference by Frontinus to an episode in Sulla's military activity near the city, in which Sulla, trapped with his army in a narrow pass, sues for peace, is denied, but escapes during the truce granted for negotiations (*Str.* 1.5.17).

32 Caesar provides an alternative response to the influx of non-combatants between city walls and besieging soldiers, at Alesia: *BGall* 7.78. On the perception of the category, see Kinsella (2011, 29–37, 53–81).

The town of Grumentum offers a useful parallel. Florus (2.6.11) included it, as with Aesernia, in his list of towns laid waste, but we are not told by whom. Appian (*B Civ.* 1.41.184) has P. Licinius Crassus, as legate, defeated by the Italian commander M. Lamponius and thence take refuge in Grumentum, which would plausibly place the town on the Roman side of the war in 90.[33] But in an anecdote related by Seneca the Younger, and attributed by him to Claudius Quadrigarius, Grumentum falls to the Romans:

> In the eighteenth book of the *Annals* Claudius Quadrigarius reports that when Grumentum was being besieged and had now reached the height of desperation, two slaves deserted to the enemy and performed worthy service for them. When, subsequently, the city had been captured and the victors were running about in all directions, the slaves ran ahead along the routes they knew to the house in which they had been slaves and then drove their mistress in front of them. To those who asked who the woman was, they proclaimed that she was their mistress, and indeed an extremely cruel one, whom they were taking off to punishment. When they had taken her outside the walls, they concealed her with the greatest of care, until the enemy's anger died down. Then, when the soldiers had been sated and quickly returned to the behavior of Romans, they too returned to theirs, and gave themselves a mistress. She immediately freed them and did not resent having received her life from those over whom she had the power of life and death.[34]

Macrobius has it thus:

> When Grumentum was under siege, slaves left their mistress and crossed over to the enemy: when the city was captured they attacked their mistress' house (this was part of their plot) and dragged her out with threatening looks, telling those they met that they had finally been given an

---

33 Appian does not give Crassus' praenomen. Frontinus repeats a brief mention of (probably) this event ("In the same way, Publius Crassus in the Social War narrowly escaped being cut off with all his forces," *P. Crassus bello sociali eodem modo prope cum copiis omnibus interceptus est*) at *Str.* 2.4.16 and 4.7.41.

34 Sen. *Ben.* 23.2–3 = Claudius Quadrigarius *FRHist* F82 (trans. Briscoe): *Claudius Quadrigarius in duodevicesimo annalium tradit, cum obsideretur Grumentum et iam ad summam desperationem ventum esset, duos servos ad hostem transfugisse et operae pretium fecisse. Deinde urbe capta passim discurrente victore, illos per nota itinera ad domum in qua servierant praecucurrisse et dominam suam ante egisse. quaerentibus quaenam esset dominam et quidem crudelissimam ad supplicium ab ipsis duci professos esse. eductam deinde extra muros summa cura celasse, donec hostilis ira consideret. deinde, ut satiatus miles cito ad Romanos mores rediit, illos quoque ad suos redisse et dominam sibi ipsos dedisse. manumisit utrumque e vestigio illa nec indignata est ab iis se vitam accepisse in quos vitae necisque potestatem habuisset.*

opportunity to repay their cruel mistress. After carrying her off, as though to punish her, they protected her and attended her devotedly.[35]

The relationship between the two versions is instructive. Macrobius, for his part, did not specify who "the enemy" is in this story. Seneca tells us only obliquely that the attackers were not Italians, but Romans. Both Seneca and Macrobius managed to convey the fearful reality of the city's fall; we are not supposed to doubt that the slaves could have had the worst of intentions, and that "those they met" were prepared to allow them to proceed. Seneca's version, however, emphasizes elements of the slaves' humanity: they do not leave the city until its plight is most desperate, they prove valuable to the attackers, and they are prepared to return to the condition of slavery despite, seemingly, having taken possession of their former owner as a prize of war. The overall effect is one of careful planning, as if the slaves left the city originally for the purpose of accomplishing this act of loyalty, which Seneca confirms in his moralizing conclusion. We are, in his continuation of the anecdote, informed that the lady became well known through the tale (*nobilis fabula*) and provided an *exemplum* for both Rome and her home town.

As with Aesernia, what we do not see matters. There is no denying the violence which sets the stage for this dramatic rescue, but it takes place out of view. Moreover, at both towns, the transfer of slaves from one side to another – a recurrence in the war, and one that must have had a particularly fraught legacy – is interpreted (or interpretable) within a larger set of objectives. Men's motives are complicated; permitting former slaves to take revenge on an especially cruel mistress, within the context of the chaos attendant a city's fall, is a rational application of violence, though it should still give us pause, as should all that is not said between *satiatus* and *cito*.[36] The point here is that the information we have concerning the two towns combines to reinforce a vague, but yet clear, narrative of the war, and the choices and loyalties that cannot be understood in isolation.

Ties that bind

Both Seneca and Macrobius followed their stories of Grumentum's slaves with the death of Vettius Scato, one of the most militarily successful leaders

---

35 Macrob. *Sat.* 1.23: *Cum premeret obsidio Grumentum, servi relicta domina ad hostes transfugerunt. capto deinde oppido impetum in domum habita conspiratione fecerunt et extraxerunt dominam vultu poenam minante ac voce obviis adserente, quod tandem sibi data esset copia crudelem dominam puniendi raptamque quasi ad supplicium obsequiis plenis pietate tutati sunt.*

36 See further Gaca (2014), on ancient expectations of sexual violence against the civilian residents of captured cities.

of the war on the Italian side.[37] Again, their details and color differ slightly. Seneca has:

> When Vettius, the praetor of the Marsians, was being conducted to the Roman general, his slave snatched a sword from the very soldier who was dragging him along, and first slew his master. Then he said: "Now that I have given my master his freedom, the time has come for me to think also of myself," and so with one blow he stabbed himself. Name to me anyone who has saved his master more gloriously.[38]

While Macrobius noted,

> Consider too a largeness of spirit – in servile circumstances – that prefers death to dishonor. When Gaius Vettius, a Paelignian of Italica, was seized by his own soldiers to be handed over to Pompey, his slave slew him and then killed himself, so that he might not survive his master.[39]

From what was this slave saving his master? Vettius Scato probably did not need to fear dishonorable treatment by his Roman captors; Cicero described in an anecdote how he, during his early military service, observed a parley between Scato and the consul Cn. Pompeius Strabo in which the men's former ties of friendship loomed large. This meeting took place in 89, the same year as Scato's death. Cicero related how Scato asked the consul's brother, Sextus, "How shall I greet you?" and received the eloquent reply, "By my will, as a guest-friend; by necessity, as an enemy." What followed was a fair parley – not without dislike, but without fear.[40] Scato, the previous year,

---

37 See Dart (2011) for the close connection between military command and the political organization of the nascent Italian confederacy.

38 Sen. *Ben.* 3.23.5: *Vettius, praetor Marsorum, ducebatur ad Romanum imperatorem; servus eius gladium militi illi ipsi, a quo trahebatur, eduxit et primum dominum occidit, deinde: "Tempus est," inquit, "me et mihi consulere! iam dominum manu misi," atque ita traiecit se uno ictu. Da mihi quemquam, qui magnificentius dominum servaverit.*

39 Macrob. *Sat.* 1.24: *Vide in hac fortuna etiam magnanimitatem exitum mortis ludibrio praeferentem. C. Vettium Pelignum Italicensem, comprehensum a cohortibus suis ut Pompeio traderetur, servus eius occidit ac se, ne domino superstes fieret, interemit.*

40 Cic. *Phil.* 12.27: *Quem cum Scato salutasset, quem te appellem, inquit. At ille voluntate hospitem, necessitate hostem. Erat in illo colloquio aequitas; nullus timor, nulla suberat suspicio, mediocre etiam odium.* ("When Scato had greeted him, he added: 'What am I to call you?' And the other replied: '"Guest-friend" by my choosing, "enemy" by necessity.' There was fair play at that parley; no covert fear, no suspicion; even the hostility was not extreme"). It is of course relevant that Cicero is contrasting this meeting of foes with his own experience of Antony; on *aequitas* in war and parley, see Brennan (2012, 485–86). On *hospitium* among Romans and Italians, see Fronda (2010, 316–20); Fronda (2011, 232–33, 237–39, 246–52), with references; on the specific relation to the Social War, Lomas (2012).

had defeated the consul, L. Julius Caesar, and also ambushed and killed the other consul, P. Rutilius Rufus, at the Tolenus River.[41] This battle cost the lives of so many notable Romans that, according to Appian (*B Civ.* 1.43.195), both Romans and Italians ceased repatriating their dead because of the effect of the funerals on morale. It is remarkable to consider Cicero's representation of honorable enemies in this light, and we might thus suspect that Vettius' death was enabled by those who wished either to spare him the experience of walking captive in a triumph or to spare his onetime hosts the sight.

We might compare Vettius' absence from a Roman triumph with the presence, in the triumph of Cn. Pompeius Strabo, of a certain P. Ventidius Bassus. We can say virtually nothing about his or his family's role during the war.[42] Bassus was paraded, along with his mother, as a captive, but went on to run a successful business providing mules to the Roman army. He gained the patronage of Julius Caesar and ultimately attained not only curule office but also a command against the Parthians (under Antony); he may even have had a descendant of the Marsic leader Poppaedius Silo as his legate there.[43] His victories in the East led him to celebrate Rome's only triumph over the Parthians. Bassus is a truly exceptional figure who benefited from a time of revolutionary change at Rome, but it is worth dwelling on the power of his story (for which we have five good sources[44]): carried in his mother's arms, a prisoner of a short, brutal, incomprehensible war, this child of defeat grew – as a Roman – into a man so honored that he was awarded a public funeral.

---

41 The defeat of Caesar is noted by Livy, *Per.* 73; App. *B Civ.* 1.41.182; Oros. 5.18.11; that of Rutilius by Livy, *Per.* 73; Ovid *Fasti* 6.563–566; Vell. Pat. 2.16.4; Flor. 2.6.11–12; App. *B Civ.* 1.43.191–94; Cass. Dio 98.1–2; Obseq. 55; Eutr. 5.3.2; Oros. 5.18.11–13. A fragment of Sisenna, quoted by Nonius (*FRHist* F11), is often taken to refer to the defeat at the Tolenus but cannot be placed with precision; see further Briscoe's commentary at *FRHist* 3. 372–3.

42 Cassius Dio suggests that this Bassus himself fought in the war, in contrast to other sources which describe him as a child at the time of the triumph (43.51.4–5, though the name requires emendation); Sisenna (F19 *FRHist*) refers to a Bassus who may be his father (*Sisenna historiarum libro iii, Bassus assiduitate indulgitate victus*), and the two may have been conflated. No Ventidius appears among lists of Italian commanders in Velleius (2.16.1) or Appian (*B Civ.* 1.40.181); see further Dart (2014, 223–24).

43 Bassus is the best known of the children of the Social War, but he does not reflect an isolated phenomenon. Compare Q. Salvidienus, consul designate for 39, whom Cassius Dio charged with humble origins as a shepherd (48.33.1–2) and whom Wiseman (1964, 130) plausibly identifies with a family known in Peltuinum and Aveia, in the territory of the Vestini.

44 On Bassus: Livy, *Per.* 127, 128; Tac. *Hist.* 5.9; Plut. *Ant.* 34; Cass. Dio 49.10.1–2, 39–41; Fest. *Brev.* 18; Eutr. 7.5; as captive in the triumphal parade of Cn. Pompeius in 89: Vell. Pat. 2.65.3; Val. Max. 6.9.9; Plin. *HN* 7.135; Gell. *NA* 15.4; Cass. Dio 21.3; further brief mentions in other sources. See also Fronda (this volume) for paraded individuals' role in the reception of Roman triumphal processions.

Conclusions

In the face of the manifold difficulties that attend the pursuit of a causally linked and chronologically coherent campaign history of the Social War, it is worth considering that our evidence gives us, more generously, something of at least equal value. In contrast to the tropes of civil war, where we, for example, find brother against brother, tales from the Social War repeatedly ask us to confront the more visceral juxtaposition of parents and children and the more socially complicated connections between slaves and owners.[45] The point is simple: violence among polities is ground-proofed in violence within families. Dominance, moreover, is not inherently defined as the ability to enact greater violence; the choice of whose lives to save is both the greatest and the worst of powers, and the most illusory conceit.

We know, as historians, that arguments *ex silentio* are weak at best. We cannot know how much the Romans wrote about the Social War, and we cannot realistically posit reticence in Livy or Sisenna when it comes to their probable approach to the horrors of the war. Even when we can read a continuous narrative of a war, though, as with Livy and the Hannibalic War or Appian and Spain in the later second century, the protagonists' motivations, the strategic causal connections, and the processes by which events were remembered and transmitted are often both elusive and fragmented. For a war like that between Rome and its putative Italian allies, in which the stakes were so high and the issues so complex, we cannot always compensate adequately for the dearth of data in pursuit of a coherent narrative. We are well served by accepting the limits of our knowledge and thence exploiting what we do have on its own terms. In this, perhaps, the *ad Herennium* is our best source, as it tells the next generation to stop trying to one-up their neighbors' dead: to stop fighting the war in the theater of memory. That was one front on which, given the hometowns of so many of Rome's leading authors and statesmen, the Romans could not win.

---

45 On the importance of fraternal relationships in Roman military thought, see Armstrong (2013b).

14 *SPQR* SNAFU

Indiscipline and internal conflict in the late Republic*

Lee L. Brice

Introduction

The period between 91 and 30 was marked by numerous conflicts, some of which were internal to Rome. It began with the broader conflict of the Social War in 91, which became a part of the civil war that erupted in 88. Internal conflicts continued after Sulla captured Rome in 82, as Lepidus revolted in 78, and the anti-Marian mopping up campaigns continued until 72. There was additional conflict between 77 and 49, but most of it was outside Italy and unconnected to civil war. The second phase of civil wars began in 49, when Caesar crossed the Rubicon, and lasted until 45 when he won at Munda. The last phase of civil wars started in 44 after Caesar's assassination and continued with varying intensity until 30. It was a busy period on and off the battlefield, full of conflict.

These years also stand out for the numerous reports of breakdowns in Roman military discipline. Indeed, despite the Roman legions' reputation for discipline,[1] ancient sources attest to seventy six incidents during the internal conflicts between 91 and 30 of both collective and individual indiscipline in Roman armies – including conspiracy, mutiny, expressions of grievances, as well as insubordination in all its varieties, especially defection.[2] These occurrences were not limited to common soldiers, but often included officers

* Versions of this chapter were presented in Sydney, Australia, and Auckland, Wellington, and Christchurch, New Zealand. I am grateful to audience members for their insightful comments. I am also thankful to Dominic Machado, Brian Turner, and the anonymous peer reviewers who examined earlier versions of this chapter and made helpful comments. The errors that remain are entirely my own. Thank you Nate for all your support and counsel over the years, and all you have done for ancient history. All dates are BC unless otherwise noted.

1 The modern reputation of the Roman legions for steady discipline in battle has not been diminished by the events of the first century. Present in popular histories in books, videos, and films, as well as numerous academic works until recently; see Brice (2015b). Messer (1920, 160–62), the starting point for examinations of indiscipline in the Republic, traces the early historiographical origins of the reputation. For Roman appeals to the soldiers' lack of discipline in combat to explain military defeats, see Rosenstein (1990, 92–114).

2 On the varieties of indiscipline see Brice (2015a) with citations.

DOI: 10.4324/9781351063500-14
This chapter has been made available under a CC-BY-NC-ND license.

at various levels. The number of cases of military disobedience for this period of just over 60 years exceeds that in any previous century.

Ancient authors made a connection between civil war and a breakdown in discipline, unsurprisingly stressing moral decline as a root cause.[3] Modern scholars, too, have posited a direct link between the civil wars (and other political aspects) and the apparent rise in indiscipline. Arthur Keaveney is typical in generalizing, "acts of disobedience ... are to be found, in the main, in the periods of civil strife."[4] Historians have linked the indiscipline of the civil war period to the political instability of the state, Marius' alleged recruiting reforms, a heightened political awareness on the part of soldiers, the emergence of "client armies," or a combination of these factors.

This chapter takes a somewhat different perspective. There is little doubt that political instability and persistent internal conflicts in the first century created a more chaotic atmosphere, which in turn contributed to the apparent increasing frequency of military indiscipline, but this recognition does not really help us understand specific reasons for these incidents of indiscipline. We need to take into account other issues – the amount of combat as a contributing factor, the varieties of incidents that occur, and the fact that the unusual number of extant sources may skew our impression of military disobedience. Closer analysis of the individual episodes of indiscipline, which are relatively well documented in the ancient sources, reveals that these outbreaks were not the product of soldiers' political awareness. Rather, they were triggered by the same forces that drove most forms of military disobedience in other periods – complex social and military problems that were more normal than the admittedly abnormal political period in which they occurred.

Before proceeding, a note on the nature of the sources. As hinted at above, it may be objected that a rise in the frequency of reported instances of indiscipline is, at least in part, the product of a distortion in the sources: we have more sources for the first century, and so it stands to reason that more notices of indiscipline for this period survive to us. This is certainly a possibility. Although indiscipline is reported in earlier periods,[5] our limited sources make it impossible to prove for certain, one way or the other, whether military disobedience was just as frequent in previous centuries or if these incidents increased (significantly) in the first century. However, we do find a similar rise in military disobedience during later periods of internal conflict.[6] Moreover, the density of references to first-century indiscipline

---

3 For example, see Sall. *Cat.* 11, App. *B Civ.* 5.17, and Plut. *Sull.* 12.7–9.

4 Keaveney (2007, 91). See also Brunt (1988, 257); Gruen (1995, 373–74); Blois (1987); Blois (2000); Wolff (2009); Machado (2017, 140–49). However, Chrissanthos (1999, 162–63), concludes, based on highly selective definitions, that there was not more indiscipline during the civil wars.

5 Livy 2.24–27; Dion. Hal. *Ant. Rom.* 6.23–24. Messer (1920) and Chrissanthos (1999, 13–27) provide some early incidents with sources.

6 For example, Wellesley (2000).

is striking, especially incidents of conspiracy and mass defection that appear on a scale simply not seen prior to the conflicts of the first century. On the whole, the impression given by the sources is probably more or less accurate, and the years 91–30 saw an increase in military indiscipline of different sorts.

Yet even if the evidence is skewed, this does not significantly impact the main threads of this chapter. I am more interested in exploring the different kinds of indiscipline, and to identify the conditions that induced such insubordination, mutiny, etc. For this, the large number of often detailed accounts of military disobedience in the first century offer a rich corpus of evidence to determine the characteristics of, and the critical factors contributing to, individual outbreaks. In other words, skewed or not, sources for the first century have much to inform us about how and why Roman soldiers disobeyed orders or engaged in indiscipline. This chapter will show that salient conditions leading to cases of collective indiscipline in the first century maps well onto broader patterns identified in comparative research on organizational behavior. In other words, the military indiscipline of the era of Roman internal conflict was not particularly distinct except in its scale and frequency.

Conditions encouraging discipline and indiscipline

Before discussing examples of indiscipline in more detail, several points need to be emphasized – what discipline is and why (generally speaking) otherwise disciplined soldiers break ranks. Discipline is more than punishments and it is certainly more than following commands.[7] It is a network of control with physical, mental, and social elements that are reinforced with a matrix of positive rewards and negative sanctions. Discipline is not just about victory. All of these components make it possible to maintain military order, allowing officers and commanders to manage and direct soldiers both during and outside of combat. Since soldiers are trained to work together under stress and in the efficient use of weapons, outbreaks of indiscipline have the potential to become violent. Organized violence outside of combat was to be avoided.[8]

A combination of Roman institutional rules and circumstances made violations of discipline less likely. An important part of these was the *sacramentum* or oath. Because soldiers took the *sacramentum*, vowing to remain loyal and obedient to commands for the duration of the conflict, any violation of discipline carried religious and legal sanctions, including the possibility of death. In practice, officers often adjusted penalties to match the severity of the offense, though there were commanders known for harsh

---

7 *Contra* Carney (2015, 28), who defined discipline too narrowly, as "obedience to commands."

8 Brice (2011, 36–41), with references; Brice (2015b, 103–4).

punishments. Supplementing these sanctions were the cultural norms and mores of Roman society, and especially of the military.[9] The fact that measures like the oath tended to be accepted by soldiers as mores worked against outbreaks of severe indiscipline.

Yet sometimes soldiers and officers still act out. Comparative research in organizational behavior has identified conditions that increase the likelihood that soldiers would join in collective indiscipline such as mutinies and mass defection. These factors include, in no particular order, homogeneity of the organization, cohesive identity, lack of opportunities to report complaints, loss of faith in upper-level leadership, and the likelihood that authorities will not punish participants.[10] These conditions, identified for modern organizations, can also be found in accounts of the Roman army. They were not prerequisites for soldiers' participation in collective indiscipline.[11] Rather, they help us understand why some soldiers participated and others did not, as collective action by all members of a unit is extremely uncommon.

Roman legions, in general, lacked homogeneity. Republican legions were made up of citizens and they were all soldiers, but that was often the limit of their homogeneity. Sometimes, legions were filled with recruits from the same city or region (for example, Picenum, Rome) but, as Brunt and Blois have shown, more often soldiers came from diverse localities, with different skill levels and varying socio-economic backgrounds (for example, *capite censi* or propertied, rural or urban). After the Social War, when all Italians became Roman citizens, those serving in the legions might now have come from diverse cultural backgrounds.[12] However, while lack of homogeneity can contribute to individuals joining in indiscipline, there is little evidence that this played a serious role in contributing to Roman military disobedience before the imperial period.[13] Indeed, since Roman legions were heterogeneous as a rule, it would be difficult to isolate this factor as decisive in any particular case of indiscipline.[14] Moreover, homogeneity was no guarantee of discipline, as ancient authors consistently report that legions raised entirely from urban recruits were unruly and disordered.[15]

Cohesive identity exists in military units (ancient and modern) in terms of "vertical" and "horizontal" forces. "Vertical cohesion" is the strength of

---

9 Phang (2008); Keaveney (2007, 71–77); Milne (2009, 8–42); Brice (2011, 36–39). See Milne in this volume on related cultural mores, such as status and the military.

10 Lammers (1969, 564–66).

11 *Contra* Chrissanthos (1999, 174).

12 Brunt (1971a); Brunt (1988); Blois (2007, 166).

13 Regional origin may have played a role in the mutiny of 49 against Caesar, but little other Republican evidence exists.

14 There is little other evidence for heterogeneity as a cause of indiscipline before the Imperial period.

15 See, for example, App. *B Civ.* 1.85, though unruliness of urban recruits may have been a literary trope: Eckstein (1995, 164–74); Fulkerson (2013, 162–63); cf. Wheeler (1996).

links between individuals and others ranked above or below them, up to the commander, while "horizontal cohesion" was the bond among soldiers in a unit. Strong vertical cohesion was helpful in maintaining control and managing the legion in battle, but it was a problem if a commander wanted to lead his army against the state. Durable horizontal cohesion may have been helpful in combat, but it could also become a problem when there was potential disobedience, as soldiers might identify more strongly with fellow soldiers in opposition to officers and thus engage in collective indiscipline.[16] Strong horizontal cohesion was not necessarily a problem, but if combined with weak vertical cohesion it increased the odds of collective indiscipline breaking out.

Loss of confidence in superior officers was connected with weak vertical cohesion and military disobedience. Surveys of modern mutinies have corroborated a pattern observed in reports of some Roman incidents of collective indiscipline: lack of confidence in the ability of a commander contributed to soldiers' decision to join indiscipline.[17] This pattern applies to officers, including centurions and tribunes, as well as soldiers. Because centurions and subalterns (for example, *optiones*) were often drawn from the ranks,[18] they were more likely than other officers to have a strong connection with soldiers – arguably a sense of horizontal cohesion, despite being officers and so above them in the hierarchy – and so were liable to join soldiers in collective indiscipline, and even take a leading role. Officers and soldiers often acted out when they suspected that a commander's poor leadership would produce unnecessary losses (for example, Caes. *BGall.* 1.39–40). Even when they seemed aware that certain commanders, like Caesar, were usually winners, they had doubts about his guidance if his leadership appeared to waver, as happened in 58 and 47.[19] Examples of this pattern abound during the late Republic, as will be discussed further below.

Roman soldiers had limited means to report complaints and seek change without breaking discipline. Elections in Rome were a way to change some unpopular officers, but this was of limited usefulness as a commander might be prorogued by the senate. Roman citizens did indeed have the customary right of free speech and were not afraid to use it with officers and commanders. Soldiers presented complaints and demands through centurions or the military tribunes, or they could also speak directly to the commander

---

16 MacMullen (1984); Lendon (2004, 445–46). Horizontal cohesion is not strictly equivalent to "social cohesion," see Armstrong (2016b, esp. 110–17), who defines the terms and addresses recent research into military cohesion and what causes it. Machado (2017, 7–75) addresses issues of homogeneity and cohesion within the context of community.

17 Lammers (1969, 565–66).

18 See McCall in this volume.

19 Caes. *BGall.* 1.39–41; App. *B Civ.* 2.92–94; see Chrissanthos (1999, 169–70). Milne (2009) analyzes this factor from the perspective of the soldiers' need for information and commanders' responses.

one-on-one if he was present.[20] The commander's assembly (*contio*) was a public venue where soldiers are supposed to have occasionally shouted views.[21] Since we tend to hear only about incidents when soldiers' demands were not met, it is difficult to know how effective these avenues were. Whatever their actual effectiveness, if the soldiers perceived that these channels were ineffective then this would have made at least some soldiers more inclined to break ranks.

Perceived unlikelihood of punishment is the last important condition found to increase participation in collective indiscipline. It has been correctly observed that during the late Republic most soldiers and officers who participated in military disobedience went unpunished, regardless of their offense.[22] It is not surprising that few punishments were imposed during times of internal conflict, when competing sides sought support and needed manpower in the conflict.[23] Among all the possible contributors to soldiers' openness to indiscipline, this one is most consistently present during the civil wars. Recognizing the instability of the period helps us understand why so many soldiers could follow orders and fight hard much of the time during the late Republic and also engage in collective military disobedience.

Episodes of military indiscipline, 91–30: a survey

Reported military indiscipline during the late Republic took several forms, the most serious of which was military conspiracy and murder by officers. What usually distinguishes these attacks from mutinies is that military conspiracies primarily involved officers. In one example, Q. Pompeius Rufus was the victim. He was sent by Sulla in 88 to take command of Cn. Pompeius Strabo's legions in Picenum. Strabo surrendered command and departed, but soon afterwards soldiers murdered Rufus during a sacrifice and Strabo then returned to the camp and resumed command. The anonymous assassins fled, and no one was ever punished.[24] This incident was murder, not mutiny. The fact that the incident occurred within a day of Rufus' arrival

---

20 For example, Caes. *BCiv.* 1.64.2 (communicating through centurions and military tribunes); Caes. *BGall.* 7.17 (direct appeal to the commander).

21 For example: Livy 40.35–6, 44.3.9; Caes. *BGall.* 3.24, 6.36; *BCiv.* 1.7, 71–72, 2.33, 3.6; see Chrissanthos (2004) for additional references; cf. Pina Polo (1995, 213–16); and Machado (2017, 144–48), with historiography. "Freedom of speech" is a red herring, however, since the state (and army) had no mechanism for limiting speech. In a military context, the commander might decide speech was contrary to discipline, as seen in the speech Livy puts in Paullus' mouth (44.34.2), but incidents where soldiers were punished for speaking out are rare: for example, Vell. Pat. 2.81.2; App. *B Civ.* 5.128–29; see Brice (2011, 49–50) for discussion.

22 Messer (1920, 159 n. 3); Chrissanthos (1999, 2, 164–65); Blois (2007, 175).

23 Ancient authors blame corruption rather than the lack of punishments for the poor discipline: App. *B Civ.* 5.17; Val. Max. 9.7.3; Plut. *Sull.* 12.7–9.

24 Livy, *Per.* 77; Vell. Pat. 2.20; App. *B Civ.* 1.63; Val. Max. 9.7.2; Chrissanthos (1999, 51–52).

means that specific frustration with Rufus had no time to mature. Strabo, an opportunist and opponent of Sulla, was able to retake command of the army immediately without resistance, as if he had been waiting for such a development. Events strongly suggest that Strabo had conspired with his officers to have Rufus murdered.[25] Plutarch (*Pomp.* 3.1) reports that Cinna conspired the following year to have Strabo and his son (the future Pompey Magnus) murdered by one of their officers, though this story is probably a fiction.[26] P. Tullius Albinovanus did not act alone in 82 when he murdered some of his new fellow officers, before defecting to Sulla.[27] Q. Sertorius was murdered in 72 as a result of a conspiracy by M. Perperna Vento and other officers in Spain.[28] Conspiracy was not limited to the early period of civil war either. Antony revealed in 40 that Octavian's legate in Gaul, Q. Salvidienus Rufus, had been conspiring in secret against him. Octavian reacted swiftly and eliminated Rufus.[29]

Mutinies were more common though, and thus a more dangerous threat to commanders. These outbreaks of collective opposition to established leadership often involved lower-level officers as well as soldiers. A mutinous or restive mood in a unit was not a mutiny. Mutinies were open resistance, and always carried the threat of real violence. Indeed, in this period, mutinies often turned violent. It is this violence, threatened or actual, that separates mutinies from mass defections or desertions. There were twenty four reported mutinies during the period of internal conflict and eight additional mutinies unconnected to the civil wars.[30] Rather than discuss them individually at length, it will suffice here to touch on each briefly to provide a sense of scale and cause.

The potential for collective violence made mutinies more dangerous and disruptive; a few even resulted in murder of the commander. A. Postumius Albinus was trying to respond to a mutiny in camp when he was stoned to death by his soldiers. They had mutinied against Albinus because either his haughty behavior had engendered their hatred (which may be shorthand for being perceived as a harsh disciplinarian) or because he was accused of disloyalty as a result of not pursuing the campaign aggressively enough.[31]

---

25 Keaveney (2007, 79); Blois' discussions (2000, 2007) of the importance of middle cadre officers is useful for understanding this murder as a conspiracy.

26 Badian (1958, 239 n. 16); Hillman (1996); *contra* Lovano (2002, 66–67); Keaveney (2007, 79).

27 App. *B Civ.* 1.91. Although Appian does not mention other conspirators, Albinovanus would have needed the assistance of his own original officers since none of them were victims and he had already lost his legions.

28 Livy, *Per.* 92; Sall. *Hist.* 3.81–83; Vell. Pat. 2.30; Plut. *Sert.* 26–27.5, *Pomp.* 20.3; App. *B Civ.* 1.113–4, *Hisp.* 101; Chrissanthos (1999, 64–65).

29 Vell. Pat. 2.76.4; App. *B Civ.* 5.66; Cass. Dio 48.33.1.

30 *Contra* Chrissanthos (1999, 162–63), who counts only ten incidents.

31 Livy, *Per.* 75; Val. Max. 9.8.3; Plut. *Sull.* 6.9; Polyaenus 8.9.1; Oros. 5.18.22; Keaveney (2007, 77–78) (who finds the accusations of treachery plausible); Gruen (1995, 373); Rosenstein (1990, 197–98).

It also has been suggested that Albinus' reputation as a poor military leader contributed to the outbreak of mutiny.[32] L. Valerius Flaccus was murdered in 86 after dismissing his popular quaestor, C. Flavius Fimbria, whom he had accused of criminality and undermining his authority.[33] Chrissanthos blamed the mutiny on Flaccus' reputation for tough discipline and weak leadership, but Lovano has shown that Flaccus was a qualified and capable leader, and that Fimbria undermined Flaccus' leadership and incited the mutiny.[34] L. Cornelius Cinna was stabbed to death while attempting to re-store discipline during a mutiny in 84. The mutiny broke out while he was in the process of taking newly levied troops from Ancona to Dalmatia, when he tried to restore order among already restive legions at Ancona. This mu-tiny seems to have been in response to the poor conditions for crossing the Adriatic, combined with the soldiers' reasonable fear of facing Sulla's expe-rienced legions.[35] C. Papirius Carbo was murdered in 80 during a mutiny caused by his attempts to impose traditional discipline.[36] But the death of a commander was an unusual outcome of mutinies in any period.

During the first, or Sullan phase, of internal conflicts, there were five ad-ditional mutinies or near-mutinies that did not involve killing a commander. The first occurred in 89 due to L. Porcius Cato's efforts to impose harsh dis-cipline on new legions.[37] A mutiny in 87 occurred when Cn. Pompeius Strabo died and the soldiers were unsure of his replacement.[38] While mopping-up in North Africa in late 82, Pompey's soldiers reportedly became mutinous when Sulla recalled him to Rome. Pompey resorted to threatening suicide if the men did not obey him.[39] In 78, during the revolt of Mam. Aemilius Lepidus, the legions in Mutina under M. Junius Brutus mutinied and forced Brutus to surrender himself to Pompey.[40] Finally, after Sertorius' death in 72, numerous soldiers mutinied and some defected from Perperna, leading to his defeat against Pompey.[41]

Six mutinies occurred during the second, or Caesarian phase, of internal conflict. In early 49, the soldiers under Cn. Domitius Ahenobarbus muti-nied, forcing their commander to surrender to Caesar.[42] Later in the same

32 Chrissanthos (1999, 50–51, 101–10).
33 Livy, *Per.* 82; Strabo 13.1.27; Vell. Pat. 2.24.1; Plut. *Luc.* 7.2; App. *Mithr.* 51–52; Oros. 6.2.9.
34 Chrissanthos (1999, 53–56); Lovano (2002, 98–99); Lintott (1971, 696–701); Blois (2007, 172–73); Wolff (2013).
35 App. *B Civ.* 1.77–78; Plut. *Pomp.* 5.1–2, *Sert.* 6.1, *Crass.* 6.1; Livy, *Per.* 83, 85; Vel. Pat. 2.25; Oros. 5.19.24; Chrissanthos (1999, 58–59); Lovano (2002, 108–10).
36 Val Max. 9.7.3; Gran. Licin. 36.8; Chrissanthos (1999); Keaveney (2007, 78).
37 Cass. Dio Fr. 100. Chrissanthos (1999, 49–50); Keaveney (2007, 76).
38 Vell. Pat. 2.21.2; Livy, *Per.* 79; Plut. *Pomp.* 3.1–3; App. *B Civ.* 1.68; Hillman (1996).
39 Plut. *Pomp.* 13; Seager (2002, 28–29) suggests that Pompey encouraged the incident.
40 Plut. *Pomp.* 16.1–6.
41 Vell. Pat. 2.30; Plut. *Sert.* 26; App. *B Civ.* 1.113–4.
42 Caes. *BCiv.* 1.20–23; Cic. *Att.* 4.8, 7.13, 23–24, 26, 8.1, 3, 6–7, 11–12; Livy, *Per.* 109; App. *B Civ.* 2.38; Cass. Dio 41.10–11; Chrissanthos (1999, 70–72).

year, Caesar's *Legio IX* mutinied in Placentia over bonuses and the continuing war.[43] The following year in Spain, a legion of local recruits mutinied and were joined first by *Legio II* and then also by parts of *Legio V*. All of these units were provisionally under the overall command of the Caesarian legate Q. Cassius Longinus, against whose leadership they mutinied separately, and then even initiated a local war before settling down when Cassius was replaced.[44] In 47, as many as eight legions in Campania mutinied over Caesar's absence and the continuing war, and only personal intervention restored some of them to order.[45] According to Cassius Dio, some of Caesar's veterans mutinied over the spoils connected with his triumph in 46, resulting in Caesar executing one soldier.[46] The legions in Spain that had mutinied previously (local, II, and V), did so again in mid-46, this time against C. Trebonius, and they subsequently defected to the Pompeians.[47]

The third period of civil war (44–30) experienced nine actual mutinies, two of which represent the largest mutinies of the late Republic. In late 44, some of M. Antonius' legions assembled in Brundisium mutinied after they were incited by Octavian's agents. Although Antonius executed some of the ringleaders, changed officers in the restive legions, and promised bounties, two legions (*Martia* and IV) later mutinied near Tibur in November and then defected to Octavian.[48] In spring 43, there was a mutiny in a legion (XIV) serving M. Junius Brutus. Cicero's correspondence suggests that this mutiny was triggered by C. Antonius' presence in Brutus' camp.[49] Later in 43, several legions under Brutus mutinied as a result of incitement by Antonius, which then required Brutus to execute some of the ringleaders and dismiss others while moving Antonius to Apollonia under lock and key.[50] After Cassius' suicide in 42, some of his soldiers were restive, though no mutiny broke out.[51] In 41, veterans, impatient for discharge, land, and bonuses, mutinied in the Campus Martius, killing a centurion before Octavian could bring them into

43 Caes. *BCiv.* 1.87, 2.22; Suet. *Iul.* 69, Plut. *Caes.* 37; App. *B Civ.* 2.47–48; Front. *Strat.* 4.5.2; Cass. Dio 41.26, 35–36; Chrissanthos (2001, 67–69); Brice (2015b, 108–10).

44 Caes. *BAlex.* 56.4; 57.1–3; Cass. Dio 43.29. 1. These are treated as two different mutinies, since cohorts of *Legio V* mutinied separately under different officers.

45 Caes. *BAfr.* 19.3, 28, 54.; Cic. *Att.* 11.10, 16, 20–22; Suet. *Iul.* 70; App. *B Civ.* 2.92–94; Plut. *Caes.* 51; Cass. Dio 42.52; Chrissanthos (2001).

46 Cass. Dio 43.24.3; this report may be a misunderstanding of events in the same year in North Africa reported in Caes. *BAfr.* 54.

47 Caes. *BHisp.* 7.5, 12.2; Cass. Dio 43.29. This mutiny is treated as one incident since they acted in concert.

48 Cic. *Att.* 14.11.2, 15.12, 16.8, *Fam.* 11.7.2, 12.23.2, *Brut.* 1.3.1, *Phil.*, 3.4, 6–7, 14, 24, 30, 4.5–6, 5.4, 22–23, 52–53, 11.20, 12.12, and 13.18–19; Nic. Dam. 139; Vell. Pat. 2.61.2; App. *B Civ.* 3.31, 39–40, 43–46; Cass. Dio 45.12–13. Because it is all closely connected, I treat these episodes as one mutiny, and not as two; see also Keaveney (2007, 85–86).

49 Cic. *Brut.* 1.2.

50 App. *B Civ.* 3.79; Plut. *Brut.* 26.3–8; Cass. Dio 47.23.2–24.2.

51 Plut. *Brut.* 45.3.

line.[52] The same year, legions marching to Spain under Salvidienus Rufus mutinied in Placentia and did not resume the march until they had extorted money from the town.[53] Two legions raised by L. Antonius in 41 mutinied in Alba and expelled their officers, but were restored to order by Antonius' promises.[54] Following the victory at Naulochus in 36, Octavian deposed Lepidus and took over his army. This was followed by an enormous mutiny of more than twenty legions, as the soldiers sought discharge and rewards.[55] During the Illyrian campaign of 34, an unidentified legion mutinied against Octavian for uncertain reasons. After he disarmed them, with the help of the rest of the army, he discharged the legion without benefits.[56] Finally, in the autumn of 31 after the victory at Actium and Octavian's simultaneous discharge of many legions, there was another large mutiny over discharges and bonuses, as some men wanted bonuses and others were disappointed at being excluded from the final campaign against Egypt.[57]

Not every report of mutiny in this phase is of equal value. During 43, Octavian and Lepidus each reported mutiny in their own army. These incidents resulted in each of them having to accept what turned out to be a beneficial settlement (i.e., Octavian became consul and Lepidus allied with M. Antonius).[58] Both of these so-called mutinies should be treated with skepticism, given that the commanders who reported them had allowed or incited them, and benefited most.

There were, between 78 and 44, eight mutinies or near-mutinies unconnected with the civil war. These incidents show similar causes and outcomes to the mutinies that occurred during the internal conflicts. M. Atilius Bulbus, an officer serving in Illyria between 78 and 76 under C. Cosconius, incited mutiny in a legion for unknown causes. Order was restored and the officer was tried in Rome.[59] In 77–75, C. Atilius Paetus Staienus, quaestor in the army of Mam. Aemilius Lepidus in Gaul incited a mutiny for unknown reasons. Lepidus was able to restore order and the quaestor was later tried successfully in Rome.[60] One of four legions serving under C. Scribonius Curio in Dyrrachium in 75 mutinied and refused to march with the

---

52 App. *B Civ.* 5.16; Cass. Dio 48.9.1–3.
53 Cass. Dio 48.10.1.
54 App. *B Civ.* 5.30.
55 Vell. Pat. 2.80.2–81.2; App. *B Civ.* 5.122–29; Cass. Dio 49.11.2–14.6. See Keaveney (2007, 88–90); Brice (2011, 45–50).
56 Cass. Dio 49.34–35; Brice (2015b, 112–14).
57 Cass. Dio 51.3.1–3, 4.2; Suet. *Aug.* 17; Brice (2011, 51–54). Keaveney (2007, 89) incorrectly asserts "there was no place for mutiny or desertion" after the events of 36, since the following wars were "foreign." The conflicts after 36 clearly took place in the context of internal conflict and civil war.
58 Octavian: App. *B Civ.* 3.86–88; Cass. Dio 46.43.6–46; Suet. *Aug* 26.1; Cic. *Brut.* 1.15, 18. Lepidus: Cic. *Fam.* 10.23, 24, 35; Plut. *Ant.* 18.2–3.
59 Cic. *Clu.* 97.
60 Cic. *Clu.* 99; Livy, *Per.* 76; Diod. Sic. 37.2.

rest of the army on campaign against the local tribes. Curio responded by breaking up the legion and dispersing its men among his other legions and then proceeded with the campaign. During the winter of 68, while pursuing Mithridates into Armenia, two of L. Licinius Lucullus' legions (II and III) mutinied and refused to continue campaigning. Lucullus withdrew to Nisibis for the winter, but the following spring the legions mutinied again rather than campaign against Mithridates near Zela. Soon afterwards Cn. Pompey replaced Lucullus in command and the mutinous soldiers were discharged without punishment.[61] When Caesar had assembled his legions at Vesontio in 58 for a first campaign against the Germans, numerous legions and officers became mutinous and threatened to refuse orders. Caesar persuaded the officers that their fears were groundless and thus resolved the near-mutiny.[62] During the retreat following the defeat at Carrhae, Crassus' remaining soldiers became mutinous and threatened him when he initially resisted a Parthian truce offer. Crassus' death, after accepting the parlay, resolved the mutiny. Two years later, A. Claudius Pulcher's legions in Cilicia mutinied in 51 due to the lack of pay.[63] This incident ended once Claudius found funds to pay the men. Finally, in 45 after the end of the civil war legions commanded by Sex. Julius Caesar in Syria mutinied and killed their commander due to incitement by Q. Caecilius Bassus, who took command.[64]

Mass desertion was another prominent form of collective indiscipline during this period, posing a problem for virtually every commander. Of course, desertion, even collectively, was not a new form of insubordination.[65] Roman soldiers either changing sides or going home had occurred in previous wars. There were so many desertions between 89 and 30 that it is necessary to cover them all in groups instead of discretely. Desertions traditionally occurred because soldiers (individuals or groups) sought a means of getting out of service or became separated from their unit for a variety of reasons. During the civil wars, there were still men who deserted to escape serving, as was the case in 82, when several legions serving under C. Marcius Censorinus went home after an ambush, and also in 49, when some Pompeian troops went home, and in 44 and 40, when some of Octavian's recruits went home after they learned they would be fighting Antonius.[66] More often,

---

61 Plut. *Luc.* 32–35; Dio 36.14; Front. *Str.* 2.1.14; Chrissanthos (1999, 110–26); Keaveney (2007, 85).

62 Caes. *BGall.* 1.39.

63 Cic. *Att.* 5.14–17; *Fam.* 3.6, 15.4.

64 Cic. *ad Fam.* 12.18.1; App. *B Civ.* 3. 77; Cass. Dio 47.26.3–27.1; Botermann (1968, 99–101); Keaveney (2007, 84).

65 Desertion is a form of indiscipline, categorized as insubordination even when it is collective, so long as it is not in connection with a mutiny; see Brice (2015a); see also Keaveney (2007, 79–92), Wolff (2009, 181–355); Machado (2017, 157–62); *contra* Chrissanthos (1999), who blends the two forms of indiscipline.

66 Censorinus: App. *B Civ.* 1.90. Pompeian deserters: Caes. *BCiv.* 1.12. Octavian in 44: App. *B Civ.* 3.42; in 40: App. *B Civ.* 5.57; Cass. Dio 48.28.1–2.

however, there was collective desertion from one side to another. Sometimes deserting troops first mutinied before deciding to switch sides, though, more typically, large numbers of soldiers changed sides without having mutinied beforehand. This usually occurred for one of two reasons: subornment or soldiers decided changing sides was preferable to fighting.

Subornment was an effective tactic used by numerous commanders. Inducements included promises of mercy, money, and loyalty. In 90, the rebel commander C. Papius Mutilus convinced the Roman garrison in Nola to join the Italian revolt by promising them mercy.[67] In 87, Cinna bribed the centurions and tribunes of A. Claudius Pulcher's legions to assist in luring the soldiers to his side, though he also appealed to the soldiers' loyalty to the *res publica*.[68] Sulla used cash offers in combination with other promises in 83 to acquire the soldiers of L. Cornelius Scipio.[69] The great effect of Caesar's offers of clemency throughout the conflict – sometimes coupled with pay – has long been recognized. Octavian used bounties and appeals to loyalty in undermining M. Antonius in 44 and against Lepidus in 36.[70] M. Antonius sent agents to suborn the troops of Decimus Brutus in 43, but they were caught (Cass. Dio 46.36.1). Appeals to soldiers' loyalty to the *res publica*, a former commander, or their fellow veterans were used by Cinna, Sulla, Caesar, M. Junius Brutus, C. Cassius Longinus, and Octavian. The success or failure of such appeals relied heavily on whether soldiers accepted the legitimacy of the commander to extend the statements. For example, Octavian and Antonius were certainly able to use their attachment to Caesar to appeal to their opponents' troops. Those who effectively used attachment to the *res publica* must have been seen by soldiers as legitimate agents for it to have worked so often.[71]

Yet, the sources more often report units changing sides on their own initiative, without subornment. This form of indiscipline occurred more often (though not exclusively) during the first phase of internal conflict, when the defecting soldiers and cavalry could not have known which side would win. Also, raw recruits and even experienced troops in smaller armies might choose not to stand against larger, more experienced forces. This dynamic is seen in numerous collective desertions, including the Fimbriani to Sulla in 86, the army of Domitius to Pompey in 81, numerous soldiers from Sertorius to Metellus in 74, Pompeian cohorts to Caesar in Italy in 49, C. Antonius' legion to Brutus in 43, the Syrian army of Bassus, Crispus,

67 Livy, *Per.* 73; App. *B Civ.* 1.42.
68 Livy, *Per.* 79; Vell. Pat. 2.20.4; App. *B Civ.* 1.65–66; Chrissanthos (1999, 52–53); Blois (2007, 171–72).
69 Livy, *Per.* 85; Vell. Pat. 2.25; App. *B Civ.* 1.85–86; Plut. *Sull.* 28, *Pomp.* 7; Chrissanthos (1999, 59); Keaveney (2007, 79).
70 Antonius: Cic. *Att.* 16.8, *Fam.* 12.23; Nic. Dam. 139; App. *B Civ.* 3.31, 39–40, 43–44. Lepidus: Livy, *Per.* 129; Vell. Pat. 2.80; App. *B Civ.* 5.124–27; Cass. Dio 49.12; Oros. 6.20.6.
71 Keaveney (2007, 37–55, 74–75); Morstein-Marx (2011); Machado (2017, 144, 154–55).

and Murcius to Cassius in 43/42, and from M. Antonius to Octavian in 31 and 30.[72] It was a problem all commanders had to remain vigilant against. In two cases, collective desertions resulted in the death of a former commander: Fimbria, by suicide in 83, and Decimus Brutus in 43, trying to evade capture later.[73] There is little evidence to show that soldiers engaged in collective desertion merely to change their commanding officer, a motivation seen in mutinies. Changing commanders was an outcome, not the cause of collective desertion. Rather, they often seem to have reckoned that switching sides gave them a better chance to survive. Broadly speaking, then, they sought better service conditions.

Another type of collective indiscipline is made up of expressions of grievances. These are non-violent, collective acts in which soldiers seek to protect their interests without resorting to mutiny or desertion.[74] There were six such expressions of grievances in our period; two of which stand out as particularly important. When L. Antonius and Fulvia provoked conflict in 41, veteran officers of both Octavian and M. Antonius tried to maintain peace (and preserve their benefits) by intervening on their own initiative to force a settlement on the leaders. There were several efforts, the last of which was called the Pact of Teanum; but each reportedly failed due to L. Antonius' aggressiveness.[75] In this case, the officers and soldiers acted out of their own interests without resorting to violence.

The most important expression of grievances occurred soon afterward, in 40, when M. Antonius returned to Brundisium and civil war between the triumvirs appeared to be imminent. Soldiers on both sides were not happy about renewed conflict and, having learned from the failure in 41, acted against their commanders' orders and refused to serve or fight. They engaged in indiscipline to protect their benefits and worked peacefully (and successfully) to force a settlement, the so-called Pact of Brundisium.[76] This was the most important of the various expressions of grievances during the three phases of civil war because of its scale and success.

---

72 Fimbriani: Livy, *Per.* 80–83; App. *Mithr.* 60; Plut. *Sull.* 25.1, Cass. Dio 31.104. Domitius: Plut. *Pomp.* 10.7. Caesar: Caes. *BCiv.* 1.12–20. C. Antonius: Cic. *Brut.* 1.2, 2.5.3, App. *B Civ.* 3.79; Cass. Dio 47.21.4–7. Syrian armies: App. *B Civ.* 4.57–59; Cass. Dio 47.28. M. Antonius: Plut. *Ant.* 76.1–2; Cass. Dio 51.5.6, 10.4. These examples do not comprise a comprehensive list. Keaveney (2007, 90) argued that there were no more desertions after 36, but Antonius' forces were still deserting in 31 and 30.

73 Fimbria: Livy, *Per.* 83; App. *Mith.* 60; Plut. *Sull.* 25.1. Brutus: Vell. Pat. 2.64; App. *B Civ.* 3.97; Cass. Dio 46.53.

74 Brice (2015a, 72–73). See also MacMullen (1984, 449–50, 454–55); Goldsworthy (1996, 147–48, 264); Rowe (2002, 155–64).

75 App. *B Civ.* 5.20–23; Cass. Dio 48.10.2–12.4; Chrissanthos (1999, 143–53). That the veterans supported Octavian rather than L. Antonius supports the assertion against the latter.

76 Vell. Pat. 2.76.2; Plut. *Ant.* 30.2–4; App. *B Civ.* 5.55–60, 63–64; Cass. Dio 48.27.4–5, 28.1–2; Chrissanthos (1999, 153–59).

Lastly, another form of collective indiscipline,[77] collective insubordination, included cowardice, failure to follow orders, fighting within a unit, and similar group disturbances. Fear was not new. It has already been pointed out that numerous desertions occurred because a unit or an army preferred defecting to a stronger opponent over death in battle. Cowardice in the midst of battle was a different kind of insubordination, since it occurred during combat. Reported cases include soldiers serving under Sulla in 86, Crassus in 71, Caesar in 48 and 45, Calvinus in 39, Antonius in 36, and Octavian in 34.[78] There were also incidents that seldom survive in sources, when whole units disobeyed orders, pillaged the communities they were supposed to protect, or fought among themselves. Commanders who encountered this kind of collective indiscipline include Cinna in 87, Pompey in 81 and 48, Sertorius in 76, Caesar in 52, and Antonius in 36.[79] There is no doubt that the comparatively extensive sources for the period have contributed to our knowledge of these kinds of insubordination, which must have been a fairly normal occurrence in other periods.

The causes of group indiscipline, 91–30

Reviewing the detailed record left by our sources, we find diverse indiscipline. There were four military conspiracies, three of which removed commanders during the first phase of the civil wars, and one revealed prematurely during the third phase. Thirty-two mutinies or near-mutinies erupted.[80] Five commanders died as a result of mutinies. There were more than forty incidents of collective desertions, which affected the equivalent of more than seventy legions. Two commanders, and several other officers, died due to desertions, only one by suicide. One of these commanders' deaths happened during the first phase and the other during the second phase of the civil wars. At least six expressions of grievances occurred, most of which happened during the third phase of the civil wars. There were also ten episodes of group anticipatory fear or cowardice in combat, and multiple reported cases of other collective insubordination. This is an impressive accumulation of disturbances, but one that stands out for its quantity, not its novelty. Indeed, when we

---

77 Besides collective indiscipline, there were numerous individual offenses, including desertion, dereliction of duty, disorderly conduct (drunk on duty or fighting), and other minor offenses. Such instances of individual insubordination rarely survive into the historical record and, given this rarity, will not be discussed in this chapter.

78 Sulla: Plut. *Sull.* 16.3, 5. Caesar: Caes. *BCiv.* 69, 74; App. *B Civ.* 2.61–63, 103. Calvinus: Vell. Pat. 2.78.3; Cass. Dio 48.42. Antonius: Plut. *Ant.* 39.7. Octavian: Cass. Dio 49.38.4; Suet. *Aug.* 24.

79 Cinna: App. *B Civ.* 1.74 and Plut. *Ser.* 3. Pompey: Plut. *Pomp.* 10.7 and 67–68. Sertorius: App. *B Civ.* 1.109. Caesar: Caes. *BGall.* 45.8–9, 47.3, 52. Antonius: Plut. *Ant.* 48.2–5.

80 Keaveney (2007, 91) argued that mutiny was rare after 80, but of these 32 mutinies, all but seven occurred after 80, and the majority of other outbreaks of indiscipline also predominate after 80.

examine the sources for these incidents, we can identify several common features. These generally match the conditions identified as common to many forms of institutional indiscipline, including military disobedience (as discussed above). Moreover, as demonstrated by the incidents unconnected to the internal conflicts, in most cases there is nothing to suggest that these factors were exacerbated by, let alone distinct to, periods of civil war.

Numerous incidents can be connected with the soldiers' perceptions of their commanders as being too harsh, or militarily weak, or both. The men under Cato, Albinus, and Carbo seem to have resented the commanders' efforts to impose more rigid disciplinary standards. Mutinies against Brutus in 78, Crassus in 53, and L. Domitius in 48, were the result of soldiers' perception of militarily weak leadership. The mutinies in Spain in 47 and 46 were connected with perceptions of poor leadership by the governor, and by extension Caesar. Soldiers claimed the mutinies in 49 and 47 were, in part, a result of doubts about Caesar's leadership – though there was a more immediate actual cause. In several cases, after commanders died (Strabo, Sertorius, and Cassius), insubordination occurred due, at least in part, to a lack of confidence in future leadership. The mutiny against Octavian in 36 was due, in part, to his leadership – having shown a willingness to subvert discipline in other men's armies, he had failed to appreciate its importance in his own. The mutiny against Cinna and many collective desertions were cases in which soldiers lacked confidence in victory against stronger armies. These were, again in part, a reflection on their leadership. Fear in (or on the eve of) battle caused several incidents of insubordination during the civil wars as well, and it was another typical cause of insubordination during periods of internal stability. Subornment, as a means of encouraging desertions, was effective in part because of the perceived weakness of the suborned units' leadership. Yet, although subornment and the incitement to mutiny or desertion was a regular issue during the civil wars, the cases of Bulbus, Staienus, and Bassus show it also occurred during more stable periods. Likewise, perception of weak or poor leadership, including uncertainty about future leaders, was a cause of indiscipline during the period of internal conflict, but as the case of Caesar in 58 demonstrates, it was not a new problem or limited to the civil wars.[81]

Cohesive identity and homogeneity also appear in some outbreaks of indiscipline. Sulla's, Pompey's, and Caesar's legions fought together for long enough to each develop a strong cohesive identity that made them resistant to subornment and desertion, but in the case of Caesar's troops it contributed to the size of the mutiny in 47. After Caesar's death, it contributed to these units' tendency to participate in mutinies and expressions of grievances

---

81 The most famous earlier example was when soldiers mutinied at Sucro in 206, while Scipio had been ill (Livy 28.24–29, 32; Polyb. 11.25–33; App. *Hisp.* 32–38). The mutinies of AD 14 were triggered by the death of Augustus and uncertainty about the new emperor, but these are just three of numerous incidents.

against Octavian and Antonius. The mutinous legions of Strabo, Albinus, Cato, Cinna (in 84), and Caesar (in 49) were largely recruited from areas that made them relatively more homogeneous, and this quality would also have encouraged the development of stronger horizontal cohesion. New units lacked time to create strong vertical cohesive bonds with commanders (for example, Cinna in 84, Domitius in 49, and L. Antonius in 41), which contributed to the likelihood of soldiers joining in indiscipline. At the same time, the wider political instability of the period may have damaged vertical cohesion within some units and so further increased the likelihood of indiscipline. But such damage could occur through casualties and was true of any serious combat and not limited to civil wars. Strong horizontal cohesion with weak vertical cohesion in a unit is potentially a bad combination for maintaining discipline. Thus, while these factors may contribute to soldiers joining in military disobedience, they may have played only a slightly more significant role in periods of civil war, when leaders' legitimacy was in question.

During this period, soldiers did not gain new avenues for complaining about conditions. The *contio* remained the only public means of communicating directly with commanders. It is little wonder, then, that assemblies were the place where mutinous soldiers stoned unpopular commanders: it was the only place where groups had an opportunity to gather with the commander. It should be stressed that the assemblies themselves did not cause mutinies: the causes of each mutiny (for example, conditions of service, incitement, exhaustion) were already present and the mobilization of participants had begun before the *contio*. As one of the few channels angry soldiers had to voice complaints, and virtually the only opportunity for them to assemble in large numbers where they knew their commander would be present, assemblies were a natural *locus* for discontent to erupt.[82] Centurions and tribunes remained another important channel for communicating with the commander, but they often joined in mutinies and desertions. The weakened vertical cohesion contributed to the failure of communication channels.

Lastly, limited punishment was the second most common feature of incidents during the internal conflict. Not surprisingly, cases of desertion were generally not punished. The only possible exception was the Fimbriani, whom Sulla left in Asia when he returned to Italy.[83] Conspirators also faced no punishment in three of the four cases identified. Out of 32 mutinies, few resulted in the punishment of mutineers. After the mutiny against Cato, a ringleader was sent to Rome for punishment. The mutiny against Caesar in 49 resulted in decimation of the ringleaders, and the mutinous behavior in connection with his triumph resulted in one soldier possibly

---

82   *Contra* Chrissanthos (2004, 359–63) and Machado (2017, 154–60), arguing for a connection between rough Roman politics and the *contio* that *caused* several mutinies (e.g., Cato, Albinus, Flaccus).
83   Badian (1958, 201).

getting executed. During the North Africa campaign, Caesar dismissed six men dishonorably for prior indiscipline. M. Antonius, in 44, executed some ringleaders of the mutiny in Brundisium, and M. Brutus, in 43/42, punished ringleaders with death. During 34, Octavian discharged a mutinous legion without benefits. All five of the cowardice episodes resulted in punishment of the unit and two of the three cases (87 and 76) of collective insubordination led to death for all participants.

Put simply, during the civil wars of the late Republic, most mutineers went unpunished. This is a strikingly different pattern than found in other periods. Indeed, a review of previous mutinies, those that occurred in this period outside the civil wars, and those later, finds that punishment was the more typical outcome, both for mutinies and for indiscipline generally.[84] Given the competition for manpower during the period of the civil wars and the unstable legitimacy of commanders' authority, they could not risk alienating soldiers. In addition, several commanders had been killed trying to assert traditional discipline. Soldiers probably knew this, since it was a period of active warfare and it is to be expected that word of responses to mutinies and desertions traveled among the soldiers who were not entirely cut off from society. This knowledge – that they could get away with some kinds of indiscipline – was strong encouragement to many who might not normally have participated, and thus stimulated further indiscipline. The likelihood that indiscipline would go unpunished appears to be one of the main conditions contributing to group indiscipline (as identified at the beginning of this chapter) that increased in the context of internal conflicts. This, then, at least partly explains the apparent increase in the frequency of military indiscipline in this period.

We might also consider other underlying causes for these episodes of military disobedience which link indiscipline to civil war. Some scholars suggest that the frequency of collective desertion and the novel employment of mutiny to eliminate commanders derive directly from political instability during the civil wars.[85] The political instability also made the most important examples of expressions of grievances, in 41 and 40, possible. While these two incidents led to truces (one unsuccessful, the other more lasting), they were nevertheless a form of indiscipline. It is clear that the political instability flowing from the civil wars contributed to a climate that encouraged much of the collective indiscipline discussed here – especially the conspiracies and desertions. We cannot conclude that similar incidents would not have occurred without civil war, but the unstable environment brought traditional legitimacy and loyalty into question, and emboldened soldiers to

---

84 Chrissanthos (1999, 126); Brice (2011, 2015b); *contra* Keaveney (2007, 91), who claimed incorrectly that "[c]ertainly in all periods mutineers went largely unpunished ...".

85 Gruen (1995, 372–73); Keaveney (2007, 71–92); Machado (2017, 144–48).

act out knowing punishment was unlikely. Such linkage – while plausible – is difficult to prove.

Thus far, we have generally discussed more structural factors or conditions that encouraged group indiscipline, but, in many cases, the immediate cause of indiscipline was unconnected with political or social instability. The stress caused by extensive combat service, for example, is a much more tangible consideration that led to a number of episodes of collective indiscipline and contributed to others. In addition to traditional explanations for military disobedience – political instability and poor conditions of service (pay, officers, quarters) – extended combat service under difficult conditions, even if soldiers are victorious, contributes significantly to the likelihood of outbreaks of indiscipline. The two mutinies against Caesar (49 and 47) were due to combat exhaustion (though the withholding of bounties provided a pretext). In both mutinies, the units involved had been fighting for eight or more years, and in both cases they complained about Caesar continuing the war.[86] Sulla's soldiers had not been fighting for as long when he marched on Rome, and then immediately took them east – but the siege and campaigns in Greece were strenuous, and there was some indiscipline reported among Sulla's troops in 87. It is also probable that the constant marching and fighting contributed directly to the defection of Fimbria's soldiers to Sulla in 85, as it also did to their participation in the mutiny against Lucullus in 68. Sertorius' army, which had been fighting intermittently in Italy and Spain for more than a decade, also experienced outbreaks of indiscipline in which exhaustion was a contributory or causal factor. It is likely that some of the other indiscipline in the second and third phases of the period (especially in 36 and 31) arose, in part, from the exhaustion of long, stressful, combat experience. Exhaustion-based military disobedience was not unique to the civil war. Since extended combat experience was a function of the amount of fighting in which soldiers engaged, periods of civil war were similar to vigorous external wars (as opposed to garrison duty) in the overall amount of combat an individual soldier saw. It shows up in numerous incidents, including the mutiny at Sucro under Scipio in 206, under Galba and Tappulus in 199, and under Lucullus in 68, each a case where soldiers involved had served for extended periods.[87] Commanders were aware of the problems created by combat exhaustion: super-annuated soldiers were retired in 46, 36, and 31, in an effort to avoid renewed indiscipline. Given the relationship between combat exhaustion and indiscipline, it is little surprise that we find more indiscipline in periods of vigorous fighting, regardless of whether it is an external war or a civil war.

Conditions of service were another typical cause of indiscipline, for example when soldiers upset at lack of pay turned to group indiscipline. Caesar (in

---

86 Chrissanthos (2001); Brice (2015b).
87 Scipio: Polyb. 11.25–33; App. *Hisp.* 32–38; Galba and Tappulus: Livy 25.5–7, 26.21, 29.24, 31.8, 14, 32.1, 8–9, 28, 34.52; summaries in Chrissanthos (1999). On imperial mutinies see Brice (2011).

48 and 47), Octavian (in 43, 41, 36, and 31), and M. Antonius (43, 41, and 34) each encountered indiscipline caused by issues connected with lack of pay or rewards. In each case, they had to raise what funds they could and promise to pay the rest later. But this too was a common cause of mutinies throughout Roman military history, and not limited to civil war: it was a cause or contributing factor in mutinies against Scipio (206), Lucullus (68 and 67), and Claudius in 51.

As should be clear, the climate of instability in the period – social, military, and political – encouraged indiscipline of all sorts, but most incidents discussed in the chapter were the result of the same forces that had always caused military unrest: service conditions, weak leadership, and limited fear of punishment. Of all of the conditions discussed, perhaps only the last (lack of fear of punishment) was distinctly more pronounced in this period because of the distinct nature of mobilization and competition for military resources in civil war.

Conclusions

Historians have suggested a variety of causes for the unusual amount of indiscipline during Rome's internal conflict in the first century; many center on the alleged politicization of the soldiers. Thus, one argument has been that the indiscipline was due to Marius' alleged reforms, which supposedly meant the regular recruitment of the poor to the legions, thus creating client armies who needed their commander for economic security. A further component of this argument is that Sulla figured out how he could use the client army for his own ends, and that by doing so he set the precedent for Pompey, Caesar, and the Triumvirs.[88] This is a seductive thesis, and it has driven much discussion, but it is problematic for several reasons. As Gauthier argues in this volume, the reforms attached to Marius are not all his responsibility and they did not usher in armies made up entirely of the desperate poor. Moreover, Gruen, Blois, and Keaveney have effectively demonstrated that the "client army thesis" is deeply flawed as an explanation for the indiscipline and the fall of the Republic, despite the continued appearance of the term in general works and textbooks on the late Republic.[89] Yet another thesis emphasizes that Roman citizen-soldiers became aware in the first century of their collective political power, and increasingly used this agency to protect their own interests – which, in the process, facilitated opportunistic leaders.[90] This argument, too, fails to explain the military indiscipline of

---

88 Premerstein (1937); Badian (1958); Gabba (1976).

89 Gruen (1995, 378–79); Blois (2000); Blois (2007); Keaveney (2007, 30–33).

90 Boterman (1968); Chrissanthos (1999); Keaveney (2007); Machado (2017). Mangiameli (2012) seems to be blending aspects of both the "political power thesis" and the "client army thesis," moving toward a synthesis. Morstein-Marx (2011) emphasizes political identity but also argues that we should see the army as a component of Roman politics and society, not a separate group.

this period. Adherents of the "political thesis" cannot demonstrate that soldiers used existing political institutions to protect their interests.[91] More importantly, their alleged political awareness cannot be shown to explain the vast majority of the numerous outbreaks of military indiscipline in the first century. Instead, the actual explanation for these episodes must be sought in more mundane forces: combat exhaustion, lack of confidence in leadership, money problems, conditions of service, and fear. An additional factor is the climate of instability arising from the internal conflict. This climate encouraged military disobedience by diminishing the likelihood of punishment and bringing legitimacy and loyalty into question. It was the cause of the unusual number of mass defections.

Finally, the extraordinary number of available sources give us an unusual view of this period. Although even this does not skew our data as much as one might suppose. Given the large number of mass defections in unique circumstances, indiscipline almost certainly occurred more often during this period than usual. But if we subtract those, the amount of military disobedience and insubordination was not so unusual. Given comparison with what we find in the Roman Empire, and in other regular militaries, we should accept that indiscipline was a regular problem throughout the history of the Roman military – an unavoidable function of the nature of military service.[92] Although it was a normal problem, that does not mean it was overwhelming or even usually large scale. Most of it was mundane – the standard varieties of insubordination, both collective and individual – and only occasionally serious. When the collective indiscipline is measured against the amount of actual combat, we recognize that most units remained disciplined – in battle, on the march, and in camp – most of the time. Indiscipline did not overwhelm the military or the state as long as it could respond under normal, and, as has been seen here, sometimes in abnormal circumstances. The Roman military's ability to absorb indiscipline and usually respond contributed to the effectiveness and resilience of Rome.

---

91 Chrissanthos and Keaveney both argue that the pacts of Teanum and Brundisium represent the high point of the soldiers' political power, but these truces were not constructed through political mechanisms or institutions; they make no sense outside of a military context, as a form of military indiscipline, the expressions of grievances.
92 On indiscipline as a normal problem for officers see: Brice (2011, 2015a, 2015b); Machado (2017).

15 From slave to citizen

The lessons of Servius Tullius*

Jack Wells

Introduction

A society's myths help it navigate the internal contradictions that develop as that society changes. In Republican Rome, war was arguably the most powerful engine of change, and perhaps the most important way that war changed Rome was by driving it to incorporate new peoples into the Roman community through conquest. This was done in a number of different ways, including granting various grades of citizenship to neighboring communities, and offering elite members of Italian allied cities Roman citizenship as an inducement to closer bonds of friendship. But the most difficult and awkward way that conquest changed the Roman social body derived from the practice of admitting freed slaves into the community through manumission. Rome obtained a significant percentage of its slaves through conquest, and so it had to find a way to incorporate its bitter enemies – who had not only been defeated in battle but also forced to endure servitude – into its own community.[1] This presented the Romans with a terrible contradiction: they despised slaves, and yet they made former slaves (including those descended from former enemies) their fellow citizens. They grappled with this contradiction and, to a certain extent, they resolved it to their liking through the vehicle of the legends they told about the origin of their city.

My purpose here is threefold. First, to explore the question of how the Roman community explained and justified, to itself and its slave population, the practice of granting citizenship to slaves after formal manumission, including those who had been enslaved after being captured in war. Second, to argue that Roman legends about the founding of the city – and in particular the stories about Rome's sixth *rex*, Servius Tullius – provided

---

* The author is very grateful to Jeremy Armstrong and Michael Fronda for their suggestions for improvement, as well as to the suggestions of two anonymous reviewers. Any remaining infelicities of thought or analysis are entirely my own. All dates are BC unless otherwise noted.
1 On the relationship between war and slavery in Rome, see Hopkins (1978, esp. 8–15). In this volume, see Fronda on appropriation (and misappropriation) of plunder, and Roselaar on the effect of spoils on the Italian economy.

DOI: 10.4324/9781351063500-15
This chapter has been made available under a CC-BY-NC-ND license.

a way for the Romans to explain why they incorporated captives and slaves into their community. The stories about Servius also taught both the Romans and their slaves the benefits and limitations of the offer of citizenship through manumission. Finally, to argue that the traditions about Servius reveal a striking ambiguity and tension in Roman attitudes toward this system – one which is reflected in Roman policy toward freedmen in the Republican period. The legends told about Servius informed slaves that after obtaining freedom they would be full members of the community but also scorned and marginalized; they might ascend to the heights of power based on their own merit, but there would be limits to what they could accomplish. In other words, they both could, and could not, be full members of the community. Ultimately, these mixed messages did exactly what the Romans would have liked: they acknowledged a strength of the Roman community, that is, its willingness to admit outsiders as citizens, but did so in a way that assuaged the anxieties of the freeborn population. Furthermore, these stories offered an incentive to freedmen to remain part of the community, yet also reminded them of the practical limits of what they were being offered.

I assume from the start that the historiography of early Rome served as a particular kind of myth.[2] This kind of myth acted, as Bruce Lincoln put it, as "ideology in narrative form,"[3] and, as Mary Midgley described, as "imaginative patterns, networks of powerful symbols that suggest particular ways of interpreting the world" which become "organic parts of our lives, cognitive and emotional habits, structures that shape our thinking."[4] Stories of early Rome communicated what it meant to be a Roman by providing good and bad examples (*exempla*), not simply by prescribing what ought to be done or not done, but by giving the framework for a discussion about the boundaries of good and bad behavior. The power of the myths to shape public opinion was magnified because the Romans themselves did not see their interpretations of Servius' stories as myth-creation, but as historical research into an actual past whose values should and did shape the attitudes of contemporary Romans. Much of the work done by Roman historians lay in the area of selecting a preferred version from among various forms and accounts of the stories that were available. The availability of variants encouraged the Romans to debate crucial communal issues that the stories tried to resolve.

Slavery and freedmen in the Roman Republic

In contrast to the rest of the ancient world, Rome was relatively open in admitting aliens to citizenship. Non-Romans could earn citizenship in many ways. It is usually assumed that, by the late Republic, if not earlier,

---

2 On myth, history, and Rome, see Scheid (1998, 148–49).
3 Lincoln (1999, 147–49). The example Lincoln gives of this phenomenon is the origins of the battle between Connaught and Ulster in the Cattle Raid of Cooley; he shows that the ideological purpose of the myth was to establish and justify the superiority of men over women.
4 Midgley (2011, 1, 5).

magistrates in Latin communities were awarded Roman citizenship upon taking their magistracy; this helped bind these elites more closely to Rome by giving Rome a base of supporters in every Latin community.[5] Most extraordinary is the Roman decision to offer citizenship to former slaves.[6] This right was hallowed, according to Roman custom; the Romans believed that it had been instituted during the regal period or very early in the Republic.[7] The traditionally minded Romans were thus very reluctant to eliminate it, even given the "snobbish" attitude toward slaves and freedman that characterized Roman society in later periods, and so it persevered throughout Roman history into the Empire.

The Roman successes in warfare during the middle and late Republic made the question of what to do about freedmen exceptionally important. During the conquests of the second century, thousands of slaves were captured and brought to Italy to work in mines, plantations, and households. Though many of the thousands of slaves taken in the wars probably died in captivity, many thousands more would have eventually been manumitted by their owners. Since the Romans admitted freed slaves into their citizen body, social tension between the newcomers and the older citizens – particularly those suffering in poverty – would be significant. The awkwardness of the situation is demonstrated by the example of Scipio Aemilianus, on the occasion when a crowd in Rome reacted angrily to his condemnation of the Gracchi. He reminded his listeners that Italy was only their stepmother and told them, when they got even angrier, that he was not going to be intimidated by those whom he had once carried off to Rome in chains as prisoners of war.[8] The logic of the retort tells us much about the assumptions that elite Romans had about the population of the city in the late second century. Aemilianus' jibe only works if the crowd he was trying to impress saw many of the inhabitants of the city as freedmen or their children, felt that these freedmen and families were engaged in un-Roman political activities, and thought that they needed to be reminded of their second-class status. Clearly, there were tensions between the two groups.

---

5 The *ius adipiscendae civitatis Romanae per magistratum* is usually assumed to have been granted to Latins c. 124 in the context of the revolt of Fregellae and M. Fulvius Flaccus' citizenship proposal, though some scholars suggest that it was instituted only in the late Republic or early Imperial period. See: Tibiletti (1953); Piper (1988); Keaveney (2005, 84–86); Coşkun (2009b), all arguing for c. 124 as the likely date. *Contra* Bradeen (1959); Mouritsen (1998, 99–108), arguing for a later date.

6 For modern discussions on the motives of manumission, see: Treggiari (1969, 11–20); Hopkins (1978, 115–32). For a review of the legal and extralegal forms of manumission, see: Duff (1928, 23–30); Treggiari (1969, 20–31); Bradley (1987, 81–112); Mouritsen (2011, 10–15). For an overview of Roman citizenship, Sherwin-White (1973) remains the central study.

7 Livy 2. 5.8–10 places this the first year of the Republic, in a very different context: a slave named, conveniently enough, Vindicius, was freed after revealing a plot against the newly formed Republic. As we will see, however, Dionysius (Dion. Hal. *Rom. Ant.* 4.23) explicitly stated that it was instituted in the reign of Servius Tullius, during the regal period.

8 Vell. Pat. 2.2.4; Val. Max. 6.2.3; [Aur. Vic.] *De vir. ill.* 58.8.

Such tensions can be also be illustrated by the difficulty that the community had in figuring out how precisely freed slaves should vote in the tribal assembly. The first reported attempt to define the matter occurred in 312, when Appius Claudius allegedly distributed freedmen among all the Roman tribes – making them politically no different from their freeborn citizens.[9] According to Livy, this innovation was repealed in 304 by the censors of that year, who restricted freedmen to the four urban tribes.[10] In a similar notice, Livy (*Per.* 20) again reports that freedmen, who had hitherto been allowed to vote in all of the tribes, were restricted to the four urban tribes, in reference to the census of 230, 225, or 220. The later report suggests that the system implemented in 304 had been modified, implying that the problem of how to enroll freedmen into voting tribes persisted throughout the third century.[11]

One of the censors of 169, Ti. Sempronius Gracchus (father of the famous Gracchi), tried to remove freedmen from the tribal assembly altogether, but was prevented by a veto of his colleague C. Claudius Pulcher. Gracchus did, however, manage to restrict their vote to a single tribe chosen by lot.[12] The timing of this needs to be noted: this was in the middle of the Third Macedonian War (171–68), famous because at the end of that war L. Aemilius Paulus enslaved an unusually large number of Epirots and brought them to Italy. It is possible that Gracchus and the senators, who supported his plan, were anticipating an influx of slaves into Italy, and the issue of citizenship status may have been on peoples' minds. Nevertheless, at some point before 88, freedmen appear to have been redistributed back into the four urban tribes.[13]

The electoral status of freedmen reemerged as an issue in the first century, with attempts to redistribute freedmen among the voting tribes recorded in 88, 66, and 52. In all three cases, the goal was to expand the number of tribes in which freedmen could vote. The first two attempts were made by tribunes: Sulpicius in 88 and Manilius in 66. The last was a proposal of Clodius, candidate for the praetorship in 52. Sulpicius seems to have been successful, but his reforms were undone by Sulla. Manilius' proposal was

---

9 Livy 9.46.10–11. Livy also condemned Appius Claudius for debasing the senate by appointing, or trying to appoint, sons of freedmen to its ranks, which again gestures to the "snobbish" attitude against freedmen.

10 Livy 9.46.12–14; see also Treggiari (1969, 38–44).

11 It is also possible that Livy's earlier report is an anachronism reflecting later squabbles. Even so, the accounts of it being placed in the early Republic suggest that the Romans saw the issue as long term, divisive, and difficult to resolve. For a full discussion of the reforms of Appius Claudius and the subsequent efforts to restrict the voting of freedmen, Oakley (2005a, 628–35).

12 Livy 45.15.1–9. The passage is lacunose and difficult to interpret; see Briscoe (2012, 648–50) for discussion. At some point before 88, freedmen seem to have been redistributed back into the four urban tribes: Livy, *Per.* 77.

13 See Treggiari (1969, 45–49).

blocked by the senate after it had been passed through violence, an indication of how controversial the matter was, and Clodius' was abandoned after his murder.[14] Both the earnestness of the attempts, and the passionate opposition they aroused, demonstrate the explosiveness of the issue. The Romans of the late Republic clearly had difficulty working out the exact electoral status of fellow citizens who had been slaves, reflecting deep divisions among both the political elite who were making the proposals and the community as a whole.

Also, during the first century, the question of the freedmen was linked to another issue relating to the integration of new citizens into the community: how to incorporate Italians newly enfranchised after the Social War (91–88). The huge influx of newly enfranchised citizens meant that the questions of who got to be a citizen, and how new citizens should behave, became vitally important public discussions. Arguably, the civil war between Marius and Sulla was triggered by this very question. When the senate was moving toward confining the newly enfranchised Italians to a restricted number of tribes to limit their voting power,[15] the tribune Sulpicius countered by trying to have them fully and fairly distributed into the 35 existing tribes. When he needed additional political support, he allied with Marius, whose price was the command against Mithridates – the issue that led to six years of civil war, further bloodshed, enslavement, and violence.[16]

The introduction of large numbers of newly enfranchised Italians into the Roman body politic caused massive disruption to the political system. It took until the census of 70 to implement a workable procedure for counting the newly enrolled citizens.[17] Even then, Roman identity remained a contested concept. Cicero (*de Leg.* 2.5) famously claimed that all Italians had two fatherlands, their hometown and Rome. Habinek notes that local, Italian, and Roman identity were in tension down to the time of Augustus, whose invocation of *tota italia/cuncta Italia* reflected an aspirational slogan rather than a reality.[18] Just as the "Romanness" of newly enfranchised

14 Sulpicius: Livy, *Per.* 77; Manilius: Asc. 64C; Clodius: Cic. *Mil.* 87. See also: Treggiari (1969, 49–51); Lintott (1999, 52); Tatum (1999, 236–39).
15 Appian (*B Civ.* 1.49) reported that the newly enfranchised Latins were to be placed in ten new tribes, while Velleius Paterculus (2.20.2) reported that the new citizens were to be placed in eight tribes, but does not specify whether these were new or existing tribes. An obscure fragment of Sisenna (*FRHist* 26 F38) mentions the proposal to create two new tribes. For discussion, see Salmon (1958, 179–84) and Mouritsen (1998, 162–63).
16 On Sulpicius, see Lintott (1971) and Powell (1990).
17 The relatively small increase in the number of citizens enrolled in the census of 86, from the previous census of 115, shows that there were serious obstacles to extending the census effectively to the entire peninsula. Yet, the significantly larger number of citizens reported for the census of 70 suggests that the logistical difficulties at least had been addressed.
18 Habinek (1998, 88–102); for example, *RGDA* 25.3–4: *Iuravit in mea verba tota Italia sponte sua et me belli, quo vici ad Actium, ducem depoposcit* ("The whole of Italy voluntarily took oath of allegiance to me and demanded me as its leader in the war in which I was victorious at Actium").

Italians remained a question until the reign of Augustus, so to the position of freedmen remained a vexing issue through the end of the Republic. Indeed, Augustus specifically addressed manumission and the status of freedmen in his legislative program. I will come back to this later. For the moment, let us turn away from grand policies and toward the child of a captive woman who made good: the story of Servius Tullius.

Servius' origins and ascension

According to tradition, Servius Tullius ruled over Rome from 578 to 534. The legends of Servius were old, and it is difficult to trace their exact origins in Roman sources. The main narrative accounts about early Rome are preserved most importantly in the works of Livy and his contemporary Dionysius of Halicarnassus, who were both writing in the late first century. Yet, these accounts derive from earlier sources from the Republican era. Servius begins appearing in literature no later than the middle of the second century[19]– that is, roughly contemporaneous with Rome's dramatic Mediterranean expansion and what is generally assumed to have been a corresponding massive growth of the slave population in Italy.[20] This is not to say he was not a fixture of Roman history at an earlier point, as of course the middle of the secondary century is also when our source tradition picks up, but he was certainly part of the narrative by this time. Furthermore, he seemed to have a very mixed reputation, being either a champion of ordinary citizens or a tyrant who seized power illegitimately, or both.[21]

The story of Servius Tullius begins with his birth and childhood in the household of Rome's fifth *rex*, Tarquinius Priscus, either as a slave or as the free child of a mother who had been captured in battle. Dionysius of Halicarnassus leaves his reader no doubt but that Servius was slave-born. He states that,

> There lived at Corniculum, a city of the Latin nation, a man of the royal family named Tullius, who was married to Ocrisia, a woman far

---

19 Gabba (1991, 164), for instance, noted that Servius was already seen as a good *rex* who worked for the best interests of his people in the play *Brutus* by the second century playwright Accius. Cf. Ogilvie (1965, 156–57); Richard (1987); Ridley (2014).

20 Brunt (1971a, 67, 124) estimated that there were 500,000 slaves in Italy in 225, and approximately 3 million by 1 BC. Rosenstein (2004, 9–12) and Scheidel (1999a) rightly note that Brunt's estimates were based on little more than guesses; see also Scheidel (2005). They suggest, plausibly, that the slave population of Italy may have been considerably lower than is usually assumed. Nevertheless, approximately 388,000 new slaves (mainly war captives) are attested in literary sources between 217 and 167, which figure surely undercounts the actual number of new slaves in those years: see Ziolkowski (1986); see also Rosenstein (2004, 171–73). The largest percentage of these new slaves would have been introduced after the notorious sack of Epirus in 167.

21 Richard (1987, 210–14).

excelling all the other women in Corniculum in both beauty and modesty. When this city was taken by the Romans, Tullius himself was slain while fighting, and Ocrisia, then with child, was selected from the spoils and taken by King Tarquinius, who gave her to his wife. She, having been informed of everything that related to this woman, freed her soon afterwards and continued to treat her with kindness and honour above all other women. While Ocrisia was yet a slave she bore a son, to whom, when he had left the nursery, she gave the name of Tullius, from his father, as his proper and family name, and also that of Servius as his common and first name, from her own condition, since she had been a slave when she had given birth to him.[22]

At the outset, then, Servius is an ambiguous figure. In Dionysius' version, he is the son of a Latin noble who died fighting for his homeland. He is also born to a captive woman, a position that the Romans would perceive as slavery.[23] Already, he does not fit neatly into categories; he is both noble and slave, foreign and Roman, which makes him an excellent person on which to build discussions of slavery and freedom.

His position in the household of the *rex* is also significant. He was the child of a captive in war, but not a captive himself, and he was raised in the *domus* of his master, not sent to work in mines or fields. Roman slaves were more likely to be freed and admitted to citizenship if they served in the household of the master, because they had more opportunity to obtain his gratitude and affection, and because their hard work was more likely to be seen by him. Servius' early life thus corresponds to the lives of many freedmen who achieved citizenship; the house slaves, working closely with their masters, were more likely both to hear stories about Servius and see him as a model for their own lives than slaves working hard labor in, for instance, the mines.[24]

The double nature of Servius is amplified when we look at the role of the gods in his origin story. Dionysius records a legend in which Servius' captive mother, Ocrisia, saw a phallus rising up out of the hearth as she was about

---

22 Dion. Hal. *Ant. Rom.* 4.1.2–3: ἐν Κορνικόλῳ πόλει τοῦ Λατίνων ἔθνους ἀνήρ τις ἐκ τοῦ βασιλείου γένους Τύλλιος ὄνομα γυναικὶ συνῆν Ὀκρισίᾳ καλλίστῃ τε καὶ σωφρονεστάτῃ τῶν ἐν Κορνικόλῳ γυναικῶν. αὐτὸς μὲν οὖν ὁ Τύλλιος, ὅθ' ἡ πόλις ὑπὸ Ῥωμαίων κατελαμβάνετο, μαχόμενος ἀποθνήσκει, τὴν δ' Ὀκρισίαν ἐγκύμονα οὖσαν ἐξαίρετον ἐκ τῶν λαφύρων λαμβάνει Ταρκύνιος ὁ βασιλεὺς καὶ δίδωσι δωρεὰν τῇ ἑαυτοῦ γυναικί. μαθοῦσα δ' ἐκείνη πάντα τὰ περὶ τὴν ἄνθρωπον οὐ πολλοῖς χρόνοις ὕστερον ἐλευθέραν αὐτὴν ἀφίησι καὶ πασῶν μάλιστα γυναικῶν ἀσπαζομένη τε καὶ τιμῶσα διετέλεσεν. ἐκ ταύτης γίνεται τῆς Ὀκρισίας ἔτι δουλευούσης παιδίον, ᾧ τίθεται τραφέντι ἡ μήτηρ τὸ μὲν ἴδιόν τε καὶ συγγενικὸν ὄνομα Τύλλιον ἐπὶ τοῦ πατρός, τὸ δὲ κοινὸν καὶ προσηγορικὸν Σερούιον ἐπὶ τῆς ἰδίας τύχης, ὅτι δουλεύουσα ἔτεκεν αὐτόν.

23 Vasaly (2015, 47).

24 Treggari (1969, 9–17); but see Bradley (1987, 95–112) on the many difficulties facing a slave seeking manumission.

to offer sacrificial cakes to the goddess Vesta. Everyone was puzzled about what to do with this until Tanaquil, Tarquin's wife, explained that whoever mated with the phallus and conceived would produce a superhuman child; as the prodigy first appeared to Ocrisia, she ought to do the honors. And so, Servius is conceived, still the child of a slave woman, but with divine parentage.[25] And so another ambiguity is introduced: Servius becomes part divine, part slave.[26]

Both Dionysius and Livy record that Servius was given another divine sign of future greatness. While the young Servius slept, a miraculous flame appeared around his head, burning brightly but leaving the sleeping child unharmed. As the entire household became alarmed by this amazing event, Tanaquil intervened to prevent a slave from dousing the child with water and announced that this child, being brought up "in such humble circumstances" (according to Livy) was being signaled out for greatness: he will be the guardian of the house of Tarquin, and therefore should be raised as a member of the family.[27] This rapid transition from low to elevated status calls attention to his unusual and ambiguous character.

Servius' ambiguities appear to have made him an awkward figure for the Romans to deal with, for whereas the Greek author Dionysius clearly asserted that Servius was born to a slave, the Roman author Livy preferred a variant of the story that denied Servius' servile status. Servius' mother, so Livy asserts, was the wife of yet another Servius Tullius, leader of Corniculum, whose city had been captured by Tarquinius Priscus. The queen was about to be sold into slavery when Tanaquil recognized her and spared her. She brought the queen into her own household, where the child was raised.[28] This version was possibly created as a result of the hostility of Roman citizens to former slaves, who were despised for their low birth and condition, particularly if these freedmen seemed to be putting on airs of social importance.[29] Livy's alternative, however, while avoiding the embarrassment of calling him a slave, does not put the Roman *rex* in a much better social position. Servius' mother was taken captive, which the Romans explicitly associated with slavery.[30] Servius was born to her after her capture and,

---

25 Dion. Hal. *Ant Rom.* 4.2.1–2; cf. Ov. *Fast.* 6.30–635; Plin. *HN* 36.70; Plut. *De fort. Rom.* 10.

26 Ogilivie (1965, 156–58); Ridley (2014); Vasaly (2015, 46–47).

27 Livy 1.39.1–4: *quem tam humili cultu educamus.* Cf. Cic. *Rep.* 2.37. Dionysius' account (*Ant. Rom.* 4.2.4) is substantially the same, though Tanaquil's prediction of his future greatness is omitted as unnecessary, as Tanaquil had just predicted this from the appearance of the fiery phallus. According to Plutarch (*De fort. Rom.* 10), Valerias Antias reported a version in which the fiery sign is given to Servius much later in life, after his wife has died. This would have the effect of deemphasizing Servius' servile status much as Livy prefers to do (see below).

28 Livy 1.39. See also the discussion in Ogilvie (1965, 159–60).

29 Most famously displayed by Horace in *Epode* 4, where Horace, himself the son of a freedman, excoriated a freedman who fancied himself as being socially important.

30 Bradley (1994, esp. 10–30).

given that status followed the mother in Roman law,[31] the questions of his exact status might remain in the mind of a reader. The emendation reveals the unease that the story of slave-born *rex* caused for many members of the Roman community.

This unease is also seen in the fragmentary discussion of Servius Tullius in Cicero's *De re publica*. Cicero reports that Servius was apparently born to a slave woman and a client of the *rex* Tarquinius Priscus.[32] This, however, does not resolve Servius' status as freeborn or freed, because Cicero does not report whether the child was born before his mother was freed – and this account, too, puts Servius in a very lowly social position. We should also note that there was yet another a completely variant tradition that identified Servius Tullius with the Etruscan hero Macstarna, cited by a speech of Claudius and preserved on the Tablet of Lyons (*ILS* 212 = *CIL* 13.1668). In this version, Servius was not servile at all, but rather a powerful warlord – although how he came to be known as (or associated with) "Servius" is conveniently left out.

The toleration of such variants made it possible for historians like Livy and Dionysius to craft narratives that reflected their own concerns, and the concerns of their generation; while not infinitely flexible, the historical tradition provided leeway. And if myth is "ideology in narrative form," as Lincoln suggested,[33] the presence of variants like this one reflects competing ideas about the role of freedmen citizens in Roman society. Thus, Cicero and Livy, members of Rome's elite, could prefer the variant that gives Servius free birth, while Dionysius' outsider, Greek perspective allowed him to use Servius' slave birth to comment on slavery and freedom in Roman society.

Servius came to power after Tarquinius Priscus was assassinated by the sons of his predecessor, Ancus Marcius. Both Livy and Dionysius report that Tanaquil took charge, urging Servius to take the throne lest the sons of Ancus seize it.[34] Tanaquil then made arrangements to pave his way. She did this by telling the crowd gathered outside the royal palace that the *rex* was only injured and would be seen in public soon. Meanwhile, Servius would take over the public functions of the *rex*. After a few days, after people had a chance to get used to seeing him in regalia, surrounded by *lictors*, and sitting on the seat of the *rex*, Tarquin's death was announced. Servius' assumption of the throne was characterized by some constitutional irregularity, according to Livy, but he had the support of the senate.[35] Dionysius too noted the constitutional irregularity of how he became *rex*, but gave

---

31 Gai. *Inst.* I.82; Bradley (1994, 33–34).
32 Cic. *Rep.* 2.37; see also Zon. 7.9.1, where Dio/Zonaras reported that the sources differ as to whether Servius was conceived before or after his mother was captured.
33 Lincoln (1999, 147).
34 Livy 1.41.3; Dion. Hal. *Ant. Rom.* 4.4; cf. Cic. *Rep.* 2.38; Zon. 7.9.
35 Livy 1.41. Servius took the throne with a strong bodyguard, which Livy seemed to indicate was irregular, and he did not seek a vote of support from the Roman people.

Servius an out: he referred his position to the Roman people, who ratified it.[36] The question of Servius' legitimacy is one more aspect of his ambiguity; he is both *rex* and not *rex*, depending on your point of view. To a favorable observer, such as (perhaps) the slaves and freedmen of Rome, this son of a captive mother would epitomize the nobility of character inherent in even the lowliest member of the Roman community. To a gruff traditionalist, his manner of coming to the throne might invoke suspicion and even the label of tyrant.[37] It is significant that the former seems to have been by far the most favored interpretation of our surviving sources[38] because the general good will the Romans seem to have felt toward the slave who became a *rex* would give impetus for fair, humane consideration of freedmen citizens.

Scholars have long noted the complexity of Servius' nature and character, but they have generally focused on the idea that the ambiguity largely emerged as Servius was distilled through late Republican sources preoccupied with the political divisions of the first century. On the one hand, Roman historians of a popular bent chose to make him a patron of popular political measures, such as expanding the franchise and giving land to the poor, while historians who leaned toward the *optimates* made him out to be a defender of the social order.[39] However, we should note that, even if this analysis is correct, late Republican historians were hanging their interpretations of Servius' political program onto an existing edifice, and the stories of Servius' humble origins, unusual path to power, and horrible death were almost certainly antecedent to the attribution of specific political programs or goals. Thus, some late Republican antiquarians and historians made Servius out to be a champion of ordinary Romans, many of whom in the first century were themselves freedmen or their descendants, because the stories of his origins had already made him out to be a freedmen, and therefore a plausible champion for other freedmen. Others seem to have reacted against the idea of Servius as a *popularis* figure by instead taking advantage of his reputation as a good and just *rex* and using his name to legitimize conservative elements in the Roman constitution.

Servius' institution of the Compitalia and manumission

Among Servius' many achievements, so the annalists say, was the creation of the census and the organization of Roman citizens into categories based on wealth. Furthermore, he was credited with establishing the centuriate assembly, which would ultimately choose Rome's highest-ranking magistrates

---

36 Dion. Hal. *Ant. Rom.* 4.12.
37 Vasaly (2015, 46).
38 Vasaly noted that Livy has constructed an account that makes his readers sympathetic to Servius, even though his taking the throne seems illegitimate.
39 Richard (1987); Gabba (1993, 164ff); Ridley (2014).

and upon which the Republic's military system was structured. He is also credited with dividing the Roman community into tribes. Space does not permit a full discussion of the other constitutional reforms attributed to Servius, nor of all of the religious innovations attributed to him, including the pan-Latin temple of Diana on the Aventine.[40] We must, however, look at two specific innovations ascribed to Servius that are particularly relevant to this chapter: the founding of the festival of the Compitalia, in which slaves could participate, and the offer of citizenship to slaves upon lawful manumission.

The festival of the Compitalia took place in late December or early January, and included rituals to honor the *lares compitales* at crossroads. Representatives from each neighborhood (*vicus*) met at crossroad shrines to conduct the rites, and these representatives included slaves and freedmen.[41] Servius founded the Compitalia by ordering shrines set up in each neighborhood at a crossroads, and by establishing a place for slaves in the ceremony. Dionysius argued that the Romans' intention in maintaining this practice was to integrate slaves into the Roman community by making the burdens of their slavery a little less onerous: the rituals themselves, being public and important, give the slaves a sense of importance to the community, and the day in which the marks of their slavery are removed also lessens the onus of their servile status.[42] His analysis here is particularly acute; ritual does, in fact serve to help reinforce and maintain a particular social order, and we can see how offering slaves and freedmen the opportunity to take part in communal ritual would help bind them to their new community.[43] We should note as well the significance of associating the ritual with Servius Tullius. Associating Servius with the Compitalia both reflected the Romans' perception of him as a role model and champion for freedmen, and further established him as a figure to whom the slaves and freedmen of the city owed their incorporation into the community.

This pattern continues with the offer of citizenship to freedmen. According to Dionysius,

> Servius Tullius permitted even manumitted slaves to enjoy equal rights of citizenship, unless they chose to return to their own countries. For he ordered these freedmen to report the value of their property in the

---

40 For a review of Servius' accomplishments as *rex* and the "Servian constitution," see Livy 1. 42–43 and Dion Hal. *Ant. Rom.* 4.16–22. For modern discussions, see: Ogilvie (1965, 166–79); Thomsen (1980, esp. 115–211); Cornell (1996, 173–97); Forsythe (2005, 102–8); Armstrong (2016c, 75–93). On his religious reforms: Vernole (2002).

41 Treggiari (1969, 198–200); Mouritsen (2011, 248–51); Flower (2017, 162–74).

42 Dion. Hal. *Ant. Rom.* 4.14.3–4.

43 See Fronda in this volume, for the suggestion that the decision to include Latins and *socii* to participate in the Roman triumph may have similarly bound those allies to the ritual community.

census with all the other free men, he distributed them among the four city tribes ... and he admitted them to participation in all public matters on equal footing with the other plebeians.[44]

Dionysius' account of the new policy for freedmen occurs within the context of the completion of the first census. Servius wanted the population of Rome to grow. And so, Servius allowed freedmen to obtain citizenship and be enrolled in the four urban tribes and share in all of the privileges of the other citizens.[45] Dionysius further expanded on Servius' motives in the next chapter. The patricians objected, and so Servius spoke in the measure's defense. The speech that Dionysius wrote for Servius is instructive about attitudes toward freedmen and manumission in the first century. Servius defended the measure by saying that the elites and the community would benefit from it in several ways. In contrast to those who might suggest that slavery was the result of a person's nature, he asserted that fortune, not nature, made slaves out of both individuals and communities. He pointed out that the Romans had admitted foreigners into their citizen body without inquiry as to their character; it would be absurd to deny citizenship to slaves who had proven good character. He pointed out that manumission would be an incentive for slaves to work harder for their masters because they had something very valuable to gain from good service. He added that the Roman military would benefit tremendously because the new citizens would have children who could be drafted. Finally, he pointed out that the elites would benefit because the newly freed citizens would turn to the older, richer families for leadership, and reward members of those families with votes and support.[46] We can see, therefore, that Dionysius was using the story of Servius to work out a debate about slavery and manumission that was taking place in his own lifetime. He made Servius the most vocal defender of freedmen's citizenship.

But then Dionysius continued with one of the most important digressions in his work, where he evaluated the effects of the policy both in the imagined past and in his own day. Somewhat counter-intuitively, to modern eyes, Dionysius argued that the policy made a lot of sense in Servius' day, as most of the slaves were captives taken honorably in war. Dionysius argued

---

44 Dion. Hal. *Ant. Rom.* 4.22.4: ὁ δὲ Τύλλιος καὶ τοῖς ἐλευθερουμένοις τῶν θεραπόντων, ἐὰν μὴ θέλωσιν εἰς τὰς ἑαυτῶν πόλεις ἀπιέναι, μετέχειν τῆς ἰσοπολιτείας ἐπέτρεψε. κελεύσας³ γὰρ ἅμα τοῖς ἄλλοις ἅπασιν ἐλευθέροις καὶ τούτους τιμήσασθαι τὰς οὐσίας, εἰς φυλὰς κατέταξεν αὐτοὺς τὰς κατὰ πόλιν τέτταρας ὑπαρχούσας...καὶ πάντων ἀπέδωκε τῶν κοινῶν αὐτοῖς μετέχειν ὧν τοῖς ἄλλοις δημοτικοῖς. We should note that Livy (2.5.9–10) has an alternative source for this innovation. He claimed that freedmen were awarded citizenship upon manumission because of the patriotic actions of a slave, Vindicius, who informed in 509 on a conspiracy against the new republic and who therefore was manumitted and given citizenship, giving his name to the process of *vindicta*.

45 Dion. Hal. *Ant. Rom.* 4.22.3–4.

46 Dion. Hal. *Ant. Rom.* 4.23.

that, if such men had demonstrated their loyalty to their masters, they could be counted on to likewise demonstrate loyalty to their own community. He contrasted this with the slaves and owners of his own day. Many of the slaves, so he argued, bought their freedom after earning money through criminal or disreputable means. Moreover, owners would free huge numbers of slaves in their wills just to make sure that their funerals were more extravagant spectacles. He even suggested some reform limiting the admission of morally inferior slaves to the citizen body.[47] Dionysius was using Servius as the vehicle through which to grapple with the issue of manumission and freedmen, and, like the Romans of his own day, he was decidedly ambiguous about the system. Furthermore, we should note that the nature of the stories of Servius lend themselves to such a discussion. Dionysius was probably building off an earlier ambiguity, not creating a new element to the story out of whole cloth.

And so, we see a supposedly historiographic discussion of Roman antiquity being turned into an examination of ideas about slavery. We can see that the Romans of this period were discussing among themselves why their system admitted supposedly lowly and degraded slaves into the citizen body. The context in which they did so was potentially a problem. The stories that attribute Republican institutions to the monarchial period were designed to give legitimacy to those institutions by imbuing them with the sacredness of antiquity. The power of the *mos maiorum* and Roman unwillingness to make radical changes in their social and communal institutions seem to have led Roman story tellers to assume that, if an institution was present by the mid-to-late Republican period when the narratives about early Rome were assembled, the institution must have an ancient origin, that is, a Regal or early Republican one. This "invention of tradition" presented the Romans with a particular problem with regards to the incorporation of freed slaves. Given two inarguable principles of Roman life, that is, first, slaves and freedmen were inferior to free-born citizens and, second, that slaves formally manumitted must be given citizenship, it is not surprising that Rome's early history presents an ambiguous attitude toward freedmen in Roman life.

As in so many other cases, it was the coming of the principate that saw the resolution, of sorts, to a Republican problem. Augustus, too, was ambiguous about how best to incorporate freed slaves into the Roman community. Most importantly, and most famously, Augustus changed the process of manumission by limiting the number of slaves who could be manumitted in wills. The *lex Fufia Caninia* of 2 set up a formula based on the number of slaves in an estate: small estates could free no more than half of the slaves, while an estate of 500 slaves or more could free no more than one hundred in any given will.[48] This was strengthened in AD 4 with the *lex Aelia Sentia*,

---

47 Dion. Hal. *Ant. Rom.* 4.24.
48 Duff (1928, 32–33); Mouritsen (2011, 34–35).

which, among other things, limited the ability of manumitted slaves under 30 years old to achieve full citizenship status, and limited the right of young slave owners to manumit.[49] To some extent, this approach was against the interest of the state treasury, as manumissions required a five percent payment to the treasury based on the worth of the slave.[50] Given the love that governments have for revenue, Augustus must have wanted to make a particularly important point by making formal manumission more difficult. A once popular explanation was that he was defending Rome against the corruption and pollution of Roman society or Roman ethnicity by preventing the citizen body from being swamped by freed foreigners. Yet, we should note that Augustus also took care to make sure to protect freedmen. The *lex Iulia de maritandis ordinibus* of 18 exempted freed fathers from their legal duties toward their patron if they fathered two free children.[51] Thus, it hardly seems that Augustus was concerned about racial purity. Rather, I think it more likely that Augustus was responding to criticisms, like those leveled by Dionysius of Halicarnasus in his digression on slavery, that too many slaves were freed for the wrong reasons.[52] Augustan policy suggests Dionysius was not alone in his perception that manumission was getting out of hand.[53]

Servius' death

Servius famously met a grisly and unfortunate end. After 44 years of rule, Tarquinius Priscus' son (in Livy's version) or grandson (in Dionysius'), Lucius Tarquinius Superbus wanted the throne, and, in a conspiracy with his second wife, Servius' daughter Tullia, got it. In both versions of the story, even after all he had accomplished, his servile ancestry came back to haunt him. Livy's account has Tarquin assemble a bodyguard, enter the forum, and sit on the throne outside the *curia* in full view of everyone. Tarquin proceeded to harangue the crowd and justified his seizure of power partly on the illegitimacy of the manner through which Servius had become *rex* but also because he had been born a slave.[54] When Servius arrived in person, Superbus repeated the charge to his face, and followed it up with the assertion that

49 Duff (1928, 32, 77); Mouritsen (2011, 33–34).
50 Duff (1928, 28–29).
51 Duff (1928, 46). Though it also set rules against senators marrying freedwomen; see Mouritsen (2011, 91).
52 Mouritsen (2011, 33–35) argued that the laws addressed the perception that many slaves were too degraded or criminal to become good citizens.
53 Augustus also transformed the way that freedmen and slaves participated in Roman religious life through the introduction of the *lares augusti* in Rome, the maintenance of whose cult was the duty of the *vicomagistri*, a group of local officials dominated by freedmen. See Dion Hal. *Ant. Rom.* 4.14.3–4; Plin. *HN* 3.66; Duff (1928, 130–37); Lott (2004, 89–98, 104–8); Flower (2017, 271–83).
54 Livy 1.47.10: *ibi Tarquinius maledicta ab stirpe ultima orsus: servum servaque natum…*

he deserved the throne more than a slave.[55] Dionysius made the confrontation longer and more dramatic. Servius realized what Tarquin was up to and called a senate meeting to chastise him. Tarquin made a case for the throne based on his hereditary right and pointed out that Servius was supposed to hold power only during the minority of the grandsons of Priscus.[56] Servius defended himself, saying that he has been just both as guardian to the Tarquinii and as *rex*. He offered to resign the position of *rex*, not to Tarquin, but to the people of Rome, who had conferred it on him.[57] When the crowd refused to accept, Servius seemed to have won his point. But Tarquin was not to be balked. When the crowds had dispersed, he gathered a bodyguard, summoned the senate, and seated himself on the throne. When again confronted by Servius, he abused him, calling him a slave and son of a slave woman taken captive. Now, Servius, having endured so many reproaches, lost his temper. The old man attacked the much younger Tarquin.[58] In both versions, Tarquin hurls Servius down the steps, where stunned, he meets his end when his own daughter's coach drives over him.[59]

We should make a particular note here of Livy's account, since, as Vasaly has pointed out, it contradicts Livy's earlier testimony. Livy, as we have seen, went out of his way to exculpate Servius from the charge of being a freed slave, but he puts exactly that charge in the mouth of Superbus, which means that Superbus was either lying or at least distorting the truth to advance his own claim to the throne.[60] That Tarquinius was lying is clear from Livy's earlier account, but the way he smeared Tullius' upbringing shows how easily Servius could be depicted as having slave birth. Intriguingly, Livy's handling of the charge makes Servius more sympathetic to the reader, and both Livy's and Dionysius' narratives seem designed to elicit compassion and sorrow because Servius had indeed ruled as a just and good *rex*. This is particularly true given that the audience knew what kind of man Superbus was, and how his reign turned out. That both Livy and Dionysius chose to remind their readers that Servius had a servile upbringing makes it almost inevitable that the readers should ponder the awkward difficulty of reconciling the fact that Servius was lowly and despicable in origin, but just and good in achievement.

Conclusions

Let me conclude by suggesting that Servius became the vehicle through which Romans explained and legitimated the unusual treatment of slaves

55 Livy 1.48.2; Cf. Ov. *Fast.* 585–609.
56 Dion. Hal. *Ant. Rom.* 4.31–32.
57 Dion. Hal. *Ant. Rom.* 4.33–37.
58 Dion. Hal. *Ant. Rom.* 4.38.
59 Livy 1.48.3–7; Dion. Hal. *Ant. Rom.* 4.39; Vasaly (2015, 50).
60 Vasaly (2015, 49–50).

and freedmen in their own society. This means that, to some extent, he ended up becoming different things to different parts of the Roman audience, which, again, is convenient given the ambiguity of the messages that the Romans sent their slaves and freedmen. As both divine and servile, foreigner and noble, he had the possibility of becoming many things to many people. So Servius was presented as a complicated character to gauge, which only strengthened the use he had as the scaffold on which the Romans could hang discussions about their system of manumitting freedmen.

So, what might free-born Romans have taken away from the stories about Servius Tullius? First, that character and birth meant something; Servius, our definitely noble (and possibly divine) slave overcame his birth status to become the Roman *rex*. He was successful in war and peace to an extent not shared by many of his fellow monarchs. But also, lurking in the back, was a perhaps uncomfortable reminder that no matter how much pride they had in their free Roman birth, Rome was a place where newcomers and foreigners had once come to make something of themselves.

What, then, were the slaves and freedmen, who comprised part of the audience of these narratives, to make of the lessons of the story of Servius Tullius? Clearly, the Romans could see Servius as an example of a low born man made good. Horace explicitly used him as an example of this, reminding Maecenas that Rome was a place where true nobility was dependent on talent, not birth.[61] A freed slave in Rome might very well conclude that their community had both potential for social advancement and fierce hostility toward that same advancement. Unlike a slave in, say, ancient Athens, Roman slaves, even *captivi*, lived in a world in which they could obtain their freedom and take a recognized place in Roman society – one that offered both an open door and a closed one. The open door was that citizenship offered the hope for a life better than endless servitude. The closed one was that their fellow citizens might very much resent the newcomers, particularly if those newcomers made efforts to surpass the older citizens in money, prestige, and power. For those already citizens, the stories provided a justification for what was probably, to many of them, a very awkward feature of Roman life: why admit slaves into the citizen body at all? The function of the legends of the regal period was to provide a coherent framework for understanding slavery and freedmen in Roman life. In addition to the specific reasons Dionysius put into Servius speech, the antiquity of Servius' supposed reforms made radical changes unthinkable, as such a change would be an assault on the *mos maiorum* and so on the entire social structure. Thus, the legends had the effect of mitigating, and to some extent ameliorating, the hostility felt toward the newcomers, which, in turn, made their assimilation into Roman society somewhat easier. These are the most important lessons of Servius Tullius.

---

61 Hor. *Sat.* 1.6.9.

16 The transformation of the Roman army in the last decades of the Republic*

François Gauthier

Introduction

For several decades, it has often been argued that the most important change in late Republican military organization was instituted by Gaius Marius, who reportedly dropped the minimum property qualification for military service and thereby supposedly created a professional army relying on volunteers coming from the lower social classes of Roman society (*proletarii* or *capite censi*).[1] This new type of army, made up of impoverished citizens rather than small landholders, was argued to be more loyal to generals than to the senate, hence its inclination to follow its leaders into civil war – that is, "client armies." However, the idea of a wide-ranging "Marian reform" that permanently abolished property qualifications for military service has recently been thoroughly rebutted.[2] In point of fact, ancient evidence attesting large numbers of very poor citizens in the legions is actually quite tenuous. On the contrary, the sources show that soldiers were usually still recruited according to their census rating even in the late Republic.[3]

Similarly – and related – Marius is also often credited with disbanding the citizen cavalry and light infantry (*velites*), replacing them with non-Roman auxiliaries.[4] Yet, this claim largely rests on an argument from silence. For the cavalry, there is no evidence stating that Marius disbanded them, and

---

* All dates are BC unless otherwise noted.
1 For example: Matthew (2010, 354–64); Gabba (1976, 1–23); Sordi (1972, 379–85); Harmand (1969, 61–74); Carney (1961, 31–33). The evidence most often cited in support of the "Marian reform" concerning recruitment is the following: Sall. *Iug.* 86, 84; Val. Max. 2.3.1; Plut. *Mar.* 9.1; Flor. 1.36.13; Gell. *NA* 16.10.14; Exup. 9–13.
2 Cadiou (2018, 35–118); Gauthier (2016b, 103–20); Aigner (1974, 11–23). Cadiou's monograph is sure to become the new reference for the army of the late Republic.
3 Cadiou (2018, 392–93).
4 Busquets Artigas (2014, 29); Rankov (2007, 32–33); Keppie (1984a, 44); Bell (1965, 404–22). Marius' alleged elimination of light infantry is seen as part and parcel of his assumed wider reform: by abolishing property qualifications, he moved to more standardized equipment, with each infantryman armed in the same way. In this model, poorer Romans would no longer serve in light infantry units but would be armed and equipped as heavy infantry, with weapons and armor paid for by the state.

DOI: 10.4324/9781351063500-16

This chapter has been made available under a CC-BY-NC-ND license.

indeed there are many references to citizen cavalry after Marius.[5] As for *velites*, those defending the traditional view of the Marian reforms some-times point out that the *velites* are supposedly last mentioned either in the Jurgurthine War or in Sulla's army in Greece.[6] To be sure, *velites* seem grad-ually to disappear from our radar during this period, but it must be said that the vocabulary used by our sources makes it hard to distinguish be-tween *velites* and non-Roman auxiliaries. Sometimes, light infantry units are named with their ethnic origin, but that is not always the case.[7] A pas-sage from the *Bellum Hispaniense* states that deserters were relegated to the light infantry and received lesser pay, without any indication as to whether these men were Roman or not.[8] In my view, it would be imprudent to as-sume that all references to light infantry in the first century must refer to non-Romans, and then to use that assumption as evidence for the so-called Marian reforms.

That being said, there is no doubt that important changes took place in the Roman army in the last century BC. In this chapter, I wish to move away from the idea of a "Marian reform," and instead look at the Social War, as well as the civil wars, as the periods of pivotal change. These conflicts nota-bly saw important changes in the financing and recruitment of the Roman army. It will be argued that the Social War greatly increased the cost of the army as a result of the enfranchisement of the *socii*. This caused a greater reliance on auxiliaries during the conflict – and probably after – since they

---

5 Cadiou (2016, 53–78). On Roman citizen cavalry in the late second and first centuries: *CIL* 1².593.1.91; Val. Max. 5.8.4; Suet. *Gramm.* 9; Plut. *Pomp.* 64.1; App. *B Civ.* 5.138. See also Crawford (1996, 384); Nicolet (1966, 52–55); McCall (2002).

6 Neue Pauly s.v. *velites*; Sall. *Iug.* 46. 7: "Accordingly, he himself led the van with the light-armed cohorts as well as a picked body of slingers and archers; his lieutenant Gaius Mar-ius with the cavalry had charge of the rear, while on both flanks he had apportioned the cavalry of the auxiliaries to the tribunes of the legions and the prefects of the cohorts. With these the light-armed troops (*velites*) were mingled." (*itaque ipse cum expeditis cohortibus, item funditorum et sagittariorum delecta manu apud primos erat; in postremo C. Marius legatus cum equitibus curabat, in utrumque latus auxiliarios equites tribunis legionum et praefectis cohortium dispertiverat, ut cum eis permixti velites.*); Frontin. *Str.* 2.3.17: "Next he arranged a triple line of infantry, leaving intervals through which to send, according to need, the light-armed troops and the cavalry, which he placed in the rear." (*triplicem deinde peditum aciem ordinavit relictis intervallis, per quae levem armaturam et equitem, quem in novissimo conlocaverat, cum res exegisset, emitteret.*); Festus *Gloss. Lat.* 274 L: "Soldiers used to fight with small bucklers. The use of which C. Marius has abolished, with Brut-tians given in their place" (*Parmulis pugnare milites soliti sunt; quarum usum sustulit C. Marius datis in vicem earum Bruttianis*). It is unclear whether *milites* in this case refers to some soldiers or all the soldiers.

7 Caes. *BGall.* 2.7, 2.10, 2.24; *BCiv.* 3.4.

8 *BHisp.* 22.7: "if any of our men deserted, they were relegated to the light infantry and did not receive more than 17 denarii" (*si qui ex nostris transfugeret, in levem armaturam coici eumque non amplius. XVII accipere*); see also *BHisp.* 23.6.

were cheaper to employ. The civil wars then led to a greater incorporation of non-Romans in the army, as well as the creation of permanent auxiliary units, laying the foundations for the standing *auxilia* of the Imperial army.

The Social War, manpower, and war finances

Rome's reserves of manpower were central to its victory over Hannibal in the Second Punic War.[9] These relied, to a large extent, on the Italian allies who fought alongside Roman soldiers. In fact, Polybius stated that the *socii* provided an equal number of foot soldiers to that of the Romans and three times as many cavalry.[10] On the other hand, Velleius Paterculus (2.15.2) recorded that, before the Social War, the Italians provided twice as many men as the Romans.[11] The plausibility of these ratios has been discussed at length by modern scholars, but it is generally agreed that the Italians provided at least half of Rome's manpower.[12]

When the Social War broke out, Rome found itself without many of its allies for the first time in more than a century. Instead of being reliable friends, many Italian communities had become fierce enemies.[13] The war thus placed a heavy burden on Roman manpower, as it had to fight several of its former allies with only its own citizens and those allies who had remained loyal.[14] Both Romans and Italians mobilized a very high proportion of their male population to wage this war. According to Appian, the rebellious allies levied an army of some 100,000 men, and the Romans were able

---

9 Brunt (1971a, 439). For a recent discussion of Rome's manpower advantages and its strategic implications in the Second Punic War, see Fronda (2010, esp. 37–50).
10 Polyb. 6.26.7–8.
11 "Every year and in every war they were furnishing a double number of men, both of cavalry and of infantry, and yet were not admitted to the rights of citizens in the state which, through their efforts, had reached so high a position that it could look down upon men of the same race and blood as foreigners and aliens." (*per omnis annos atque omnia bella duplici numero se militum equitumque fungi neque in eius civitatis ius recipi, quae per eos in id ipsum pervenisset fastigium, per quod homines eiusdem et gentis et sanguinis ut externos alienosque fastidire posset*).
12 For modern estimates and analyses: Ilari (1976, 171) (estimating on average that allies supplied about 60% of the Roman armies, soldiers between 200 and 168); Kendall (2012, 105–22); Mouritsen (2008, 481); Erdkamp (2006b, 44); Baronowski (1984, 248–52); Baronowski (1993, 181–202); Shochat (1980, 93–94); Brunt (1962, 74, 1971, 677–86); Afzelius (1944); Lo Cascio (1991–94, 309–28).
13 Lists of *socii* who revolted: Livy, *Per.* 72; Diod. Sic. 37.2; App. *B Civ.* 1.39; Vell. Pat. 2.15 states "all Italy took up arms against the Romans" (*universa Italia [...] arma adversus Romanos cepit*). This is obviously an exaggeration. On the sources for the Social War, see Clark's contribution in this volume. On Roman internal war in the first century, see Brice's chapter in this volume.
14 On the causes of the Social War and the Italians' motives: Dart (2014); Kendall (2013); Keaveney (2005); Mouritsen (1998); Brunt (1965, 90–109). On troop numbers: Brunt (1971a, 435–40).

to muster a comparable number.[15] Since the record of the census for 115/4, the last available before the war, gives the figure of 394,336 Roman citizens, this rate of mobilization would have been extremely high.[16] The fact that freedmen had to be enrolled to guard coastal areas strongly suggests that Roman manpower was stretched to the limit.[17]

In addition to manpower, the defection of many allied communities caused an important financial challenge for Rome. The *socii* not only represented a strong reserve of manpower but also their use was cheap for Rome, since the Italian allies paid for the troops they were providing for the Romans themselves.[18] The Romans were content to give free rations to the *socii* – something that was a lot cheaper than providing a *stipendium* for the allied contingents.[19] But in the Social War, Rome would have had to pay the *stipendium* to a far greater proportion of her soldiers than usual, putting immense strain on the system, and indeed Livy's *periocha* states that in 89 "the state was laboring under the burden of debts." [20]

Under these circumstances, Rome's most pressing issue was probably cavalry – both for military and financial reasons. Traditionally, the *socii* had provided 75% of the cavalry for a Roman consular army.[21] In terms of numbers, this means that every Roman legion was usually accompanied by 900 Italian cavalrymen, in addition to its complement of 300 Roman horsemen.[22] Cavalry was also the most expensive component of the army to field. According to Polybius, Roman cavalrymen received a *denarius* per

---

15 App. *B Civ.* 1.39. See also: Rich (1983, 328); Brunt (1971a, 441–45).

16 Even allowing for substantial under-registration in the census, this still represents a very important percentage of the citizen population. See Scheidel (2008, 17–70) and Rosenstein (2002). Debate about Roman census figures and demography: Hin (2013); De Ligt (2012); Launaro (2011).

17 App. *B Civ.* 1.49; Livy, *Per.* 74. Similar measures were taken during the Second Punic War when slaves were enlisted to compensate the terrible losses suffered at the battles of Trebia, Lake Trasimene, and Cannae, cf. Livy 22.2, 22.61.2.

18 Polyb. 6.21.4–5: "At the same time the consuls send their orders to the magistrates in the allied cities in Italy which they wish to contribute troops, stating the numbers required and the day and place at which the men selected must present themselves. The cities, choosing the men and administering the oath in the manner above described, send them off, appointing a commander and a paymaster." (Κατὰ δὲ τοὺς αὐτοὺς καιροὺς οἱ τὰς ὑπάτους ἀρχὰς ἔχοντες παραγγέλλουσι τοῖς ἄρχουσι τοῖς ἀπὸ τῶν συμμαχίδων πόλεων τῶν ἐκ τῆς Ἰταλίας, ἐξ ὧν ἂν βούλωνται συστρατεύειν τοὺς συμμάχους, διασαφοῦντες τὸ πλῆθος καὶ τὴν ἡμέραν καὶ τὸν τόπον, εἰς ὃν δεήσει παρεῖναι τοὺς κεκριμένους. αἱ δὲ πόλεις παραπλησίαν ποιησάμεναι τῇ προειρημένῃ τὴν ἐκλογὴν καὶ τὸν ὅρκον ἐκπέμπουσιν, ἄρχοντα συστήσασαι καὶ μισθοδότην). See Nicolet (1978); Kendall (2012, 116–17); Martin (2014, 118–19); see also Tan and Roselaar in this volume.

19 Polyb. 6.39.15. Price of wheat: Polyb. 2.15.1, 34.8.7–8; Cic. *Verr.* 2.3.163, 188–9.

20 Livy, *Per.* 74: *Cum aere alieno oppressa esset civitas.*

21 Polyb. 6.26.7.

22 Polyb. 6.20.9; McCall (2002, 100).

day – three times more than foot soldiers.[23] Even though cavalrymen were supposed to provide their own equipment, just like all other Roman soldiers, their pay was provided by the state.[24] The pay for 1,200 *Roman* horsemen nearly equaled the pay for all of the foot-soldiers in a legion.[25] To be sure, asking the allies to send three times the number of cavalry compared to the Romans was a way for the senate to spare itself the cost of fielding the bulk of this expansive arm. Since this was no longer possible, the recourse to non-Roman manpower must have represented a practical expedient to bolster Rome's cavalry, and likely other supporting troops as well. According to Cicero, referring to a later period, the peoples from outside of Italy who provided soldiers for the Roman army usually paid for them just like the Italian *socii* had before the Social War.[26] As we will see, this was not the first time the Romans called upon foreigners, but the seriousness of the Social War probably caused a significant increase in their use.

Auxiliaries as substitutes for the *Socii*?

Prag has rightly pointed out that there is plenty of evidence for the use of auxiliaries before the Social War.[27] In the middle Republic, auxiliaries tended to be deployed in the theatre of operations where they were recruited. For example, the Romans were quick to make use of local manpower to bolster their forces in Spain and Africa while they almost exclusively employed

---

23 Polyb. 6.39.12. Polybius provides figures for legionary pay in obols and drachmae. It is assumed that Polybius equated one *drachma* with one *denarius*; see Walbank (1957, 722).

24 Polyb. 6.20.9, 6.26.1, 6.39.12, 6.39.15; Livy 4.59.11.

25 According to Polybius (6.39.12) cavalry pay was one *drachma* (= one *denarius*) per day. Thus, the annual pay for 1200 cavalrymen would be 432,000 *denarii* (= 1200 × 360 *denarii*). This would have been equivalent to the pay of 3,600 foot soldiers paid at the rate of two obols per day (according to Polybius).

26 Cic. *Verr.* 5.60: "It had been the regular practice that each state should provide for its naval expenditure on provisions, pay, and all other such matters, by furnishing its own commander with the sum needed. [...] This, I repeat, was the invariable practice, and not in Sicily only, but in all our provinces, and even for the pay and maintenance of the Italian allies and Latins in the days when they supplied us with auxiliary troops." (*Sumptum omnem in classem frumento stipendio ceterisque rebus suo quaeque nauarcho civitas semper dare solebat.* [...] *Erat hoc, ut dico, factitatum semper, nec solum in Sicilia sed in omnibus provinciis, etiam in sociorum et Latinorum stipendio ac sumptu, tum cum illorum auxiliis uti solebamus*); Cic. *Font.* 13; 26. See also Martin (2014, 117–38). On the definition of *auxilia externa*, see also Fest. 16 L: "These are, in times of war, the allies of the Romans provided by foreign nations." (*Auxiliares dicuntur in bello socii Romanorum exterarum nationum*); Varro *Ling.* 5.90: "*Auxilium* ('auxiliaries') was so called from *auctus* ('increase'), when those foreigners who were intended to give help had added themselves to the fighters." (*Auxilium appellatum ab auctu, cum accesserant ei qui adiumento essent alienigenae*); Livy 22.37.7–8; Tac. *Hist.* 1.68.

27 Prag (2007, 70).

Sicilian troops to garrison the island after the Second Punic War.[28] In similar fashion, Flamininus' auxiliaries at the battle of Cynoscephalae were mostly Greek.[29] Only in major crises were auxiliaries deployed far away from their homeland – such as when foreign troops were sent from Spain, Africa, and Sicily to Italy in 207, in anticipation of the clash against Hasdrubal.[30] Evidence for auxiliaries is more plentiful for the late Republic. This reflects not merely an accident of the surviving evidence, but rather that Rome had more provinces than in the third century, and thus a much bigger and diverse reserve of manpower to draw from. Thus, for instance, Mauretanians, Bithyhians, Thessalians, and Acarnanians were sent to Sicily during the Second Servile War (104–100).[31]

By the time of the Social War, several foreign communities had become more closely incorporated into Rome's military and social structures through a long tradition of service alongside Roman troops. Perhaps the most famous example of this, in the context of the Social War, is indicated by the so-called bronze of Ascoli – an inscription recording the rewards granted to a body of Spanish horsemen, the *turma Salluitana*. This unit had successfully fought for Rome against the rebellious Italians, notably at Asculum. For their distinguished service, they were granted Roman citizenship – a reward that would eventually become quite common in the early Empire.[32] These Iberians were hardly the only auxiliaries involved in the conflict, as Gauls, Numidians, and Mauritanians are also attested in the literary tradition.[33] Of course, these examples do not necessarily prove an increase in the use of auxiliaries overall, but they nonetheless clearly indicate their presence in significant numbers during the conflict, and in a theatre of operations far away from their homeland. In the light of the aforementioned evidence on the increased use of auxiliaries in times of crisis, and given the magnitude of the war, it makes sense that the Romans relied on auxiliary forces during the Social War, probably in even greater numbers than usual.

The end of the Social War ultimately brought about the enfranchisement of the Italians, which created yet another important change in Roman military finances. It meant that Rome's former allies were no longer obliged to finance their own troops: they now would have been paid the *stipendium* while the newly enfranchised surely also benefited from the exemption from the war-tax (*tributum*), which had been suspended for Roman citizens

---

28 Spain: Cadiou (2008, 611–84); Africa: Hamdoune (1999); Sicily: Prag (2007, 68–100).
29 Plut. *Flam.* 7; Livy 32.3; Keppie (1984a, 121–25).
30 Livy 27.38.11.
31 Diod. Sic. 36.5.4, 36.8.1.
32 *CIL* 1².709. On the inscription, see Criniti (1970) and Pina Polo (2003, 197–204). See also Haynes (2013, 31–34) and Cadiou (2016, 58).
33 App. *B Civ.* 1.42, 1.50.

in 167.[34] Each legion was now composed entirely of Roman citizens, who were all paid from the *aerarium*, which would have more than doubled the cost of paying an army. For example, before the Social War, a legion comprising Roman citizens, supported by twice the number of allied infantrymen (using Velleius' ratio, discussed above), would have cost a little over 500,000 *denarii* per year in *stipendium*.[35] After the Social War the same number of men, equal now to about three legions of citizens, all paid by the Roman state, would have cost over one and a half million *denarii* in *stipendium*. As pointed out above, paying a number of cavalrymen equivalent to the contingent provided by the allies before the Social war was also quite expensive. Additionally, it should be noted that these numbers only concern the *stipendium*, the only data that can be calculated with any degree of accuracy. To be sure, the total cost of the military would be much more as it included transport, supplies, fleets, and various other expenses. Moreover, Rome hardly had any time to breathe before it had to deal with the Mithridatic War and the first series of civil wars. That Sulla, in 88, had to sell sacred property to the value of 9,000 pounds of gold in order to finance the war against Mithridates is a testimony to the sorry state of the Roman treasury after the Social War.[36] Furthermore, Sulla and his enemies raised gigantic armies during the civil war, and their cost must have been crippling.[37] Sulla's infamous proscriptions raised the huge sum of 350,000,000 sesterces, which he surely used to try to cover military expenses.[38]

Although such numbers are impossible to assess, I think it is reasonable to argue that crisis situations, like those during and immediately after the Social War, would have created an increased demand for auxiliaries, who were called upon to fill the role once played by the Italian *socii*: to provide soldiers whose pay Rome did not have to assume directly. This helped compensate for the huge cost of the enfranchisement of the *socii* and the demands of civil wars. This hypothesis provides a logical solution for the survival of Roman cavalry in the first century and an increased use of auxiliary horsemen and

---

34 Nicolet (1978, 10–11). Suspension of *tributum*: Plin. *HN* 33.17.56. Origins of the *tributum*, see Tan in this volume.

35 As mentioned above, each infantryman was paid two obols per day (120 *drachmae*/year), according to Polybius. Centurions received double pay (thus 240 *drachmae*/year). Assuming, as above, one Polybian *drachma* = one *denarius*: 4,200 infantrymen per legion × 120 *denarii*/year + 60 centurions × 240 *denarii*/year = 518,400 *denarii* according to the numbers given by Polybius (6.20.8, 6.39.12). A legion's strength could be brought up to 5,000 in times of emergency: Polyb. 6.20.8. The numbers provided by Pliny (*HN* 33.17) concerning what was in the *aerarium* (six million sesterces in 157 and 30 million in 49) seem quite low compared to the cost proposed here for a mere four legions. Pliny adds an important amount of gold and silver ingots that could presumably be exchanged or melted and coined.

36 App. *Mith.* 22.

37 Livy, *Per.* 89; App. *B Civ.* 1.82, 1.100; Plut. *Sull.* 27; Vell. Pat. 2.24.3. On the plausibility of the figures: Brunt (1971a, 445).

38 Livy, *Per.* 89; App. *B Civ.* 1.98–100; Plut. *Sull.* 33.

other troops to match what was once provided by the *socii*. In summary, the Social War caused an enormous increase in the cost of the Roman military, which was further exacerbated by the outbreak of civil war. I propose that a greater reliance on auxiliary cavalry, and perhaps other types of non-Roman troops, was an expedient to compensate for this.

The mid-first century and the armies of the last civil wars

After the end of the civil war between Sulla and his enemies, the rate of mobilization decreased – although important military commitments continued until the outbreak of the civil war between Pompey and Caesar.[39] The evidence does not allow us to tell whether the Social War and the first round of civil war had a lasting effect on auxiliary recruitment for the period between 80 and 50. However, since the Romans had a tradition of relying on local manpower to garrison provinces where there were no major wars, as well as to support their own legions, it is reasonable to suppose that these practices, at least, continued. Following this long-established custom of using non-Roman manpower, the civil wars that led to the end of the Republic further integrated provincials into Roman military structures. The Romans levied them in larger numbers than before, recruited them into the legions, and established units that remained in service for a long time – setting the stage for the reforms of Augustus.

The nature of the sources in the mid-first century can give the impression that a lot of things suddenly changed in the Roman army. This is, to a certain extent, due to Caesar's *Commentarii* and the nature of the sources covering the 70s and 60s. In his writings, Caesar seems to have recruited auxiliaries as he saw fit – even raising an entire legion composed of Transalpine Gauls who did not have Roman citizenship.[40] This unit was not recruited for one

---

39 Brunt (1971a, 449).
40 Suet. *Caes.* 24.2: "one [legion] actually composed of men of Transalpine Gaul and bearing a Gallic name too (for it was called *Alauda*), which he trained in the Roman tactics and equipped with Roman arms, and later on he gave every man of it citizenship." (*unam etiam ex Transalpinis conscriptam, vocabulo quoque Gallico —Alauda enim appellabatur—, quam disciplina cultuque Romano institutam et ornatam postea universam civitate donavit*); Plin. *HN* 11.121: "the small bird that was formerly named from this peculiarity the crested lark and subsequently was called by the Gallic word *alauda* and gave that name also to the legion so entitled." (*praeterea parvae avi quae, ab illo galerita appellata quondam, postea Gallico vocabulo etiam legioni nomen dederat alaudae*). One could see a precedent for this when Marius granted citizenship to two cohorts of allies from Camerinum at the battle of Vercellae in 101. However, these units were made up of Italians who had been part of the Roman military system for centuries. Cic. *Balb.* 46; Plut. *Mar.* 28.3. Sertorius could also be cited as an example since he relied on non-Romans to a certain extent and armed some of them with Roman weapons, but the sources are not precise about the composition of his legions. See Plut. *Sert.* 12.2, 14.1; Cadiou (2008, 627–30); Cadiou (2004, 297–314).

campaign then disbanded, as was often the case for levies of non-Romans during the Republic. Rather, this legion, eventually known as the fifth *Alaudae*, is attested until the late first century AD.[41] Additionally, Caesar's army also contained numerous auxiliaries, such as Numidians, Spaniards, Balears, Cretans, Germans, and of course large numbers of Gauls, which he used to wage war, not only in Gaul but also against Pompey.[42] Caesar's actions were not entirely new, as Roman generals had a lot of freedom to recruit auxiliaries in provinces, without necessarily needing the permission of the senate.[43] What seems new is the scale of non-Roman recruitment during the civil wars, and not only for Caesar's army.[44] Moreover, recent research has stressed not only the scale but also the great diversity in the recruitment of *auxilia*.[45]

The army raised in Greece by Pompey to confront Caesar was even more diverse than his rival's, and it reflected the regions he was effectively controlling at the time. Caesar's account of Pompey's army is very precise. He was perhaps able to acquire documents detailing its origins when Pompey's camp was captured after the battle of Pharsalus.[46] According to Caesar, Pompey had nine legions of Roman citizens; five of these were recruited in Italy, one in Cilicia, one in Crete and Macedon, and two in Asia. He was also expecting two additional legions to come from Syria. To keep all these units at full strength, Pompey had to incorporate large numbers of local inhabitants into them – no doubt because there were not enough Roman citizens living in the provinces he controlled. Therefore, he recruited Thessalians, Boeotians, Achaeans, Epirotes, Syrians, and various other peoples as legionaries.[47] Moreover, Pompey enlisted auxiliaries trained in specialized fighting styles and weapons, such as Cretan, Lacedaemonian, Pontic, and Syrian archers and slingers, as well as Galatian, Cappadocian,

---

41 Tac. *Hist.* 1.61; 2.43. Its name still exists in modern French as *alouette* (lark); Keppie (1984a, 70): "The new legions were raised by virtue, it would seem, of a proconsul's right to call out local forces in defence of his province."

42 Caes. *BGall.* 1.39, 1.49, 1.51, 2.7, 2.10, 2.19.4, 2.24, 3.6.5, 3.12, 3.18, 3.20, 3.25, 5.5, 5.58, 6.4, 6.5; 6.7, 6.53, 7.13, 7.37, 7.65, 7.67, 7.70, 7.80, 8.5, 8.10, 8.11, 8.13, 8.18, 8.25, 8.36 Note especially 1.51: "in full view of the enemy, he posted all the allied troops." (*omnes alarios in conspectu hostium pro castris minoribus constituit*). Before the Social War, *ala* used to refer to the detachments provided by the *socii*. It later referred to auxiliary units, even if they were not always posted on the wings. In the Imperial period, the term came to mean a unit of auxiliary cavalry.

43 Prag (2015, 281–94).

44 Cadiou (2008, 684).

45 Prag (2015, 285); Prag (2011b, 101–13); Prag (2011a, 15–28); Pina Polo (2008, 453).

46 Something similar happened after the battle of Bibracte in 58 when documents were found in the Helvetians' camp detailing their numbers: Caes. *BGall.* 1.29.

47 Caes. *BCiv.* 3.4; Plut. *Pomp.* 64.2; Cass. Dio 41.61 exaggerated and offered another literary *topos* by affirming that Pompey's army was mostly made up of untrained Ἀσιανοί: Saddington (1982, 193).

Thracian, Macedonian, Gallic, and German cavalry, along with additional troops from various other regions, such as Cappadocia and Dardania. In total, it has been calculated that there were 33 different ethnicities in Pompey's army.[48]

Pompey's legates also recruited extensively among native populations. For example, Afranius and Petreius levied auxiliaries in Celtiberia, from among the Cantabrians, and the peoples bordering the Atlantic Ocean. These forces were considerable, amounting to around 30 cohorts of infantry (some 15,000 men) and 5,000 cavalry, levied exclusively from among non-Romans.[49] Considering that Afranius and Petreius also had five legions at their disposal, these natives formed roughly half of their total forces, which is approximately the ratio of troops once provided by the *socii*.

There are also grounds to think that even these five legions were not entirely made up of Roman citizens. Indeed, Caesar makes an interesting comment that Afranius' soldiers fought like Lusitanians because (Caesar posits) they had grown accustomed to encounter these people in battle. Caesar's own men were at first troubled by their opponents' tactics, which they had never encountered before. This could indicate that some of Afranius' men were actually Lusitanians and Celtiberians themselves, locally recruited and incorporated in the legions. Meanwhile, Varro, another of Pompey's legates, levied two legions and 30 auxiliary cohorts in Further Spain, along with a legion composed of provincial natives.[50] Pompey himself later raised two more *legiones vernaculae*.[51] In 47, two of the legions mobilized by Caesar's lieutenants against Pharnaces had been raised by the Galatian king Deiotarus and equipped in the Roman fashion. The remains of these two legions were later merged into one unit, whose existence is attested until the early second century AD as the *Legio XXII Deiotariana*.[52]

Non-Romans continued to play a prominent role during the next round of civil war between Caesar's assassins and the triumvirs. Cassius and Brutus were forced to recruit non-Romans into their legions, since there were not enough Roman citizens in the provinces that they controlled to fill the ranks of their 19 legions. Brutus levied two legions composed entirely of Macedonians, whom he then trained to fight in the Roman fashion.[53] There were also, presumably, large numbers of non-Romans recruited in other

---

48 Caes. *BCiv.* 3.4; App. *B Civ.* 2.38. Appian (*B Civ.* 2.49) gave Caesar 10 legions and Pompey 11 legions of Italian troops, and at 2.97 he gave 80,000 men to Caesar, a figure that seems inflated. Further comparison of both armies in 2.70. See Yoshimura (1961, 477–79).

49 Caes. *BCiv.* 1.38, 1.39.

50 Caes. *BCiv.* 2.18.1, 2.20. On the *legio vernacular*, see Keppie (1984a, 121); Cadiou (2008, 612 ff.), who thought that this unit was actually composed of Roman citizens living in the province.

51 *BHisp.* 7.4.

52 *BAlex.* 34, 39–40; *ILP* 86; *ILS* 1434; Parker (1928, 89).

53 App. *B Civ.* 3.79.

legions as well to bring them up to strength, as Pompey had done before. Cassius and Brutus also commanded at least 17,000 cavalrymen from many regions of the Mediterranean, which included Gauls, Lusitanians, Thracians, Illyrians, Parthians, Thessalians, Spaniards, Arabs, and Medes.[54] The final showdown of the civil wars at Actium offers a similar picture, as non-Romans are attested in very large numbers.[55] Mark Antony was supported by Libyans, Cilicians, Cappadocians, Paphlagonians, Commagenians, Thracians, Pontics, Arabs, Jews, Galatians, and Medes.[56] Octavian's propaganda famously accused his opponent of relying on the support of foreigners, but his own army included many of them as well.[57]

Considering the extent of this mobilization of non-Romans, it is no wonder that several studies on the origins of the Imperial *auxilia* start in the late Republic.[58] Looking more closely, however, it is clear that the late Republic represents a distinct phase. Modern historians, using Imperial norms, frequently draw a clear line between legions and *auxilia* as being two distinct entities.[59] In contrast, in the last decades of the Republic, this dividing line was often blurred. Legions of non-citizens were raised, whereas others contained a mix of Romans and non-Romans. Whereas legionary service had traditionally been the preserve of Roman citizens who served according to their census rating, the period of the civil wars transformed the Roman army into a more undifferentiated, pan-Mediterranean force.

Haynes has pointed out that there is no explicit evidence for the survival of a particular Republican auxiliary unit into the Principate.[60] That being said, there is no doubt that some of the auxiliary units raised during the civil wars became part of the permanent Augustan army.[61] In the *Res Gestae*, Augustus mentions only citizens who swore the military oath to him, with no reference to auxiliaries.[62] However, Tacitus makes it clear that auxiliaries were stationed alongside the legions and played a vital role in frontier defense.[63] There are good grounds to believe that several units had their origins in the late Republic. For example, the *cohortes Ituraeorum* of the early Imperial army may have been raised in the late Republic, since units

---

54 App. *B Civ.* 4.88 (17,000 cavalry), 4.108 (20,000 cavalry).

55 Speidel (2016, 84): "It seems perfectly justifiable, therefore, to classify the war of 32 – 30 B.C. not merely as a Roman civil war, but indeed as a true World War."

56 Plut. *Ant.* 61; Cass. Dio 50.13.5–9.

57 Hor. *Epod.* 9.17; Plut. *Ant.* 61; 63; 67; Cass. Dio 50.6.4; 13.5; 51.2.1–3; 7.4.

58 Speidel (2016, 79–95); Haynes (2013); Saddington (1982).

59 As pointed out by Martin (2014, 130).

60 Haynes (2013, 38).

61 Suet. *Aug.* 49: "Of his military forces he assigned legions and auxiliaries to the various provinces" (*Ex militaribus copiis legiones et auxilia provinciatim distribuit*); Tac. *Hist.* 4. 48; Speidel (2016, 88). On the transition from Republican to Imperial auxiliary units: Speidel (2016, 79–95); Haynes (2013, 32–50).

62 *RGDA* 3.3.

63 Tac. *Ann.* 4.5.

of Iturians from Syria are attested in Caesar's and Mark Antony's armies.[64] Similarly, the *ala Atectorigiana*, named after its (probable) founder, the Gallic commander Atectorix, was likely created under Caesar and continued to serve in the Augustan army.[65]

This greater reliance on non-Romans during the civil wars can be explained by two reasons. First, there was, of course, the need to match or outnumber one's adversary in wars of unprecedented magnitude. Since generals and legates sometimes operated without access to Italy's manpower, local recruitment was often the only way not only to raise new auxiliary units but also to reinforce or even form new legions. Second, using contingents sent by provincial peoples allowed generals to levy more soldiers without "breaking the bank." Although, as argued above, Sulla and other Roman generals and politicians found various ways to bolster Rome's finances during this period, the military budget likely remained on a knife's edge for much of the first century.[66] These auxiliary troops, which were usually supplied and paid by their own communities, would have been useful in this context.

It is worth noting here as well, perhaps, that things may not be quite as clear as they appear on the surface. For instance, there is some evidence that auxiliaries were paid by Roman generals in the late Republic, often using local coinage.[67] This can, perhaps, be understood as a desperate attempt to attract more auxiliaries while coping with the high cost of military expenditure. It is impossible to tell which part of auxiliary units was paid by Roman officials, or how much. Nevertheless, the appearance of at least some auxiliary units receiving pay from Rome was a trend that started before the reign of Augustus.[68] Still, the sheer size of the armies of the civil wars probably made it impossible for Roman generals to pay for *all* auxiliary units, since the upkeep of so many legions was already a huge burden. For example, in 40, Octavian was in command of forty legions, a far larger number than were in service in the early Imperial period.[69] The cost of maintaining such a force would have been enormous, which explains why exceptional measures were taken to raise huge sums of money such as proscriptions, new taxes, and exceptional cash requisitions.[70]

---

64 *BAfr.* 20.7; Cic. *Phil.* 2.44.122; Dabrowa (1986, 221).
65 *CIL* 13.1041; Birley (1978, 257–73).
66 According to Dio (46.31.3), some form of *tributum* was reinstituted after the death of Caesar, when the senate declared war on Anthony. Given that Augustus funded the *aerarium militare* with other revenue streams, this was likely only a temporary measure – but it does suggest that the financial pressure of Rome's armies was a constant worry during this period.
67 Caes. *B Civ.* 3.59–60; *BAfr.* 6.1, 8.5. For the use of local coinage: Cadiou (2008, 524–43); Busquets Artigas (2014, 68–97, 138–46, 204–23, 292–339; Martin (2014, 130–32).
68 Speidel (2016, 93–95).
69 App. *B Civ.* 5.53.
70 Proscriptions, new taxes and reintroduction of the *tributum*: Cic. *ad Fam.* 12.30; Cass. Dio 46.31.3; 47.14.2; App. *B Civ.* 4.34. Money taken from temples and from rich women: App. *B Civ.* 4.3; 4.5; 4.32–34, 5.13, 5.22, 5.24, 5.27; Cass. Dio 48.12.4; Mark Antony collecting money in Asia: App. *B Civ.* 5.3, 5.5, 5.6, 5.13, 5.15, 5.22; Plut. *Ant.* 24.4–5.

So, to summarize, the *socii* had provided a cheap and powerful addition to the Roman army during the Republic. However, the enfranchisement of the Italians after the Social War meant the loss of this cheap reservoir of manpower. All of these soldiers, although still serving in Rome's armies, now had to be paid by the Roman treasury. As a result, although auxiliaries were already a common feature before the Social War, their use became even more important after the loss of the *socii* because they were also relatively cheap – and also, conveniently, plentiful. Whereas *socii* had to provide troops for the Roman army and pay them as per the terms of their treaty, auxiliaries could be obtained through a wide range of systems. For example, a Roman commander could conduct a levy or ask local chieftains for military help. In both cases, the natives usually paid for the troops they were providing. The civil wars then led to a "recruitment race" in the provinces during which natives could serve as legionaries or auxiliaries. Generals eventually paid some of their auxiliaries, presumably in order to help win the recruitment battle. All of this set the stage for the reforms of Augustus and the formal establishment of a professional army funded by a military treasury.

Conclusion

Historians have often attributed the increased prominence of auxiliaries in sources for the first century to reforms allegedly implemented by Gaius Marius. His supposed abrogation of the property qualifications was also, according to the argument, accompanied by the disbandment of citizen cavalry and light infantry, the latter replaced by auxiliary units. Yet, this theory actually finds little support in the sources while there is plenty of late Republican evidence militating against this.

In contrast to this standard line, I would suggest that far more emphasis should be accorded to the Social War and its impact on the Roman army. The outbreak of the Social War created a profound manpower problem for Rome, as it was deprived of many allies who normally supplied at least half its infantry and the vast majority of its cavalry. The subsequent enfranchisement of the Italians meant that soldiers, who had previously been recruited as *socii*, were now paid the *stipendium* while they were (like all Roman citizens at the time) dispensed from paying *tributum*. The recourse to auxiliaries was thus a logical expedient both for military and financial reasons, as these troops not only compensated for the loss of much of the Italian manpower during the Social War itself, but they were also (like the *socii* in previous times) paid for by the communities sending them.

Finally, the civil wars of the end of the Republic saw a tremendous mobilization of Roman manpower that was matched by an unprecedented use of foreign manpower, both in auxiliary units and in the legions. This was motivated by the scale of the wars fought at the time but also because military dynasts often did not have access to the main reservoir of Roman citizens – Italy. Non-Romans were thus called upon more than ever before.

Furthermore, raising auxiliary units typically allowed generals to save a great deal of money. Yet, the veritable arms race of the civil wars, as the competing dynasts tried to out-recruit each other, increased even more the value of auxiliary and other non-Roman levies. Thus, Roman commanders seem to have begun to pay auxiliaries, at least on occasion.

The measures taken in recruitment during the civil wars had a lasting impact on the composition of the Roman army, as many units were kept under arms and continued to serve in the permanent army established by Octavian (later Augustus). After his victory over Mark Antony and Cleopatra, Octavian found himself in command of some 60 legions plus a considerable number of auxiliary units: a formidable force that was far too expensive and dangerous to maintain.[71] He demobilized many legions and auxiliary units, and stationed the rest in the provinces. He recognized that the Republican policy of *laissez-faire,* concerning recruitment for governors, was dangerous for political and financial stability. He thus chose to forbid them from levying additional troops and funds without proper authorization.[72] He also instituted more standardized pay for soldiers, both those in the legions and auxiliaries, thus formalizing a practice that occurred irregularly during the civil wars.[73] By formally integrating auxiliaries into the Roman army, Augustus fully acknowledged their importance. Furthermore, by regularizing the length of service, pay, and rewards on discharge for legions and auxiliary units,[74] Augustus completed the transformation of the Roman army that began in the Social War. Yet, no transformation is ever truly complete, and the more sophisticated military structures that Augustus instituted and formalized continued to evolve over the next centuries.

71 Keppie (1984a, 126).
72 Cass. Dio 53.15.6. Examples of auxiliaries raised in the provinces: Tac. *Ann.* 1.56.1, 3.41.3, 12.49.1; *Hist.* 4.17.1, 4.24.1, 4.71.2.
73 Haynes (2013, 48) estimated that an auxiliary infantryman was paid 750 HS/year and a cavalryman 1050 HS/year under Augustus, against 900 HS/year for a legionary infantryman and 900 HS/year for a cavalryman.
74 Cass. Dio 53.15.6; 54.25.5–6; 55.23.1; Suet. *Aug.* 49.2–3; *RGDA* 17.2. Haynes (2013, 38–50); Martin (2014, 135): "La création de corps auxiliaires permanents n'a pas entraîné la disparition des pratiques républicaines et des arrangements *ad hoc.*"

17 Epilogue

Nathan Rosenstein

No student of Republican Rome can doubt the centrality of warfare to its history. That fact has long been patent. What this important collection of papers convincingly demonstrates is just how much its study still has to contribute to our understanding of that history. Its chapters encompass nearly every aspect of the Republic: its *Staatsrecht*, politics, and economy; the motivation of its citizens and allies to go to war; religion, collective memory, and cultural prejudice; as well as the venerable topics of military operations and the evolution of the Roman army. To sum them all up succinctly presents an epilogist with a considerable challenge. Nevertheless, a few key themes emerge, the most prominent being how the army and its leadership responded to the changing nature of the military challenges Rome faced.

Fred Drogula explores the development of military leadership in the early Republic, between the fall of the monarchy and the middle of the fourth century. Like a number of recent scholars, he predicates his analysis on the theory that Rome's population in this era was divided between those living in the city and the various *gentes* that occupied tracts of the surrounding countryside. Warfare was principally a private affair, carried out by warbands drawn from the *gentes* and commanded by their leaders. Only when it required the efforts of both the *gentes* and the city dwellers was a praetor chosen through auspication from among the leaders of the *gentes*. As military challenges increased, however, the warbands sought additional manpower from the city populace, giving rise first to tribunes of the plebs to protect the latter's inhabitants from abuse and then to a demand that leaders seek authority to exercise command via a grant of *imperium* from the *comitia curiata*. Yet, these were not formal magistrates, merely publicly authorized war leaders; hence, the fluctuating number of them appointed annually, whose names have come down to us under the rubric "military tribunes with consular power." Only with the passage of the Licinio-Sextian legislation c. 367 was this irregular appointment of commanders replaced with the annual election of three praetors, two of whom would eventually come to be termed consuls.

DOI: 10.4324/9781351063500-17
This chapter has been made available under a CC-BY-NC-ND license.

Drogula's reconstruction of the evolution of military command has much to commend it, and represents a refinement and elaboration of arguments offered in his 2015 monograph, *Commanders and Command in the Roman Republic and Early Empire* on the origins of *imperium*. The suggestion that the appointments of military tribunes with consular power simply represent instances of war leaders securing public endorsement of their private war-making is particularly helpful: finally, this puzzling practice begins to make some sense. His study contributes to the recent renewed interest in Roman constitutional law and institutions, as well as to the current wholesale revision to our picture of early Rome.

Recent archaeological work, a closer understanding of the nature of the *gentes*, and the rejection of much of the narrative offered by Livy and the annalistic tradition, have been fundamental to this revisionism, and nowhere has this been more true than in the work of Jeremy Armstrong, whose 2016 monograph, *War and Society in Early Rome: From Warlords to Generals*, challenges nearly everything we thought we knew about early Roman warfare. No surprise, therefore, that his chapter offers a radically different take on the development of the early Roman army. Starting from the position that Rome never possessed a phalanx-style army (certainly correct) and building on his own previous studies of the centrality of warbands in early Roman warfare, Armstrong makes the very shrewd observation that little differentiates a warband from a maniple. He sees the latter developing organically out of the former. From the discovery of this "proto-manipular" early Roman army, he is led to wonder whether the manipular army itself, as famously described by Polybius in Book 6, ever in fact existed. He finds Polybius' account overly schematic and idealizing; hence, he rejects it in favor of Livy's description of the Roman army in Book 8, where only one line of soldiers is organized in maniples. Even Polybius' description of the manner in which recruits were sorted into maniples is challenged: the mid-Republican legions, Armstrong argues, remained a patchwork of various contingents mustered on the basis of clan or ethnic ties, much as the *socii* were.

There is more to this bold and challenging chapter than can easily be summarized here, and it is sure to arouse controversy. One wonders how many will be persuaded by his rejection of Polybius – who was, after all, not only a contemporary of the army he describes but also in a position to see it in action and could well have observed its recruitment – in favor of Livy's description of Rome's army c. 340, which is usually seen as "no more than an antiquarian construct."[1] It is worth noting that Polybius' description of Roman castrametation in Book 6 was for all its schematism and idealization long ago confirmed in its essence by Schulten's excavations of the Roman camps at Numantia (now restudied and updated in Michael Dobson's important study[2]).

---

1 Oakley (1998, 463); cf. Rawson (1971a, 26–31).
2 Dobson (2008); see also McCall in this volume.

Still, there is much to commend in this iconoclastic study; progress is rarely made by simply repeating conventional wisdom. In particular, if Armstrong is right to see maniples as growing organically out of the *gentes'* earlier warbands and yet Polybius' account of the sorting of recruits into their maniples is also accurate, then an important problem emerges: at what point and how did the Republic acquire sufficient control over its citizens to be able to force them to go to into combat and risk their lives not beside their friends, relatives, and fellow clansmen, but strangers, to whom only a shared citizenship linked them? At Athens, as van Wees has shown, this critical step seems to have occurred around the turn of the sixth century and marked a major step forward in the formation of the Athenian state.[3] Was the same true at Rome?

The question of the citizen's relationship to the state lies at the center of Peter VanDerPuy's study of warfare and debt in early Rome. In order to understand the turmoil surrounding debt in this period, he argues, we must understand the fundamental change in its nature. The debt small farmers incurred in the sixth and fifth centuries was "embedded" in their relations of dependence on locally powerful figures and was repaid in kind. It was a social bond as much as an economic exchange, and so uncontroversial. The cluster of notices of debt problems in our sources for the fourth and early third centuries reflects a change in the nature of the debt from a largely social to a primarily economic transaction, one repaid in money. Small farmers' need for cash, in turn, grew out of a need to buy weapons and equipment to participate in the wars that Rome was now recruiting them for, as well as to pay the *tributum* to support those who fought them. Cash, too, was needed to finance the farms on the colonial allotments that the victories in those wars produced. All of this arose as a consequence of the growing strength of same state structures in the fourth century that may have brought about the method of selecting recruits for the maniples that Polybius describes. And while the military consequences may have improved the legions' combat effectiveness, the ecological consequences, VanDerPuy argues, were devastating: a need for cash led to intensification of agriculture, destabilization of the landscape, the growing unsustainability of many farms, a demand for new settlements, and ultimately more conquests to win land on which to establish colonies and begin the cycle all over again. One wonders how the Republic managed to pull out of what appears to have been a downward spiral?

James Tan's incisive analysis of the fiscal considerations behind Rome's imposition of citizenship on those it conquered may supply a large part of the answer. Early fourth century complaints about the burdens of *tributum,* in his view, led the senate to make new citizens out of a number of recently conquered enemies: enlarging the number of taxpayers diminished the tax any one of them had to pay. Expanding the *dilectus-tributum* system to the

---

3 Van Wees (2004, esp. 80–81).

whole of Italy, however, proved impractical. Employing a rational choice model, Tan analyzes the costs and benefits of the senate's options. Creating Latin colonies, in his view, represented a win-win for Rome. They contributed contingents to Rome's armies, but not only was the Republic under no obligation to pay these soldiers, no financial down-side resulted from allowing Romans to join them. Because *tributum* was (principally) a tax on land, the farms the colonists had once owned continued to yield *tributum* – only now paid by the citizens who became their new owners. Equally cheap to run were unpaid contingents of *socii* drawn from areas the senate considered too poor to exploit for anything but manpower. And the senators found, in the creation of citizenship without the vote, a way to have their fiscal cake and eat it, too. *Cives sine suffragio,* like the Campanians, were subject to the *dilectus* as well as payment of *tributum*, but the *patres*, while happily availing themselves of the latter, could keep their poor-quality infantry in separate units and utilize them or not as circumstances dictated.

The picture Tan paints accords well with our impression of the mid-Republican senators that Polybius knew: hard-headed and tight-fisted where money was concerned (e.g. 31.26.9–27.11). Yet, Tan recognizes that Rome could not in every case impose the status most fiscally advantageous to itself upon those it subjugated. Other considerations could supervene and, as the camps at Numatia demonstrate in concrete terms, the Romans adapted well-established systems and frameworks to facts on the ground. However, the *stipendium* side of the *dilectus-tributum* balance remains uncertain: expanding the number of taxpayers only lightens each one's burden if the number of troops to be paid remains relatively constant. That seems to have been largely the case down to the eve of the Hannibalic War – with notable exceptions, but more investigation is called for.

How Rome and its allies profited from its conquests also forms the focus of Saskia Roselaar's survey of the acquisition and distribution of one of the most important kinds of war spoils: land mulcted from Rome's defeated enemies. Warbands in the early Republic seized cattle, easily transportable valuables, and slaves, but as the character of Rome's wars changed, so did the booty. Land became the principal prize of war. Yet, the Romans never established formal procedures for distributing spoils, raising the danger that a general might use them to win popularity. So, the *patres* regularly dispatched colonies but hesitated over *viritim* allotments, reluctant to increase the power of the man who handed them out. The soldiers' share of the spoils, however, constituted a critical element in the glue that bound the Italians to Rome, for allied contingents fought in the expectation of material rewards. *Socii* received cash and precious objects, but land in colonies, Roselaar argues, went only to Latin allies, no others, at least until the second century. The shock provided by the Second Punic War and the subsequent era of rapid overseas expansion brought significant changes and challenges. The Romans modified some of the practices with respect to *ager publicus*: some new viritim allotments were made, and a few of a new type of Roman colony were founded. Yet, land distribution dried up by the second quarter of the second century. On the one hand, Roselaar argues, this was because Roman elite were not interested in sharing out more public land, either because of political jealousy, or because they did not see the strategic need. On the other hand, oversees expansion put greater emphasis again on the acquisition of moveable booty, as well as encouraging the development of new kinds of exploitation such as taxes and indemnities. The unwillingness of the Roman state to share more *ager publicus* bred discontent, which exploded when the Gracchi brothers attempted not only to redistribute *ager publicus* but also to found new colonies, including overseas. The economic and political importance of land ownership, and so therefore the significance of access to land taken from Rome's enemies, continued through the first century and was a major subject of debate resolved only with the fall of the Republic.

Financial benefits and economic imperatives are likewise at the center of François Gauthier's study of the developments that shaped the Roman army of the late Republic. He starts from a position that has become increasingly accepted among scholars (although unfortunately not among popular writers), namely that Marius was not responsible for the key changes that distinguished first-century legions from their mid-Republican predecessors: the elimination of citizen cavalry and light infantry, and the replacement of maniples with cohorts. He highlights the evidence for Roman citizens continuing to serve as cavalrymen in the first century and notes that evidence for the abandonment of citizen *velites* is inconclusive. However, there is no disputing that auxiliary cavalry and light infantry become ubiquitous in first-century Roman armies. The explanation, Gauthier argues, lies in the senate's pressing need for troops during the Social War. *Proletarii* had to be pressed into the legions and so were unavailable for light infantry service while fiscal realities led the *patres* to shift the principal burden of mounted combat to auxiliary cavalry. A citizen cavalryman's *stipendium* cost three times an infantryman's while the treasury paid nothing for auxiliaries. Hence, more bang for the buck, and the settlement of the Social War did nothing to make those savings any less attractive. Once all *socii* became citizens, the recruits they had previously sent to Rome's armies at the allied communities' expense now became legionaries that the treasury had to pay without having the *tributum* to tap for the funds to do so.

Rome's increased reliance on foreign *auxilia* reached its logical conclusion, in Gauthier's view, during the civil wars, when commanders desperate for troops enrolled provincials into their legions. Caesar's *Alaudae* anticipated Pompey's generals' recruitment of non-Romans in the run-up to Pharsalus as well as in Spain and elsewhere, and the practice continued in the struggle between Octavian and Anthony, ultimately transforming Roman armies into "a mix of various peoples from all over the Mediterranean world." Perhaps so, but Brunt presumes those same legions comprised exclusively citizens; indeed, they provide what he sees as the only sound basis for estimating the number of citizens domiciled overseas during the first century and so an important element in his calculation about the size of the Italian population at that time.[4] Gauthier will need to confront Brunt's arguments head-on before his claim that the Roman army became "globalized" during the civil wars can carry conviction.

The Republic's elite and their Italian allies were not the only ones concerned with the costs and benefits of war, according to Marian Helm. Ordinary recruits, in his view, saw their service in terms of a return on investment. Because the cost of equipment was higher for soldiers in the first census class than other *manipularii* and cheapest for *velites*, equal shares of booty produced unequal "Return on Investments" (ROIs). Accordingly, motivation to serve varied: the poor, who served primarily as light infantry, saw their low-cost military service as potentially yielding a high ROI and an (albeit risky) avenue to social mobility. Wealthy citizens serving at a higher cost to themselves, by contrast, had less to gain from undergoing the dangers of combat and so served in less exposed parts of the battle line. Hence, the Republic's way of war conduced to political stability, protecting its more valuable citizens while offering poorer ones a chance to better their lot. Young Roman aristocrats were equally calculating when considering military service, but their accounting focused on political rather than financial gains. War offered them an opportunity to display valor and advertise their protection of ordinary citizens, assets that, spread abroad by the men who served under them, would pay dividends as these aristocrats made their way up the *cursus honorum*. This connection between elite and commons within the legions also fostered social cohesion and underwrote political stability.

Yet, Kathryn Milne reminds us that humans are complex creatures and their motivations not easily reducible to a simple financial spreadsheet. Opportunities to win honor in battle, in her analysis, were a principal spur for recruits to go to war. The decorations and awards they won were acknowledged and their experiences validated, approved, and celebrated in triumphs, which, in turn, incited others to seek similar esteem. Indeed, she argues, triumphs were as much about public acknowledgment of the soldiers' valor as the general's leadership. The prestige reflected in the decorations soldiers displayed as they marched, and the spoils they hung outside their houses, stemmed both from a common standard for judging what deeds merited them and a sense that generals who awarded them did so appropriately and fairly. Adherence to those high standards, in turn, attested to a general's discernment and enhanced the prestige of the decorations he bestowed. Soldiers who celebrated their commanders and officers with songs during a triumph thus underscored the legitimacy of their own awards. Even soldiers whose legions never marched in triumph in Rome had their service commemorated via spoils, memorials, temples, and paintings in Rome, and often in the public spaces of their *municipia* as well as outside

4 Brunt (1971a, 227–33), cf. 118.

of private homes. These things constituted a kind of narrative of valor and its benefits that perpetuated enthusiasm for military service.

Helm and Milne are both right. People act from a variety of often complicated motives. When 4,000 or 5,000 men came together to form a Roman legion, it would be surprising if many different reasons did not impel each of them to take up arms. What Lee Brice shows us is how the discipline that was also a key factor in soldiers' willingness to fight could collapse under a variety of circumstances. Indiscipline might take a variety of forms, ranging from a collective expression of grievances to a refusal to obey orders, desertion, mutiny, and even outright cowardice in the face of the enemy. Most instances occurred during the internal conflicts and political turmoil of the first century, but as Brice notes the impression that acts of disobedience were especially frequent in the first century may simply be an artifact of the abundant sources for this period. And while political turmoil could create conditions in which insubordination could occur, other factors – in particular unit cohesiveness, loss of confidence in officers, a lack of opportunities to register complaints or lack of fear of punishment – all increased its likelihood.

Even in far less tumultuous times, the ways the Republic conducted its wars profoundly affected politics, as Jeremiah McCall demonstrates. Beginning from the fact that Roman generals were politicians elected to these posts without necessarily having much if any experience in command of armies, he asks an obvious question: what role did commanders actually play in the success or failure of their armies? As he rightly notes, Roman manipular armies were collections of systems; hence, he tackles the problem in systemic terms: where were the fundamental command decision points and who made those decisions? The senate assigned armies, designated *provinciae*, and funded campaigns. Subordinate officers, usually not selected by their general, recruited the troops, organized them into their maniples, exercised day-to-day command, and played an important role in managing logistics. The order of march, layout of marching camps, and deployment of units in the battle line were all largely dictated by established protocol. It emerges therefore that a general's decisions were critical at a relatively limited number of points. Selecting a route to the enemy and where and when to offer battle were the most important. But a general rarely made these decisions alone. His *consilium* offered guidance and a check on rash decisions. Once combat began, tribunes and centurions maintained unit cohesion and kept the soldiers fighting. Cavalry typically launched flank attacks on their own initiative while generals rarely ordered infantry maneuvers more complex than bringing up the reserves if the front lines wavered. Their role, rather, was mainly to inspire their men and order the pursuit of a defeated enemy. "And so," McCall concludes, "Roman aristocrats could happily compete for the consulship, knowing that...they would not generally require any special qualifications, other than simply being an aristocrat, to avoid disaster and probably even secure some level of victory in their year of command." And so as well, the system underwrote the practice of allocating military commands to many different annually elected aristocrats, offering each a chance at military glory and the aristocracy collectively a reason to support the political status quo.

Nowhere however did the effects of war more dramatically reshape that political status quo than in the early years of the Hannibalic War. By teasing out the implications of M. Fabius Buteo's adlection of 177 new senators in the wake of Cannae, Cary Barber exposes how a "lost generation" of *patres* altered the distribution of power in the *curia*. Recognizing the limitations of the evidence preserved in our narrative sources, he turns to demography and a nuanced examination of three models of high mortality populations in order to estimate the senate's age structure after the battle. In each, the number of *iuniores* is very close to the 177 dead senators whose places the senators chosen in 215 took. That these dead senators perished in the war's early battles, Barber argues, seems certain based on the nearly total mobilization of senate's *iuniores* as well as equestrians reported by Livy while few *seniores* among the *patres* are likely to have accompanied them into battle. What emerges from his analysis is a picture of a radical distortion of the senate's normal age structure: few men under 45 and a preponderance of senators over that age. Buteo's enrollment of new senators redressed that imbalance, but these men will all have been either much younger than usual or candidates who had seen their progress up the *cursus honorum* stall before they reached offices that would have brought them into the senate. As a group, their collective *auctoritas* was therefore minimal, and that state of affairs opened the way for a few very senior *patres*, notably Q. Fabius Maximus, M. Claudius Marcellus, and Q. Fulvius Flaccus, to dominate the senate and control the course of the war. Barber's findings shed important new light on the senate's internal dynamics during the war as well as offer intriguing hints about how the normal balance of age and *auctoritas* therein was maintained.

One of the men who most benefited from the political backlash that long dominance provoked was T. Quinctius Flamininus, the conqueror of Philip of Macedon. While his rapid rise and remarkable career have not lacked scrutiny, Michael Fronda focuses on a neglected – but as he shows, quite significant – incident, his victorious army's "triumph-like" march in 194 from Brundisium to Rome. The procession finds its context in the scaled-up competition in display among those generals whose spectacular victories established Rome's second-century dominion over the Hellenistic East. The rich hauls of booty these conquests produced fueled ever more elaborate triumphs, games, and monumental building intended to foster popularity, enhance prestige, and lay the groundwork for future political power. Flamininus' aim was no different and, as Fronda shows, intended to rival in quite specific ways the magnificent triumph Scipio Africanus celebrated for his conquest of Hannibal and Carthage. The political challenge Flamininus' spectacle posed to Scipio's predominance was met later in 194, when the two faced off, backing rival candidates for the following year's consulate. Flamininus' triumph, so Livy avers, helped sway the voters to elevate his candidate over Scipio's. Yet, the audience Flamininus' procession addressed extended far beyond Rome. Traveling up (probably) the *Via Appia* it passed through areas allotted to Scipio's veterans as well as territories inhabited by Rome's allies, many of whom will have lined the route to witness the spectacle. What is more, Flamininus' procession included the contingents of *socii* that had fought alongside the legions, whose participation in Roman triumphs up to that point had been, in Fronda's view, rare if not lacking altogether. Flamininus' triumph thus set a precedent. It formed part of a larger Roman effort to conciliate formerly rebellious subjects upon whose increasing participation in its wars the Republic depended in the years following the Hannibalic War.

Perhaps so, yet in the absence of Livy's second decade, it is impossible to know what was innovative in the second century and what simply followed pre-Hannibalic practice. The risks of *ex silentio* arguments are obvious. Whichever the case, however, there is no doubt that the *socii* followed Rome to war for a share of the spoils and a cut of the profits as well as a portion of the glory, all of which went a long way toward reconciling the Italians for many years to Rome's hegemony.

Yet in the end, conciliation proved unattainable in the face of irreconcilable conflicts that finally eventuated in revolt. Untangling course of the Social War continues to frustrate historians and, Jessica Clark argues, we should not even try. No ancient source offers us a linear narrative of events, nor did the war itself form a single, distinct campaign. Several different conflicts played out simultaneously. Attempts to form a sequential account of what happened must work against the grain of how events unfolded. Clark therefore offers a vision of the war that, in her view, would have been "recognizable, and meaningful" to the generation that followed by focusing on the inconsistencies, uncertainties, and unknowns contained in three anecdotes from the war. In unpacking these events, she exposes how memories of the war shaped events to fit the needs of those who came afterwards. In the stories of events at Pinna, Aesernia, and Grumentum, inhabitants of these Italian and Roman towns, when forced to make difficult choices, acted (by their lights) honorably. Yet those choices cannot be understood in isolation; they must be set in contexts of terrible violence. And in the story of P. Ventidius Bassus, who was a baby in his mother's arms when she marched, a captive, in a Roman triumph but went on to gain the consulate and a triumph of his own and in the epitaph of Sergius, a freedman's son who died in a Roman defeat, we see parents confront the fates that sons suffer in war. Wars, she reminds us, are not fought only on the battlefield; they waged in the theater of memory as well.

Those memories, as Jack Wells reminds us, can reach far back into the past, even into the realm of myth. He addresses one of the Republic's more surprising contradictions: on the one hand, the Romans despised freed slaves, many of whom they had captured in war; on the other, manumission made them full citizens of the Republic. This paradox, he argues, was focalized and resolved via the legend of Servius Tullius, Rome's sixth king, which presented "ideology in narrative form." Servius, as the Romans remembered him, was a great king, responsible for many of Rome's foundational institutions. Yet, his origins were also servile: born a slave or – perhaps – a free infant, born to a once war-captive mother. Servius thereby became a vehicle for freeborn Romans of the late Republic to express their ambiguous feelings about slavery and the practice of manumission. Once long ago, in their view, slaves had earned their freedom through loyalty and personal character and their admission to the citizen body strengthened Rome. Now, however, disreputable slaves simply bought their freedom or masters freed large numbers by will to glorify themselves. For slaves and freedmen, on the other hand, the message Servius' story conveyed was ambiguous. While the king inaugurated the practice of granting freed slaves citizenship and established the festival of the Compitalia, giving slaves and freedmen a place within the community, his servile origins were never forgotten and played a central role in his downfall. This aspect of his story told freedmen that despite their legal equality as full citizens, socially there were always going to be limits that they would do well not to forget. But the fact that a slave had become king at Rome also told them that they could make their way in spite of the prejudice against them.

Ideological dimensions of war are also central to John Serrati's examination of the relationship between Rome's two ancient war gods, Mars and Bellona. Curiously, however, while Mars – like all other martial deities at Rome – was masculine, Bellona was indisputably a goddess, the only one associated with war. The reason, Serrati argues, is to be found in the ideology associated with warfare at Rome. While, for Roman soldiers, war was about loot and personal glory, the fetial rituals carried out at Bellona's temple that marked the opening of a war demonstrate that the Romans always conceived of themselves as the victims of enemy aggression. Having been refused a rightful recompense, the *fetiales* proclaimed, Rome undertook to wage war for a just revenge. And vengeance, Serrati notes, is particularly associated with women in Roman literature. Hence, a female divinity was the appropriate embodiment of Roman vengeance, whose primary function was the pursuit of justice through warfare. In that role, the Romans also associated Bellona with the chaos, fury, and sounds of battle itself. When civil war erupted between Sulla and his Marian enemies, however, warfare could no longer be associated with Roman vengeance. Bellona, a particular favorite of Sulla's, lost that aspect of her persona and was left with only battlefield chaos. This negative association carried over into the Augustan era, when she came to represent Romans spilling one another's blood.

To have had this impressive collection of papers by friends, some of whom I have had the privilege of teaching, dedicated to me is an unexpected and deeply heartwarming honor, and it is a great pleasure to express my sincere thanks to its editors, who also organized the session of the Celtic Classical

Conference in Montreal in 2017. Several of the talks presented there appear here as revised and expanded chapters. Some take up questions my own work on Roman warfare has touched on in one way or another over the years, and in every case, they have advanced the discussion and brought fresh and important perspectives to the debate. Others strike out in new directions, surprising and impressing with their boldness and novel approaches. All represent ways forward in the study of Republican warfare and reveal how much more there is to be said on the ways the Romans waged war and the ways those wars affected Rome.

Notably, however, no study here examines a particular battle, analyzes a campaign, or traces the strategies pursued by the belligerents in a war – the traditional stuff of military history. This is not entirely surprising. Although warfare ancient and modern continues to attract a substantial readership among the general public and draw healthy undergraduate enrollments, academic historians of Roman warfare, at least, tend to shy away from these topics. Possibly, this is the fault of our sources: apart from Caesar and the remains of Polybius, we lack reliable contemporary (or near contemporary) historians like Herodotus or Thucydides whose detailed accounts could serve as the basis for scholarly study of the events comprised in the Republic's many wars. But I think there is more to it than that. Among scholars there is a sense that "drums and trumpets" military history – narrative accounts of battles and campaigns – is not the stuff of "serious history," that is, work that will command the respect of academic peers in other disciplines (and one suspects contribute toward their positive votes in favor of tenure and/or promotion at institutions of higher learning). This is a pity. War is too important a subject to be left to popularizers, whose knowledge too often is a generation or two out of date and whose ideas about how the Romans waged war do a disservice to the realities involved. As someone who has enjoyed and benefited greatly from the blessings of tenure and promotion at a major university, I would never dismiss the concerns of those who are working to attain them. But to ignore the popular audience for Roman military history – or any other field of history for that matter – does a disservice both to the public and to our profession. It withholds the fruits of current scholarship and the best studies from interested non-specialists who, in many cases, are the taxpayers who fund the institutions that employ us. At the same time, it contributes to a tendency among the public to dismiss what we academic historians do as unimportant and irrelevant and – at worst – not worth supporting with those tax dollars. So, perhaps some of us, and especially those who have reached the stage of their careers where we no longer have to worry about tenure, or promotion votes, or even merit raises any longer, should turn our hands to the sorts of narrative military history that can engage, in a constructive way, with the interests of a general readership. One cannot pretend this will be a panacea for all the challenges facing academic history and the liberal arts today, but it may contribute in a small way to fighting those battles.

Bibliography

Adamo, M. (2018), "*Sedes et Rura*: Landownership and the Roman Peasantry in the Late Republic," DPhil dissertation, University of Oxford, Oxford.

Adcock, F. E. (1957), "Consular Tribunes and their Successors," *JRS* 47, 9–14.

Afzelius, A. (1944), *Die römische Kriegsmacht während der Auseinandersetzung mit den hellenistischen Grossmächten: Studien über die römische Expansion II*. Aarhus.

Afzelius, A. (1975), *Two Studies on Roman Expansion*. New York.

Aigner, H. (1974), "Gedanken zur sogenannten Heeresreform des Marius," in F. Hampl, and I. Weiler (eds.), *Kritische und vergleichende Studien zur Alten Geschichte und Universalgeschichte*. 11–23. Innsbruck.

Alexiou, M. (2002), *The Ritual Lament in Greek Tradition*[2]. Lanham, MD.

Alföldi, A. (1965), *Early Rome and the Latins*. Ann Arbor, MI.

Álvarez Pérez-Sostoa, D. (2015), "*Clementia* o 'visión diplomática': devolución voluntaria de los cautivos en la república romana," in B. Grass and G. Stouder (eds.), *La diplomatie romaine sous la République*. 107–25. Besançon.

Anders, A. (2015), "The 'Face of Roman Skirmishing,'" *Historia* 64, 263–300.

Anderson, G. (2005), "Before *Turannoi* Were Tyrants: Rethinking a Chapter of Early Greek History," *ClAnt* 24, 173–222.

Anderson, G. (2009), "The Personality of the Greek State," *JHS* 129, 1–22.

Ando, C. (2009), *The Matter of the Gods: Religion and the Roman Empire*. Berkeley, CA.

Ando, C. (2016), "Making Romans: Citizens, Subjects and Subjectivity in Republican Empire," in M. Lavan, R. E. Payne, and J. Weisweiler (eds.), *Cosmopolitanism and Empire: Universal Rulers, Local Elites, and Cultural Integration in the Ancient Near East and Mediterranean*. 169–86. Oxford.

Arlacchi, P. (1983), *Mafia, Peasants, and Great Estates: Society in Traditional Calabria*. Translated by Jonathan Steinberg. Cambridge.

Armstrong, J. (2008), "Breaking the Rules: Recruitment in the Early Roman Army (509–450 BC)," in E. Bragg, L. Hau, and E. Macaulay-Lewis (eds.), *Beyond the Battlefields: New Studies on Warfare and Society in the Graeco-Roman World*. 47–66. Cambridge.

Armstrong, J. (2013a), "Claiming Victory: The Early Roman Triumph," in A. J. Spalinger, and J. Armstrong (eds.), *Culture and History of the Ancient Near East: Rituals of Triumph in the Mediterranean World*. 7–22. Leiden.

Armstrong, J. (2013b), "'Bands of Brothers': Warfare and Fraternity in Early Rome," *JAH* 1, 53–69.

Armstrong, J. (2016a), *Early Roman Warfare: From the Regal Period to the First Punic War*. Barnsley.

Armstrong, J. (2016b), "The Ties That Bind: Military Cohesion in Archaic Rome," in Armstrong (2016d), 101–19.

Armstrong, J. (2016c), *War and Society in Early Rome: From Warlords to Generals*. Cambridge.

Armstrong J. (ed.) (2016d), *Circum Mare: Themes in Ancient Warfare*. Leiden.

Armstrong, J. (2017a), "The Consulship in 367 BC and the Evolution of Roman Military Command," *Antichthon* 51, 124–48.

Armstrong, J. (2017b), "The Origins of the Roman *Pilum* Revisited," *JRMES* 18, 65–74.

Astin, A. E. (1978), *Cato the Censor*. Oxford.

Attema, P., T. de Haas, and M. Termeer (2014), "Early Colonization in the Pontine Region (Central Italy)," in Stek and Pelgrom (2014), 211–32.

Badian, E. (1958), *Foreign Clientelae (264–70 B.C.)*. Oxford.

Badian, E. (1968), *Roman Imperialism in the Late Republic*. Ithaca, NY.

Balmaceda, C. (2017), *Virtus Romana: Politics and Morality in the Roman Historians*. Chapel Hill, NC.

Bannon, C. J. (2009), *Gardens and Neighbors: Private Water Rights in Roman Italy*. Ann Arbor, MI.

Barber, C. (2016), *The Lost Generation of the Roman Republic: Elite Losses and the Senate of the Hannibalic War*, PhD dissertation, The Ohio State University, Columbus, OH.

Barber, C. (2020), "Quibus patet curia: Livy 23.23.6 and the Mid-Republican Aristocracy of Office," *Historia* 69 (forthcoming).

Baronowski, D. W. (1984), "The 'Formula Togatorum,'" *Historia* 33, 248–52.

Baronowski, D. W. (1993), "Roman Military Forces in 225 B.C. (Polybius 2.23–4)," *Historia* 42, 181–202.

Bastien, J.-L. (2007), *Le triomphe romain et son utilisation politique à Rome aux trois derniers siècles de la République*. Rome.

Beacham, R. C. (1991), *The Roman Theatre and its Audience*. London.

Beard, M. (2007), *The Roman Triumph*. Cambridge, MA.

Beard, M., J. North, and S. R. F. Price (1998a), *Religions of Rome, Vol. I: A History*. Cambridge.

Beard, M., J. North, and S. R. F. Price (1998b), *Religions of Rome, Vol. II: A Sourcebook*. Cambridge.

Beck, H. (2005), *Karriere und Hierarchie: Die römische Aristokratie und die Anfänge des cursus honorum in der mittleren Republik*. Berlin.

Beck, H. (2011a), "Consular Power and the Roman Constitution: The Case of *Imperium* Reconsidered," in Beck, Duplá, Jehne, and Pina Polo (2011), 77–96.

Beck, H. (2011b), "The Reasons for the War," in Hoyos (2011), 225–41.

Beck, H. (2011c), "Ineditum Vaticanum(839)," in *BNJ* (doi:10.1163/1873-5363_bnj_a839).

Beck, H. (2015), "Beyond 'Foreign Clienteles' and 'Foreign Clans': Some Remarks on the Intermarriage between Roman and Italian Elites," in Jehne and Pina Polo (2015), 57–72.

Beck, H., A. Duplá, M. Jehne, and F. Pina Polo (eds.) (2011), *Consuls and Res Publica. Holding High Office in the Roman Republic*. Cambridge.

Beck, H., M. Jehne, and J. Serrati (eds.) (2016), *Money and Power in the Roman Republic*. Brussels.

Beck, H., and U. Walter (2001), *Die Frühen Römischen Historiker I: Von Fabius Pictor bis Cn. Gellius.* Darmstadt.

Beck, H., and U. Walter (2004), *Die Frühen Römischen Historiker II: Von Coelius Antipater bis Pomponius Atticus.* Darmstadt.

Belier, W. W. (1991), *Decayed Gods: Origin and Development of Georges Dumézil's "Idéologie Tripartie".* Leiden.

Bell, M. J. V. (1965), "Tactical Reform in the Roman Republican Army," *Historia* 14, 404–22.

Beloch, J. (1886). *Die Bevölkerung der griechisch-römischen Welt.* Leipzig.

Beloch, K. J. (1926), *Römische Geschichte bis zum Beginn der punischen Kriege.* Berlin.

Berrendonner, C. (2001), "La formation de la tradition sur M. Curius Dentatus et C. Fabricius Luscinus: un homme nouveau peut-il être un grand homme?," in M. Coudry, and T. Späth (eds.), *L' invention des grands hommes de la Rome antique: Actes du colloque du Collegium Beatus Rhenanus, 16–18 septembre 1999.* 97–116. Paris.

Bergk, A. (2011), "The Development of the Praetorship in the Third Century BC," in Beck, Duplá, Jehne, and Pina Polo (2011), 61–74.

Bernard, S. (2016), "Debt, Land, and Labor in the Early Republican Economy," *Phoenix* 70, 317–38.

Berry, W. (2015), *The Unsettling of America: Culture and Agriculture.* Updated edition. Berkeley, CA.

Billows, R. (1989), "Legal Fiction and Political Reform at Rome in the Early Second Century B.C.," *Phoenix* 43, 112–33.

Billows, R. (2000), "Polybius and Alexander Historiography," in A. Bosworth and E. Baynham (eds.), *Alexander the Great in Fact and Fiction.* 286–306. Oxford.

Birley, E. (1978), "*Alae* Named After their Commanders," *AncSoc* 9, 257–73.

Bishop, M. C. (2016), *The Gladius: The Roman Short Sword.* Oxford.

Bishop, M. C. and J. C. N. Coulston (1993), *Roman Military Equipment: From the Punic Wars to the Fall of Rome.* London. Second, Revised Edition 2006.

Bispham, E. (2006a), "Literary Sources," in Rosenstein and Morstein-Marx (2006), 29–50.

Bispham, E. (2006b), "Coloniam Deducere: How Roman was Roman Colonization during the Middle Republic?" in Wilson and Bradley (2006), 73–160.

Bispham, E. (2007), *From Asculum to Actium: The Municipalization of Italy from the Social War to Augustus.* Oxford.

Bizzarri, E. (1973), "Titolo Mummiano a Fabrateria Nova," *Epigraphica* 35: 140–2.

Bleckmann, B. (2002), *Die römische Nobilität im Ersten Punischen Krieg: Untersuchungen zur aristokratischen Konkurrenz in der Republik.* Berlin.

Bleckmann, B. (2016), "Roman War Finances in the Age of the Punic Wars," in Beck, Jehne, and Serrati (2016), 82–96.

Blois, L. de (1987), *The Army and Politics in the First Century B.C.* Amsterdam.

Blois, L. de (2000), "Army and Society in the Late Roman Republic: Professionalism and the Role of the Military Middle Cadre," in G. Alföldy, B. Dobson, and W. Eck (eds.), *Kaiser, Heer und Gesellschaft in der Römischen Kaiserzeit.* 11–31. Stuttgart.

Blois, L. de (2007), "Army and General in the Late Roman Republic," in Erdkamp (2007c), 164–79.

Boddington, A. (1959), "The Original Nature of the Consular Tribunate," *Historia* 8, 356–64.

Bolle, H.-J. (2003), "Climate, Climate Variability, and Impacts in the Mediterranean Area: An Overview," in H.-J. Bolle, M. Menenti, and I. Rasool (eds.), *Mediterranean Climate: Variability and Trends*. 5–86. Berlin.

Boren, H. C. (1980), "Rome, Republican Disintegration, Augustan Reintegration: Focus on the Army," *Thought* 55, 51–64.

Botermann, H. (1968), *Die Soldaten und die römische Politik in der Zeit von Caesars Tod bis zur Begründung des Zweiten Triumvirats*. Munich.

Botsford, G. W. (1909), *The Roman Assemblies from Their Origin to the End of the Republic*. New York.

Bourdin, S. (2012), *Les Peuples de l'Italie Préromaine: Identités, Territoires et Relations Inter-Ethniques En Italie Centrale et Septentrionale (VIIIe–Ier s. Av. J.-C.)*. Rome.

Bradeen, D. W. (1959) "Roman Citizenship *per magistratum*," *CJ* 54, 221–28.

Bradley, G. (2014), "The Nature of Roman Strategy in Mid-Republican Colonization and Road Building," in Stek and Pelgrom (2014), 60–72.

Bradley, K. (1987), *Slaves and Masters in the Roman World: A Study in Social Control*. Oxford.

Bradley, K. (1994), *Slavery and Society at Rome*. Cambridge.

Breitenberger, B. M. (2007), *Aphrodite and Eros: The Development of Erotic Mythology in Early Greek Poetry and Cult*. London.

Bremmer, J. N. (1982), "The *Suodales* of Poplios Valesios,"·*ZPE* 47, 133–47.

Brennan, T. C. (1996), "Triumphus in Monte Albano," in Wallace and Harris (1996), 315–37.

Brennan, T. C. (2000), *The Praetorship in the Roman Republic*. Two vols. Oxford.

Brennan, T. C. (2012), "Roman Legal Ideology in the Military Sphere: Insights on *aequitas* from the Case of the Caudine Forks (321 BC)," in J.-L. Ferrary (ed.), *Leges publicae. La legge nell'esperienza giuridica romana*. 475–88. Pavia.

Brennan, T. C. (2014), "Power and Process under the Republican 'Constitution,'" in Flower (2014), 19–53.

Brice, L. L. (2011), "Disciplining Octavian: An Aspect of Roman Military Culture during the Triumviral Wars, 44–30 BCE," in W. E. Lee (ed.), *Warfare and Culture in World History*. 35–60. New York.

Brice, L. L. (2015a), "Military Unrest in the Age of Philip and Alexander of Macedon: Defining the Terms of Debate," in T. Howe, E. E. Garvin, and G. Wrightson (eds.), *Greece, Macedon and Persia: Studies in Social, Political and Military History in Honour of Waldemar Heckel*. 69–76. Oxford.

Brice, L. L. (2015b), "Second Chance for Valor: Restoration of Order after Mutinies and Indiscipline," in L. L. Brice and D. Slootjes (eds.), *Aspects of Ancient Institutions and Geography: Studies in Honor of Richard J.A. Talbert*. 103–21. Leiden.

Briquel, D. (2000), "Le tournant du IVe siècle," in F. Hinard (ed.), *Histoire romaine: Tome I: Des origines à Auguste*. (In collaboration with G. Brizzi). 203–243. Paris.

Briquel, D. (2010), "Une alternative à la vision romaine: l'Italie dans le monnayage des insurgés de la guerre sociale," in A. Colombo, S. Pittia, and M.T. Schettino (eds.), *Mémoires d'Italie*. 83–100. Como.

Briscoe, J. A. (1973), *A Commentary on Livy, Books XXXI–XXXIII*. Oxford.

Briscoe, J. A. (1981), *A Commentary on Livy, Books XXXIV–XXXVII*. Oxford.

Briscoe, J. A. (2007), *A Commentary on Livy, Books 38–40*. Oxford.

Briscoe, J. A. (2012), *A Commentary on Livy, Books 41–45*. Oxford.

Broadhead, W. (2008), "Migration and Hegemony: Fixity and Mobility in Second-century Italy," in de Ligt and Northwood (2008), 451–470.

Broughton, T. R. S. (1951), *The Magistrates of the Roman Republic Volume 1: 509 B.C.–100 B.C.* New York.

Broughton, T. R. S. (1952), *The Magistrates of the Roman Republic Volume 2: 99 B.C.–31 B.C.* New York.

Brunt, P. A. (1962), "The Army and the Land in the Roman Revolution," *JRS* 52, 69–86. Revised and reprinted in *The Fall of the Roman Republic and Related Essays* (1988). 240–275. Oxford.

Brunt, P. A. (1965), "Italian Aims at the Time of the Social War," *JRS* 55, 90–109.

Brunt, P. A. (1971a), *Italian Manpower: 225 B.C.–A.D. 14*. Oxford.

Brunt, P. A. (1971b), *Social Conflicts in the Roman Republic*. London.

Brunt, P. A. (1974), "Conscription and Volunteering in the Roman Imperial Army," *SCI* 90–115. Reprinted in *Roman Imperial Themes* (1990). 188–215. Oxford.

Brunt, P. A. (1988), *The Fall of the Roman Republic and Related Essays*. Oxford.

Bruun, C. (ed.) (2000), *Roman Middle Republic: Politics, Religion, and Historiography*. Rome.

Bucher, G. (2000), "The Origins, Program, and Composition of Appian's Roman History," *TAPhA* 130, 411–58.

Budin, S. L. (2002), "Creating a Goddess of Sex," in D. L Bolger, and N. Serwint (eds.), *Engendering Aphrodite: Women and Society in Ancient Cyprus*. 315–24. Boston.

Bunse, R. (1998), *Das römische Oberamt in der frühen Republik und das Problem der "Konsulartribunen"*. Trier.

Buonocore, M., A. R. Staffa, and L. Franchi dell'Orto (eds.) (2010), *Pinna Vestinorum. La città romana*. Rome.

Burnett, A. (2012), "Early Roman Coinage and its Italian Context," in W. E. Metcalf (ed.), *The Oxford Handbook of Greek and Roman Coinage*. 297–314. Oxford.

Burns, M. T. (2003), "The Homogenisation of Military Equipment under the Roman Republic," in *"Romanization"? Digressus Supplement* 1, 60–85.

Burns, M. T. (2006), "The Cultural and Military Significance of the South Italic Warrior's Panoply from the 5th to the 3rd Centuries BC," PhD dissertation, Institute of Archaeology, University College London, London.

Burton, P. J. (2011), *Friendship and Empire: Roman Diplomacy and Imperialism in the Middle Republic (353–146 BC)*. Cambridge.

Burton, P. J. (2017), *Rome and the Third Macedonian War*. Cambridge.

Busquets Artigas, S. (2014), "Los externa auxilia en el siglo final de la República romana (133–27 a.C.)," PhD dissertation, Universitat Autònoma de Barcelona, Barcelona.

Cadario, M. (2014), "Preparing for Triumph: *Graecae Artes* as Roman Booty in L. Mummius' Campaign (146 BC)," in Lange and Vervaet (2014), 83–101.

Cadiou, F. (2004), "Sertorius et la guérilla," in C. Auliard, and L. Bodiou (eds.), *Au jardin des Hespérides: Histoire, société et épigraphie des mondes anciens: Mélanges offerts à Alain Tranoy*. 297–314. Rennes.

Cadiou, F. (2008), *Hibera in terra miles: Les armées romaines et la conquête de l'Hispanie sous la république (218–45 av. J.-C.)*. Madrid.

Cadiou, F. (2009), "Le service militaire et son impact sur la société à la fin de l'époque républicaine: un état des recherches récentes," *Cahiers du Centre G. Glotz* 20, 157–71.

Cadiou, F. (2016), "Cavalerie auxiliaire et cavalerie légionnaire dans l'armée romaine au Ier s. av. J.-C.," in Wolff and Faure (2016), 53–78.

Cadiou, F. (2018), *L'armée imaginaire: Les soldats prolétaires dans les légions romaines au dernier siècle de la République*. Paris.

Cagniart, P. F. (1989), "L. Cornelius Sulla's Quarrel with C. Marius at the Time of the Germanic Invasions (104 – 101 BC)," *Athenaeum* 67, 139–49.

Cagniart, P. (2007), "The Late Republican Army (146-30 BC)," in Erdkamp (2007c), 80–95.

Campbell, B. (1978), "The Marriage of Soldiers under the Empire," *JRS* 68, 153–66.

Campbell, B., and L. A. Tritle (eds.) (2013), *Oxford Handbook of Warfare in the Classical World*. Oxford.

Campbell, D. B. (2013), "Arming Romans for Battle," in Campbell and Tritle (2013), 419–37.

Capogrossi Colognesi, L. (2014), *Law and Power in the Making of the Roman Commonwealth*. Translated by Laura Kopp. Cambridge.

Carney, E. D. (2015), "Macedonians and Mutiny: Discipline and Indiscipline in the Army of Philip and Alexander," in E. D. Carney (ed.), *King and Court in Ancient Macedonia: Rivalry, Treason and Conspiracy*. 27–59. Swansea.

Carney, T. (1961). *A Biography of C. Marius*. Assen.

Carter, M. J. (2006), "Buttons and Wooden Swords: Polybius 10.20.3, Livy 26.51, and the *Rudis*," *CPhil.* 101, 153–60.

Cavaignac, E. (1932), "Le Sénat de 220: Étude Démographique," *REL* 10, 458–68.

Cecchet, L. (2017), "Greek and Roman Citizenship: State of Research and Open Questions" in L. Cecchet and A. Busetto (eds.) *Citizens in the Graeco-Roman World: Aspects of Citizenship from the Archaic Period to AD 212*. 1–30. Leiden.

Chagnon, N. A. (1988), "Life Histories, Blood Revenge, and Warfare in a Tribal Population," *Science* 239, 985–92.

Chambers II, J. W. (2003), "S.L.A. Marshall's *Men against Fire*: New Evidence Regarding Fire Ratios," *Parameters* 33, 113–121.

Champion, C. B. (2004), *Cultural Politics in Polybius's Histories*. Berkeley, CA.

Champion, C. B. (2017), *The Peace of the Gods: Elite Religious Practices in the Middle Roman Republic*. Princeton, NJ.

Cheesman, G. L. (1914), *The Auxilia of the Roman Imperial Army*. Oxford.

Chrissanthos, S. G. (1999), "*Seditio*: Mutiny in the Roman Army, 90–40 B.C.," PhD dissertation, University of Southern California, Los Angeles.

Chrissanthos, S. G. (2001), "Caesar and the Mutiny of 47 B.C.," *JRS* 91, 63–75.

Chrissanthos, S. G. (2004), "Freedom of Speech and the Roman Republican Army," in I. Sluiter and R. Rosen (eds.), *Free Speech in Classical Antiquity*. 341–67. Leiden.

Churchill, J. B. (1999), "*Ex Qua Quod Vellent Facerent*: Roman Magistrates' Authority Over *Praeda* and *Manubiae*," *TAPA* 129, 85–116.

Clark, J. H. (2014), *Triumph in Defeat: Military Loss in the Roman Republic*. Oxford.

Clark, J. H. (2016), "Were *Tribuni Militum* First Elected in 362 or 311 BCE?" *Historia* 65, 275–97.

Clark, J.H., and B. Turner (eds.) (2018), *Brill's Companion to Military Defeat in Ancient Mediterranean Society*. Leiden.

Clemente, G. (1981), "Le leggi sul lusso e la società romana tra III e II secolo a.C.," in A. Giardina and A. Schiavone (eds.), *Società romana e produzione schiavistica: Modelli etici, diritto e trasformazioni sociali*. 1–14. Rome.

Coale, A. J. and P. Demeny (1983), *Regional Model Life Tables and Stable Populations*[2]. New York.

Coarelli, F. (1972), "Il sepolcro degli Scipioni," *DArch* 6, 36–106.

Coli, U. (1951), "Regnum I–VII," *Stud. Doc. Hist. Iur.* 17, 2–168.

Cooley, A. E. (2006), "Beyond Rome and Latium: Roman Religion in the Age of Augustus," in C. E. Schultz and P. B. Harvey (eds.), *Religion in Republican Italy.* 228–52. Cambridge.

Cornell, T. J. (1982), "Review of Wiseman, *Clio's Cosmetics* (1979)," *JRS* 72, 203–6.

Cornell, T. J. (1988), "La Guerra e lo stato in Roma arcaica (VII–V sec.)," in E. Campanile (ed.), *Alle Origini di Roma.* 89–100. Pisa.

Cornell, T. J. (1989), "The Conquest of Italy," *CAH*² 7.2, 351–419.

Cornell, T. J. (1995), *The Beginnings of Rome: Italy and Rome from the Bronze Age to the Punic Wars (c. 1000–264 BC).* London.

Cornell, T. J. (2000), "The *Lex Ovinia* and the Emancipation of the Senate," in Bruun (2000), 69–89.

Cornell, T. J. (2005), "The Value of the Literary Tradition Concerning Archaic Rome," in Raaflaub (2005b), 47–74.

Cornell, T. J. (ed.) (2013), *The Fragments of the Roman Historians, Vol. I–III: Introduction.* Oxford.

Coşkun, A. (2009a), *Bürgerrechtsentzug oder Fremdenausweisung?: Studien zu den Rechten von Latinern und weiteren Fremden sowie zum Bürgerrechtswechsel in der Römischen Republik (5. bis frühes 1. Jh. v.Chr.).* Stuttgart.

Coşkun, A. (2009b), "Zu den Bedingungen des Bürgerrechtserwerbs per magistratum in der späten römischen Republik," *Historia* 58, 225–241.

Coudry, M. (2009), "Partage et gestion du butin dans la Rome républicaine: procédures et enjeux," in M. Coudry and M. Humm (eds.), *Praeda: Butin de guerre et société dans la Rome républicaine / Kriegsbeute und Gesellschaft im republikanischen Rom.* 21–79. Stuttgart.

Couissin, P. (1926), *Les armes romaines: essai sur les origins et l'evolution des armes individuelles du legionnaire romain.* Paris.

Courtney, E. (1999), *Archaic Latin Prose.* Atlanta.

Crawford, M. H. (1974), *Roman Republican Coinage.* Cambridge.

Crawford, M. H. (1985), *Coinage and Money under the Roman Republic: Italy and the Mediterranean Economy.* Berkeley, CA.

Crawford, M. H. (ed.) (1996), *Roman Statutes.* 2 vols. London.

Crawford, M. H. (2006), "Aerarius," in *BNP* (doi: 10.1163/1574-9347_bnp_e105900).

Crawford, M. H. (2008), "States Waiting in the Wings: Population Distribution and the End of the Roman Republic," in de Ligt and Northwood (2008), 631–43.

Crawford, M. H. (forthcoming) "Caso Cantouio: Archaic Warfare and up-to-date Literacy."

Criniti, N. (1970), *L'Epigrafe di Asculum di Gn. Pompeo Strabone.* Milan.

Cristofani, M. (1986), "C. Genucius Cleusina pretore a Caere," *Archeologia nella Tuscia* 2, 24–26.

Cristofani, M. (1990), *La grande Roma dei Tarquini.* Rome.

Crone, P. (1989), *Pre-Industrial Societies: New Perspectives on the Past.* Oxford.

Culham, P. (1989), "Chance, Command, and Chaos in Ancient Military Engagements," *World Futures* 27, 191–205.

Culham, P. (2014), "Women in the Roman Republic," in Flower (2014), 127–48.

Dabrowa, E. (1986), "Cohortes Ituraeorum," *ZPE* 63, 221–30.

Daly, G. (2002), *Cannae: The Experience of Battle in the Second Punic War.* London.

Dart, C. J. (2011) "The 'Italian Constitution' in the Social War: A Reassessment (91 to 88 BCE)," *Historia* 58, 215–24.

Dart, C. J. (2014), *The Social War, 91 to 88 BCE: A History of the Italian Insurgency Against the Roman Republic.* London.

Dawson, D. (1996), *The Origins of Western Warfare: Militarism and Morality in the Ancient Greek World.* Boulder, CO.

Daly, G. (2002), *Cannae: the Experience of Battle in the Second Punic War.* London.

Delbrück, H. (1975), *History of the Art of War within the Framework of Political History, Vol. I: Antiquity.* Translated by Walter J. Renfroe, Jr. Westport.

Degrassi, A. (1954), *Fasti capitolini.* Turin.

de Ligt, L. (2007), "Roman Manpower and Recruitment During the Middle Republic," in Erdkamp (2007c), 114–31.

de Ligt, L. (2012), *Peasants, Citizens and Soldiers: Studies in the Demographic History of Roman Italy 225 BC – AD 100.* Cambridge.

de Ligt, L. (2014), "Livy 27.38 and the *Vacatio Militiae* of the Maritime Colonies," in Stek and Pelgrom (2014), 106–21.

de Ligt, L., and S. Northwood (eds.) (2008), *People, Land, and Politics: Demographic Developments and the Transformation of Roman Italy, 300 BC–AD 14.* Leiden.

De Martino, F. (1972), *Storia della costituzione romana.* Two vols. Naples.

Dench, E. (2005), *Romulus' Asylum: Roman Identities from the Age of Alexander to the Age of Hadrian.* Oxford.

De Pauw, L. G. (1998), *Battle Cries and Lullabies: Women in War from Prehistory to the Present.* Norman, OK.

de Vaan, M. (2008), *Etymological Dictionary of Latin and Other Italic Languages.* Leiden.

Dickert, S., D. Västfjäll, J. Kleber, and P. Slovic (2012), "Valuations of Human Lives: Normative Expectations and Psychological Mechanisms of (Ir)rationality," *Synthese* 189, 95–105.

Dillon, S., and K. E. Welch (eds.) (2006), *Representations of War in Ancient Rome.* Cambridge.

Dobson, M. J. (2008), *The Army of the Roman Republic: The Second Century BC, Polybius, and the Camps at Numantia, Spain.* Oxford.

Drogula, F. K. (2007), "*Imperium, Potestas,* and the *Pomerium* in the Roman Republic," *Historia* 56, 419–52.

Drogula, F. K. (2015), *Commanders and Command in the Roman Republic and Early Empire.* Chapel Hill, NC.

Drogula, F. K. (2017), "Plebeian Tribunes and the Government of Early Rome," *Antichthon* 51, 101–23.

Duff, A. (1928), *Freedmen in the Early Roman Empire.* Oxford

Dumézil, G. (1970), *Archaic Roman Religion.* 2 vols. Translated by Philip Krapp. Chicago.

Duncan-Jones, R. (1977), "Age-rounding, Illiteracy and Social Differentiation in the Roman Empire," *Chiron* 7, 333–53.

Duncan-Jones, R. (2002), *Structure and Scale in the Roman Economy.* Cambridge.

Earl, D. C. (1960), "Political Terminology in Plautus," *Historia* 9, 235–43.

Earle, E. M. (ed.) (1971), *Makers of Modern Strategy: Military Thought from Machiavelli to Hitler.* Princeton, NJ.

Earnshaw-Brown, L. (2009), "The Limits of Knowledge, Demography and the Republic," in A. Keaveney and L. Earnshaw-Brown (eds.) *The Italians on the Land: Changing Perspectives on Republican Italy Then and Now.* 123–36. Cambridge.

Eckstein, A. M. (1987), *Senate and General: Individual Decision-Making and Roman Foreign Relations, 264–194 B.C.* Berkeley, CA.

Eckstein, A. M. (1995), *Moral Vision in the Histories of Polybius.* Berkeley, CA.

Eckstein, A. M. (1997), "*Physis* and *Nomos*: Polybius, the Romans, and Cato the Elder," in P. Cartledge, P. Garnsey, and E. S. Gruen (eds.), *Hellenistic Constructs: Essays in Culture, History, and Historiography.* 175–98. Berkeley, CA.

Eckstein, A. M. (2006), *Mediterranean Anarchy, Interstate War, and the Rise of Rome.* Berkeley, CA.

Eckstein, A. M. (2008), *Rome Enters the Greek East: From Anarchy to Hierarchy in the Hellenistic Mediterranean, 230–170 BC.* Malden, MA.

Eder, W. (ed.) (1990), *Staat und Staatlichkeit in der frühen römischen Republik.* Stuttgart.

Ehlers, W. (1939), "Triumphus," *RE* 2/7A, 493–511.

Eichner, H. (2002), "Lateinisch *hostia, hostus, hostīre* und die stellvertretende Tiertötung der Hethiter," in M. Fritz, and S. Zeilfelder (eds.), *Novalis Indogermanica: Festschrift für Günther Neumann zum 80. Geburtstag.* 101–56. Graz.

Elshtain, J. B. (1995), *Women and War.* Chicago.

Engels, D. (1984), "The Use of Historical Demography in Ancient History," *CQ* 34, 386–93.

Erdkamp, P. (1998), *Hunger and the Sword: Warfare and Food Supply in Roman Republican Wars (264–30 B.C.).* Amsterdam.

Erdkamp, P. (2006a), "Late-Annalistic Battle Scenes in Livy (Books 21–44)," *Mnemosyne* 59, 525–63.

Erdkamp, P. (2006b), "The Transformation of the Roman Army in the Second Century BC," in T. Ñaco del Hoyo, and I. Arrayás (eds.), *War and Territory in the Roman World: Guerra y territorio en el mundo romano.* 41–51. Oxford.

Erdkamp, P. (2007a), "Polybius and Livy on the Allies in the Roman Army," in L. de Blois and E. Lo Cascio (eds.), *The Impact of the Roman Army (200 BC-AD 476): Economic, Social, Political, Religious, and Cultural Aspects.* 47–74. Leiden.

Erdkamp, P. (2007b), "War and State Formation in the Roman Republic," in Erdkamp (2007c), 96–113.

Erdkamp P. (ed.) (2007c) *A Companion to the Roman Army.* Oxford.

Erdkamp, P. (2011a), "Manpower and Food Supply in the First and Second Punic Wars," in Hoyos (2011), 58–76.

Erdkamp, P. (2011b), "Soldiers, Roman citizens, and Latin Colonists in Mid-Republican Italy," *AncSoc* 41, 108–146.

Errington, R. M. (1990), *A History of Macedonia.* Berkeley, CA.

Evans, J. K. (1980), "*Plebs Rustica*: The Peasantry of Classical Italy," *AJAH* 5: 19–47 and 134–73.

Evans, R. J. (1994). *Gaius Marius: A Political Biography.* Pretoria.

Fantham, E. (2005), "Liberty and the People in Republican Rome," *TAPA* 135, 209–29.

Fantham, E., H. P. Foley, N. B. Kampen, S. B. Pomeroy, and H. A. Shapiro (1994), *Women in the Classical World: Image and Text.* Oxford.

Farney, G. D. (2007), *Ethnic Identity and Aristocratic Competition in Republican Rome.* Cambridge.

Feig Vishnia, R. (1996), *State, Society and Popular Leaders in Mid-Republican Rome 241–167 BC.* Routledge.

Feldherr, A. (2009a), "Introduction," in Feldherr (2009b), 1–8.

Feldherr, A. (ed.) (2009b), *The Cambridge Companion to the Roman Historians*. Cambridge.

Ferrary, J.-L. (1995), "*Ius fetiale* et diplomatie," in E. Frézouls, and A. Jacquemin (eds.), *Les relations internationales: actes du colloque de Strasbourg, 15–17 Juin 1993*. 411–32. Strasbourg.

Finley, M. I. (1973), *The Ancient Economy*. Berkeley, CA.

Finley, M. I. (1981), *Economy and Society in Ancient Greece*. Edited with an Introduction by Brent D. Shaw, and Richard P. Saller. New York.

Flower, H. I. (1996), *Ancestor Masks and Aristocratic Power in Roman Culture*. Oxford.

Flower, H. I. (2010), *Roman Republics*. Princeton and Oxford.

Flower, H. I. (ed.) (2014), *The Cambridge Companion to the Roman Republic*[2]. Cambridge.

Flower, H. I. (2017), *The Dancing Lares and the Serpent in the Garden: Religion at the Roman Street Corner*. Princeton, NJ.

Foley, H. P. (2001), *Female Acts in Greek Tragedy*. Princeton, NJ.

Fontana, A. and R. Rosenheck (1994), "Posttraumatic Stress Disorder among Vietnam Theater Veterans: A Causal Model of Etiology in a Community Sample," *The Journal of Nervous and Mental Disease* 182, 677–84.

Forni, G. (1953), "Manio Curio Dentate uomo democrático," *Athenaeum* 31, 170–239.

Forni, G. (1989), "Questioni di storia agraria pre-romana: Le quattro fasi dell'agricoltura etrusca," in G. Maetzke (ed.), *Atti del II Congresso Internazionale Etrusco* 3, 1501–15. Rome.

Forsythe, G. (1994), *The Historian L. Calpurnius Piso Frugi and the Roman Annalistic Tradition*. Lanham, MD.

Forsythe, G. (2005), *A Critical History of Early Rome: From Prehistory to the First Punic War*. Berkeley, CA.

Forsythe, G. (2007), "The Army and Centuriate Organization in Early Rome," in Erdkamp (2007c), 24–41.

Forsythe, G. (2012), *Time in Roman Religion: One Thousand Years of Religious History*. London.

Foxhall, L. (1990), "The Dependent Tenant: Land Leasing and Labour in Italy and Greece," *JRS* 80, 97–114.

Foxhall, L. (1997), "A View from the Top: Evaluating the Solonian Property Classes," in L. G. Mitchell, and P.J. Rhodes (eds.), *The Development of the Polis in Archaic Greece*. 113–36. London.

Franchi dell'Orto, L., S. Agostino, and M. Buonocore (eds.) (2010), *Pinna Vestinorum e il popolo dei Vestini*. Rome.

Frank, D., P. Slovic, D. Västfjäll, D. Vasfjall (2011), "'Statistics Don't Bleed': Rhetorical Psychology, Presence, and Psychic Numbing in Genocide Pedagogy," *JAC* 31, 609–24.

Frank, T. (1919), "Agriculture in Early Latium," *American Economic Review* 9, 267–76.

Frank, T. (ed.) (1933), *An Economic Survey of Ancient Rome, Vol. 1: Rome and Italy of the Republic*. Baltimore, MD.

Frayn, J. M. (1975), "Wild and Cultivated Plants: A Note on the Peasant Economy of Roman Italy," *JRS* 65, 32–9.

Frederiksen, M. W. (1984), *Campania*. Edited by Nicholas Purcell. Rome.

Frey, B. S. (2006), "Giving and Receiving Awards," *Perspectives on Psychological Science* 1, 377–88.

Frey, B. S. (2007), "Awards as Compensation," *European Management Review* 4, 6–14.

Frey, B. S., and S. Neckermann (2008), "Awards: A View from Psychological Economics," *Journal of Psychology* 216 (4), 198–208.

Frier, B. W. (1979), *Libri Annales Pontificum Maximorum: The Origins of the Annalistic Tradition*. Rome.

Frier, B. W. (1982), "Roman Life-expectancy: Ulpian's Evidence," *HSPh* 86, 212–51.

Fronda, M. P. (2010), *Between Rome and Carthage: Southern Italy During the Second Punic War*. Cambridge.

Fronda, M. P. (2011), "Privata hospitia, beneficia publica? Consul(ar)s, local elite and Roman Rule in Italy," in Beck, Duplá, Jehne, and Pina Polo (2011), 232–55.

Fronda, M. P., and F. Gauthier (2017), "Italy and Sicily in the Second Punic War: Multipolarity, Minor Powers, and Local Military Entrepreneurialism," T. Ñaco del Hoyo and F. López Sánchez (eds.), *War, Warlords, and Interstate Relations in the Ancient Mediterranean*. 308–25. Leiden.

Fulkerson, L. (2013), *No Regrets: Remorse in Classical Antiquity*. Oxford.

Fulminante, F. (2014), *The Urbanisation of Rome and Latium Vetus: From the Bronze Age to the Archaic Era*. Cambridge.

Gabba, E. (1967), *Appiani Bellorum Civilium Liber Primus*. Florence.

Gabba, E. (1976), *Republican Rome, the Army, and the Allies*. Translated by P. J. Cuff. Berkeley, CA.

Gabba, E. (1991), *Dionysius and The History of Archaic Rome*. Berkeley, CA.

Gabrielli, C. (2003), "Lucius Postumius Megellus at Gabii: A New Fragment of Livy," *CQ* 53, 247–59.

Gaca, K. L. (2014), "Martial Rape, Pulsating Fear, and the Sexual Maltreatment of Girls (παῖδες), Virgins (παρθένοι), and Women (γυναῖκες) in Antiquity," *AJP* 135, 303–51.

Gaertner, J. F. (2008), "Livy's Camillus and the Political Discourse of the Late Republic," *JRS* 98, 27–52.

Gallant, T. W. (1991), *Risk and Survival in Ancient Greece: Reconstructing the Rural Domestic Economy*. Stanford, CA.

Galsterer, H. (1976), *Herrschaft und Verwaltung im republikanischen Italien: Die Beziehungen Roms zu den italischen Gemeinden vom Latinerfrieden 338 v. Chr. bis zum Bundesgenossenkrieg 91 v. Chr.* Munich.

Garlan, Y. (1975), *War in the Ancient World: A Social History*. Translated by Janet Lloyd. New York.

Garnsey, P. (1988), *Famine and Food Supply in the Graeco-Roman World. Responses to Risk and Crisis*. Cambridge.

Gauthier, F. (2016a), "Financing War in the Roman Republic, 201 BCE–14 CE," PhD dissertation, McGill University, Montreal.

Gauthier, F. (2016b), "The Changing Composition of the Roman Army in the Late Republic and the So-Called 'Marian-Reforms,'" *AHB* 30, 103–20.

Gauthier, F. (forthcoming), "Plunder, Common Soldiers, and Military Service in the Middle and Late Republic," in Roselaar and Helm (forthcoming).

Giovannini, A. (1983), *Consulare Imperium*. Basel.

Glinister, F. (2006), "Kingship and Tyranny in Archaic Rome," in Lewis (2006), 17–32.

Goldschmidt, W. (1986), "Personal Motivation and Institutionalized Conflict," in M. L. Foster and R. A. Rubinstein (eds.), *Peace and War: Cross-Cultural Perspectives*. 3–14. New Brunswick, NJ.

Goldstein, J. S. (2001), *War and Gender: How Gender Shapes the War System and Vice Versa*. Cambridge.

Goldsworthy, A. K. (1996), *The Roman Army at War: 100 BC - AD 200*. Oxford.

Goukowsky, P. (2014), *Diodore de Sicile, Bibliothèque Historique: Fragments, Livres XXXIII–XL*. Paris.

Gratwick, A. S. (2002), "A Matter of Substance: Cato's Preface to the *De Agri Cultura*," *Mnemosyne* 55, 41–72.

Grethlein, J. (2007), "The Poetics of the Bath in the *Iliad*," *HSPh* 103, 25–49.

Grieve, L. J. (1983). "Tabulae Caeritum," in C. Deroux (ed.), *Studies in Latin Literature and Roman History*. Vol. 3. 26–43. Brussels.

Grossman, D. (2009), *On Killing: The Psychological Cost of Learning to Kill in War and Society*. New York.

Grossman, D., and L. W. Christensen (2008), *On Combat: The Psychology and Physiology of Deadly Conflict in War and in Peace*[3]. Mascoutah, IL.

Grosso, G. (1969), *Le servitù prediali nel diritto romano*. Turin.

Gruen, E. S. (1984a), "Material Rewards and the Drive for Empire," in Harris (1984), 59–82.

Gruen, E. S. (1984b), *The Hellenistic World and the Coming of Rome*. Berkeley, CA.

Gruen, E. S. (1992), *Culture and National Identity in Republican Rome*. Ithaca, NY.

Gruen, E. (1995), *The Last Generation of the Roman Republic*[2]. Berkeley, CA.

Habinek, T. N. (1998), *The Poetics of Latin Literature: Writing, Identity and Empire in Ancient Rome*. Princeton, NJ.

Habinek, T. N. (2005), *The World of Roman Song: From Ritualized Speech to Social Order*. Baltimore, MD.

Halstead, P. (2014), *Two Oxen Ahead: Pre-Mechanized Farming in the Mediterranean*. Oxford.

Hamdoune, C. (1999), *Les* auxilia externa *africains des armées romaines, IIIe siècle av. J.-C. – IVe siècle ap. J.-C.* Montpellier.

Hammond, N. G. L. (1989), *The Macedonian State: The Origins, Institutions, and History*. Oxford.

Hanson, V. D. (1983), *Warfare and Agriculture in Classical Greece*. Berkeley, CA.

Hanson, V. D. (1989), *The Western Way of War: Infantry Battle in Classical Greece*. New York.

Hanson, V. D. (ed.) (1991), *Hoplites: The Classical Greek Battle Experience*. London.

Hantos, T. (1983), *Das römische Bundesgenossensystem in Italien*. Munich.

Harmand, J. (1967), *L'Armée et le soldat à Rome: de 107 à 50 avant notre ère*. Paris.

Harmand, J. (1969), "Le prolétariat dans la légion de Marius à la veille du second *bellum civile*," in J.-P. Brisson (ed.), *Problèmes de la guerre à Rome*. 61–74. Paris.

Harris, W. V. (1971), "On War and Greed in the Second Century B.C.," *AHR* 76, 1371–85.

Harris, W. V. (1979), *War and Imperialism in Republican Rome, 327–70 B.C.* Oxford.

Harris, W. V. (ed.) (1984), *The Imperialism of Mid-Republican Rome*. Rome.

Harris, W. V. (1990), "Roman Warfare in the Economic and Social Context of the Fourth Century B.C.," in Eder (1990), 494–510.

Harris, W. V. (1999), "Demography, Geography and the Sources of Roman Slaves," *JRS* 91, 1–26.

Harris, W. V. (2016), *Roman Power: A Thousand Years of Empire*. Cambridge.

Harris, W. V. (2017), "Rome at Sea: The Beginnings of Roman Naval Power," *G&R* 64, 14–26.

Hatzopoulos, M. B. (1996), *Macedonian Institutions under the Kings*, 2 Vols. Athens.

Haynes I. P. (2013), *Blood of the Provinces: The Roman Auxilia and the Making of Provincial Society from Augustus to the Severans*. Oxford.

Helm, M. (2017), "A Troubled Beginning: Rome and its Reluctant Allies in the Fourth Century BC," *Antichthon* 51, 202–26.

Hermon, E. (2007), "Des communautés distinctes sur le même territoire: quelle fut la réalité des '*incolae*'?" in R. Compatangelo-Soussignan and C.-G. Schwentzel (eds.) *Étrangers dans la cité romaine*, 25–42. Rennes.

Hesberg, H. von (2008), "Die Hierarchie der Räume. Straßen und Plätze in Städten und Militärlagern zur Zeit der Republik," in D. Mertens (ed.), *Stadtverkehr in der antiken Welt, Internationales Kolloquium zur 175-Jahrfeier des Deutschen Archäologischen Instituts Rom, 21. bis 23. April 2004*. 71–86. Wiesbaden.

Heurgon, J. (1969), "Inscriptions étrusques de Tunisie," *CRAI* 113, 526–551.

Heuss, A. (1944), "Zur Entwicklung des Imperiums der römischen Oberbeamten," *ZRG* 64, 57–133.

Hickson-Hahn, F. (2004), "The Politics of Thanksgiving," in C. F. Konrad (ed.), *Augusto Augurio: Rerum Humanarum et Divinarum Commentationes in Honorem Jerzy Linderski*. 31–51. Stuttgart.

Hilder, J. C. (2015), "Recontextualising the *Rhetorica ad Herennium*," PhD dissertation, University of Glasgow, Glasgow.

Hill, H. (1952), *The Roman Middle Class in the Republican Period*. Oxford.

Hillman, T. P. (1996), "Cinna, Strabo's Army, and Strabo's Death in 87 B.C.," *L'Antiquité Classique* 65, 81–9.

Hin, S. (2013), *The Demography of Roman Italy: Population Dynamics in an Ancient Conquest Society, 201 BCE –14 CE*. Cambridge.

Hinard, F. (ed.) (2000), *Histoire Romaine, Tome I, Des Origines à Auguste*. Paris.

Hofmann, F. (1847), *Der römische Senat zur Zeit der Republik nach seiner Zusammensetzung und inneren Verfassung betrachtet*. Berlin.

Hölkeskamp, K.-J. (1987), *Die Entstehung der Nobilität: Studien zur sozialen und politischen Geschichte der Römischen Republik im 4. Jhdt. v. Chr.* Stuttgart.

Hölkeskamp, K.-J. (1993), "Conquest, Competition and Consensus: Roman Expansion in Italy and the Rise of the 'Nobilitas,'" *Historia* 42, 12–39.

Hölkeskamp, K.-J. (2004), *Senatus Populusque Romanus. Die politische Kultur der Republik—Dimensionen und Deutungen*. Stuttgart.

Hölkeskamp, K.-J. (2010), Reconstructing the Roman Republic: An Ancient Political Culture and Modern Research. Translated by Henry Heitmann-Gordon. Princeton, NJ.

Hölkeskamp, K.-J. (2017), *Libera Res Publica: Die politische Kultur des antiken Rom – Positionen und Perspektiven*. Stuttgart.

Holland, L. L. (2012), "Women and Roman Religion," in S. L. James and S. Dillon (eds.), *A Companion to Women in the Ancient World*. 204–14. Oxford.

Hollingsworth, T. H. (1969), *Historical Demography*. London.

Holloway, R. R. (1994), *The Archaeology of Early Rome and Latium*. London.

Holloway, R. R. (2008), "Who Were the *Tribuni Militum Consulari Potestate?*" *L'Antiquité Classique* 77, 107–25.

Holloway, R. R. (2009), "Praetor Maximus and Consul," in C. Marangio and G. Laudizi (eds.), *Palaia philia: Studi di topografia antica in onore di Giovanni Uggeri.* 71–5. Milan.

Hölscher, T. (2001), "Die Alten vor Augen: Politische Denkmäler und öffentliches Gedächtnis im republikanischen Rom," in G. Melville (ed.), *Institutionalität und Symbolisierung: Verstetigungen kultureller Ordnungsmuster in Vergangenheit und Gegenwart.* 183–211. Cologne.

Hölscher, T. (2004), *The Language of Images in Roman Art.* Cambridge.

Hölscher, T. (2006), "The Transformation of Victory into Power: From Event to Structure," in Dillon and Welch (2006), 27–48.

Hopkins, K. (1966), "On the Probable Age-structure of the Roman Population," *Population Studies* 20, 245–64.

Hopkins, K. (1978), *Conquerors and Slaves.* Cambridge.

Hopkins, K. (1983), *Death and Renewal.* Cambridge.

Horden, P., and N. Purcell (2000), *The Corrupting Sea: A Study of Mediterranean History.* Oxford.

Hornblower, J. (1981), *Hieronymus of Cardia.* Oxford.

Hoyer, D. C. (2012), "Samnite Economy and the Competitive Environment of Italy in the Fifth to Third Centuries BC," in Roselaar (2012), 179–196.

Hoyos, D. C. (ed) (2011), *A Companion to the Punic Wars.* Oxford.

Hoyos, D. C. (ed.) (2013), *A Companion to Roman Imperialism.* Leiden.

Hughes, J. D. (2014), *Environmental Problems of the Greeks and Romans: Ecology in the Ancient Mediterranean².* Baltimore, MD.

Humbert, M. (1978), *Municipium et civitas sine suffragio: L'organisation de la conquête jusqu'à la guerre sociale.* Rome.

Humm, M. (2006), "Tribus et citoyenneté : extension de la citoyenneté romaine et expansion territoriale," in Jehne and Pfeilschifter (2006), 39–64.

Humm, M. (2011), "The Curiate Law and the Religious Nature of the Power of Roman Magistrates," in Tellegen-Couperus (2011), 57–84.

Ilari, V. (1974), *Gli Italici nelle strutture militari romane.* Milan.

Isayev, E. (2011), "Corfinium and Rome: Changing Place in the Social War," in M. Gleba and H. Horsnaes (eds.), *Communicating identity in Italic Iron Age communities.* 210–22. Oxford.

Isayev, E. (2017), *Migration, Mobility and Place in Ancient Italy.* Cambridge.

Itgenshorst, T. (2005), *Tota illa pompa: Der Triumph in der römischen Republik.* Göttingen.

Jaia, A. M. (2013), "Le colonie di diritto Romano: Considerazioni sul sistema difensivo costiero tra IV e III secolo a.C," *Scienze dell'Antichità* 19, 475–89.

Jaia, A. M., and M. C. Molinari (2011), "Two Deposits of Aes Grave from the Sanctuary of Sol Indiges," *The Numismatic Chronicle* 171, 87–97.

James, S. (2002), "Writing the Legions: The Development and Future of Roman Military Studies in Britain," *Archaeological Journal* 159, 1–58.

Jefferson, E. (2012), "Problems and Audience in Cato's *Origines,*" in Roselaar (2012), 311–26.

Jehne, M. (2002), "Die Geltung der Provocation und die Konstruktion der römischen Republik als Freiheitsgemeinschaft," in G. Melville and H. Vorländer (eds.),

Geltungsgeschichten: Über die Stabilisierung und Legitimierung institutioneller Ordnungen. 55–74. Cologne.

Jehne, M. (2006), "Römer, Latiner und Bundesgenossen im Krieg: Zu Formen und Ausmaß der Integration in der republikanischen Armee," in Jehne and Pfeilschifter (2006), 243–68.

Jehne, M. (2011), "*The Rise of the Consular as a Social Type in the Third and Second Centuries BC*," in Beck, Duplá, Jehne, and Pina Polo (2011), 211–31.

Jehne, M., and F. Pina Polo (eds.) (2015), *Foreign Clientelae in the Roman Empire: A Reconsideration.* Stuttgart.

Jehne, M., and R. Pfeilschifter (eds.) (2006), *Herrschaft ohne Integration?: Rom und Italien in republikanischer Zeit.* Frankfurt.

Johnson, T., and C. Dandeker (1989), "Patronage: Relation and System," in Wallace-Hadrill (1989b), 219–42.

Jung, J. H. (1982), "Das Eherecht der römischen Soldaten," *ANRW* 2, 302–346.

Kay P. (2014), *Rome's Economic Revolution.* Oxford.

Keaveney, A. (2005), *Rome and the Unification of Italy*². Exeter.

Keaveney, A. (2007), *The Army in the Roman Revolution.* London.

Keegan, J. (1976), *The Face of Battle: A Study of Agincourt, Waterloo, and the Somme.* London.

Keegan, J. (1993), *A History of Warfare.* London.

Keith, A. (2016), "City Lament in *Augustan* Epic: Antitypes of Rome from Troy to Alba Longa," in M. R. Bachvarova, D. Dutsch, and A. Suter (eds.), *The Fall of Cities in the Mediterranean: Commemoration in Literature, Folk-Song, and Liturgy.* 156–82. Cambridge.

Kendall, S. (2012), "Appian, Allied Ambassadors, and the Rejection of 91: Why the Romans Chose to Fight the *Bellum Sociale*," in Roselaar (2012), 105–22.

Kendall, S. (2013), *The Struggle for Roman Citizenship: Romans, Allies, and the Wars of 91–77 BCE.* Piscataway, NJ.

Kent, P. A. (2012a), "Reconsidering *Socii* in Roman Armies Before the Punic Wars," in Roselaar (2012), 71–83.

Kent, P. A. (2012b), "The Roman Army's Emergence from its Italian Origins," PhD dissertation, University of North Carolina at Chapel Hill, Chapel Hill, NC.

Keppie, L. J. F. (1984a), *The Making of the Roman Army: From Republic to Empire.* London.

Keppie, L. J. F. (1984b). "Colonisation and Veteran Settlement in Italy in the First Century A.D.," *PBSR* 52, 77–114.

Keppie, L. J. F. (2001), "Army and Society in the Late Republic and Early Empire," in T. Bekker-Nielsen and L. Hannestad (eds.), *War as a Cultural and Social Force: Essays on Warfare in Antiquity.* 130–6. Copenhagen.

Kinsella, H. M. (2011), *The Image before the Weapon: A Critical History of the Distinction between Combatant and Civilian.* Ithaca, NY.

Klar, L. S. (2006), "The Origins of the Roman *Scaenae Frons* and the Architecture of Triumphal Games in the Second Century B.C.," in Dillon and Welch (2006), 162–83.

Koon, S. (2011), "Phalanx and Legion: The 'Face' of Punic War Battle," in Hoyos (2011), 77–94.

Koortbojian, M. (2002), "A Painted *Exemplum* at Rome's Temple of Liberty," *JRS* 92, 33–48.

Kornemann, E. (1933), "Municipium," *RE* 16, 570–638.

Kromayer, J. and G. Veith (1963), *Heerwesen und Kriegführung der Griechen und Römer*. Munich.

Kron, J. G. (2008), "The Much Maligned Peasant: Comparative Perspectives on the Productivity of the Small Farmer in Classical Antiquity," in de Ligt and Northwood (2008), 71–119.

Künzl, E. (1988), *Der römische Triumph: Siegesfeiern im antiken Rom*. Munich.

Lajoye, P. (2010), "Quirinus, un ancien dieu tonnant? Nouvelles hypothèses sur son étymologie et sa nature primitive," *Revue de l'histoire des religions* 227, 175–94.

Lammers, C. J. (1969), "Strikes and Mutinies: A Comparative Study of Organizational Conflicts between Rulers and Ruled," *Administrative Science Quarterly* 14, 558–72.

Lange, C. H. (2014), "The Triumph Outside the City: Voices of Protest in the Middle Republic," in Lange and Vervaet (2014), 67–81.

Lange, C. H. (2016), *Triumphs in the Age of Civil War: The Late Republic and the Adaptability of Triumphal Tradition*. London.

Lange, C. H., and F. J. Vervaet (eds.) (2014), *The Roman Republican Triumph: Beyond the Spectacle*. Rome.

Langlands, R. (2006), *Sexual Morality in Ancient Rome*. Cambridge.

Launaro, A. (2011), *Peasants and Slaves: The Rural Population of Roman Italy (200 BC to AD 100)*. Cambridge.

Lazenby, J. (1978), *Hannibal's War: A Military History of the Second Punic War*. Norman, OK.

Le Bohec, Y. (1989), *L'armée romaine sous le haut-empire*. Paris. (= *The Roman Imperial Army*. London 1994.)

Lee, A. D. (1996), "Morale and the Roman Experience of Battle," in A. B. Lloyd (ed.), *Battle in Antiquity*. 199–217. London.

Lee, J. W. I. (2013), "The Classical Greek Experience," in Campbell and Tritle (2013), 143–61.

Leigh, M. (2004), *Comedy and the Rise of Rome*. Oxford.

Lendon, J. E. (2004), "The Roman Army Now," *CJ* 99(4), 441–49.

Lendon, J. E. (2007), "War and Society," in P. Sabin, H. van Wees, and M. Whitby (eds.), *The Cambridge History of Greek and Roman Warfare, Vol. I: Greece, the Hellenistic World and the Rise of Rome*. 498–516. Cambridge.

Lennon, J. J. (2013), *Pollution and Religion in Ancient Rome*. Cambridge.

Levene, D. S. (2010), *Livy on the Hannibalic War*. Oxford.

Levithan, J. (2013), *Roman Siege Warfare*. Ann Arbor, MI.

Lewis, S. (ed.) (2006), *Ancient Tyranny*. Edinburgh.

Linderski, J. (1986), "The Augural Law," *ANRW* 2.16, 2146–312.

Lincoln, B. (1999), *Theorizing Myth: Narrative, Ideology, and Scholarship*. Chicago.

Linke, B. (2014), "Die Väter und der Staat: Die Grundlagen der aggressiven Subsidiarität in der römischen Gesellschaft," in C. Lundgreen (ed.), *Staatlichkeit in Rom?: Diskurse und Praxis (in) der römischen Republik*. 65–90. Stuttgart.

Linke, B. (2016), "Die Republik und das Meer: Seerüstung und römische Innenpolitik zur Zeit der Punischen Kriege," in E. Baltrusch, C. Wendt, and H. Kopp (eds.), *Seemacht, Seeherrschaft und die Antike*. 163–85. Stuttgart.

Linke, B. (2017), "Die Nobilität und der Sieg: Eine komplizierte Beziehung," in M. Haake, and A.-C. Harders (eds.), *Politische Kultur und soziale Struktur der römischen Republik: Bilanzen und Perspektiven*. 381–99. Stuttgart.

Linke, B. (2018), "Die unruhige Republik: Kollektive Gewaltausübung und Friedens-
bereitschaft im republikanischen Rom," in A. Lichtenberger, H.-H. Nieswandt,
and D. Salzmann (eds.), *Eirene / Pax: Frieden in der Antike*. 105–14. Münster.

Lintott, A. W. (1971), "The Offices of C. Flavius Fimbria in 86-5 B.C.," *Historia* 20,
696–701.

Lintott, A. W. (1994), "Political History, 146–95 B.C.," CAH^2 9, 40–103.

Lintott, A. W. (1999), *Violence in Republican Rome*[2]. Oxford.

Lipka, M. (2009), *Roman Gods: A Conceptual Approach*. Leiden.

Lo Cascio, E. (1991–94), "I togati della 'formula togatorum,'" *Annali Dell' Istituto
Italiano per Gli Studi Storici* 12, 309–28.

Lo Cascio, E. (2001), "Il census a Roma e la sua evoluzione dall'età 'serviana' alla
prima età imperial," *MÉFRA* 113, 565–603.

Lodge, G. (1962), *Lexicon Plautinum*. 2 vols. Hildesheim.

Lomas, K. (2012), "The Weakest Link: Elite Social Networks in Republican Italy,"
in Roselaar (2012), 197–213.

Lott, J. B. (2004), *The Neighborhoods of Augustan Rome*. Cambridge.

Lovano, M. (2002), *The Age of Cinna: Crucible of Late Republican Rome*. Stuttgart.

Lundgreen, C. (2014), "Rules for Obtaining a Triumph—the *ius triumphandi* Once
More," in Lange and Vervaet (2014), 17–32.

Machado, D. M. (2017), "Community and Collective Action in the Roman Republican Army (218–44 BCE)," PhD dissertation, Brown University, Providence, RI.

MacKay, C. S. (2004), *Ancient Rome: A Military and Political History*. Cambridge.

MacMullen, R. (1984), "The Legion as a Society," *Historia* 33, 440–56.

Magdelain, A. (1964), "Note sur la loi curiate et les auspices des magistrates," *RHD*
62, 198–203.

Magdelain, A. (1968), *Recherches sur l'"Imperium": la loi curiate et les auspices d'in-
vestiture*. Paris.

Mangiameli, R. (2012), *Tra* duces *e* milites*: Forme di comunicazione politica al
tramonto della Repubblica*. Trieste.

Marchetti, P. (1977), "À propos du *tributum* romain: impôt de quotité ou de répar-
tition?" in A. Chastagnol, C. Nicolet, and H. Van Effentere (eds.), *Armées et fiscalité dans le monde antique: Actes du Colloque national du CNRS*. 107–31. Paris.

Marco Simón, F., F. Pina Polo, and J. Remesal Rodriguez (eds.) (2012), *Vae Victis.
Perdedores en el Mundo Antiguo*. Barcelona.

Marcovich, M. (1996), "From Ishtar to Aphrodite," *JAE* 30, 43–59.

Marincola, J. (2010), "Eros and Empire: Virgil and the Historians on Civil War,"
in C. Kraus, J. Marincola, and C. Pelling (eds.), *Ancient Historiography and its
Contexts. Studies in Honor of A.J. Woodman*. 184–204. Oxford.

Marinova, S. V., H. Moon, and L. Van Dyne (2010), "Are All Good Soldier Behaviors the Same? Supporting Multidimensionality of Organizational Citizenship
Behaviors Based on Rewards and Roles," *Human Relations* 63(10), 1463–85.

Marsden, E. W. (1974), "Polybius as Military Historian," in E. Gabba (ed.), *Po-
lybe: neuf exposés suivis de discussions. Entretiens sur l'Antiquité Classique, XX.*
267–301. Vandoeuvres-Geneva.

Marshall, S. L. A. (2000), *Men Against Fire: The Problem of Battle Command*.
Norman, OK.

Martin, S. (2014), "*Auxiliaria stipendia merere*: La solde des auxiliaires de la fin de
la guerre sociale à la fin du Ier s. p.C.," in M. Reddé (ed.), *De l'or pour les braves!
Soldes, armées et circulation monétaire dans le monde romain*. 117–38. Paris.

Marzano, A. (2007), *Roman Villas in Central Italy: A Social and Economic History*. Leiden.

Mason, G. G. (1992), "The Agrarian Role of Coloniae Maritimae: 338–241 B.C.," *Historia* 41, 75–87.

Matthew, C. A. (2010), *On the Wings of Eagles: The Reforms of Gaius Marius and the Creation of Rome's First Professional Soldiers*. Newcastle.

Maxfield, V. A. (1981), *The Military Decorations of the Roman Army*. Berkeley, CA.

McCall, J. (2002), *The Cavalry of the Roman Republic: Cavalry Combat and Elite Reputations in the Middle and Late Republic*. London.

McCall, J. (forthcoming), "Digital Legionaries: Video Game Simulations of the Face of Battle in the Republic," in C. Rollinger (ed.), *Playing with the Ancient World: Representations of Classical Antiquity in Video Games*. London.

McDonnell, M. (2006a), "Roman Aesthetics and the Spoils of Syracuse," in Dillon and Welch (2006), 68–90.

McDonnell, M. (2006b), *Roman Manliness: Virtus and the Roman Republic*. Cambridge.

Mellor, R. (1999), *The Roman Historians*. London.

Mersing, K. M. (2007), "The War-tax (*Tributum*) of the Roman Republic: A Reconsideration," *Classica et Mediaevalia* 58, 215–35.

Messer, W. S. (1920), "Mutiny in the Roman Army: The Republic," *CPhil.* 15, 158–75.

Mickel, A. E., and L. A. Barron (2008), "Getting 'More Bang for the Buck': Symbolic Value of Monetary Rewards in Organizations," *Journal of Management Inquiry* 17, 329–38.

Midgely, M. (2011), *The Myths We Live By*. New York.

Millar, F. (1989), "Political Power in Mid-Republican Rome: Curia or Comitium?" *JRS* 79, 138–50.

Milne, K. H. (2009), "The Republican Soldier: Historiographical Representations and Human Realities," PhD dissertation, University of Pennsylvania, Philadelphia, PA.

Milne, K. H. (2012), "Family Paradigms in the Roman Republican Military," *Intertexts* 16, 25–41.

Miltsios, N. (2013), *The Shaping of Narrative in Polybius*. Berlin.

Miltsios, N. and M. Tamiolaki (eds.) (2018), *Polybius and His Legacy*. Berlin.

Mitchell, R. E. (1990), *Patricians and Plebeians: The Origin of the Roman State*. Ithaca, NY.

Momigliano, A. D. (1983), "Premesse per una discussione su Georges Dumézil," *Opus* 2, 329–42.

Momigliano, A. D. (1994), *A.D. Momigliano: Studies on Modern Scholarship*. Edited by G. W. Bowersock, and T. Cornell. Translated by Tim Cornell. Berkeley, CA.

Mommsen, T. (1854–56), *Römische Geschichte*[3]. 3 vols. Leipzig.

Mommsen, T. (1887/88), *Römisches Staatsrecht*[3]. 3 vols. Leipzig.

Moore, T. J. (1998), *The Theater of Plautus: Playing to the Audience*. Austin, TX.

Morel, J.-P. (2007), "Early Rome and Italy," in W. Scheidel, I. Morris, and R. P. Saller (eds.), *The Cambridge Economic History of the Greco-Roman World*. 485–510. Cambridge.

Morell, K. (2015), "Appian and the Judiciary Law of M. Livius Drusus (tr. pl. 91)," in Welch (2015), 235–55.

Mouritsen, H. (1998), *Italian Unification: A Study in Ancient and Modern Historiography*. London.

Mouritsen, H. (2007), "The Civitas Sine Suffragio: Ancient Concepts and Modern Ideology," *Historia* 56, 141–58.

Mouritsen, H. (2008), "The Gracchi, The Latins, and the Italians Allies," in de Ligt and Northwood (2008), 471–83.

Mouritsen, H. (2011), *The Freedman in the Roman World*. Cambridge.

Mouritsen, H. (2017), *Politics in the Roman Republic*. Cambridge.

Neesen, L. (1980), *Untersuchungen zu den direkten Staatsabgaben der römischen Kaiserzeit 27 v. Chr. bis 284 n. Chr.* Bonn.

Neil, S. (2012), "Identity Construction and Boundaries: Hellenistic Perugia," in Roselaar (2012), 51–70.

Nicolet, C. (1966), *L'Ordre équestre à l'époque républicaine (312–43 av. J.-C.), T. 1: Définitions juridiques et structures sociales*. Paris.

Nicolet, C. (1976), *Tributum: Recherches sur la fiscalité directe sous la république romaine*. Bonn.

Nicolet, C. (1978), "Le *stipendium* des alliés italiens avant la guerre sociale," *PBSR* 46, 1–11.

Nicolet, C. (1980), *The World of the Citizen in Republican Rome*. Translated by P. S. Falla. Berkeley, CA.

Nicolet, C. (2000), *Censeurs et publicains: Économie et fiscalité dans la Rome antique*. Paris.

Nijboer, A. J. (1998), *From Household Production to Workshops: Archaeological Evidence for Economic Transformations, Pre-Monetary Exchange and Urbanisation in Central Italy from 800 to 400 BC*. Groningen.

North, J. A. (1981), "The Development of Roman Imperialism," *JRS* 71, 1–9.

Nuber, H. U. (1972), "Zwei bronzene Besitzmarken aus Frankfurt/M.- Heddernheim," *Chiron* 2, 483–507.

Oakley, S. P. (1985), "Single Combat in the Roman Republic," *CQ* 35, 392–410.

Oakley, S. P. (1997), *A Commentary on Livy: Books VI–X, Vol. I: Introduction and Book VI*. Oxford.

Oakley, S. P. (1998), *A Commentary on Livy: Books VI–X, Vol. II: Books VII and VIII*. Oxford.

Oakley, S. P. (2005a), *A Commentary on Livy: Books VI–X, Vol. III: Book IX*. Oxford.

Oakley, S. P. (2005b), *A Commentary on Livy: Books VI–X, Vol. IV: Book X*. Oxford.

Ogilvie, R. M. (1965), *A Commentary on Livy: Books 1–5*. Oxford.

Ogilvie, R. M. (1976), *Early Rome and the Etruscans*. Trowbridge.

O'Neill, P. (2003), "Triumph Songs, Reversal and Plautus' Amphitruo," *Ramus* 32, 1–38.

Orlin, E. M. (1997), *Temples, Religion, and Politics in the Roman Republic*. Leiden.

Östenberg, I. (2009), *Staging the World: Spoils, Captives, and Representations in the Roman Triumphal Procession*. Oxford.

Östenberg, I. (2014a), "Animals and Triumphs," in G. L. Campbell (ed.), *The Oxford Handbook of Animals in Classical Thought and Life*. 491–506. Oxford.

Östenberg, I. (2014b), "War and remembrance. Memories of defeat in ancient Rome," in B. Alroth and C. Scheffer (eds.), *Attitudes Towards the Past in Antiquity: Creating Identities. Proceedings of an International Conference Held at Stockholm University, 15–17 May 2009*. 255–65. Stockholm.

Palmer, R. E. A. (1970), *The Archaic Community of the Romans*. Cambridge.

Parker, H. M. D. (1928), *The Roman Legions*. Oxford.

Parkin, T. (1992), *Demography and Roman Society*. Baltimore.

Patterson, J. R. (2006a), "The Relationship of the Italian Ruling Classes with Rome: Friendship, Family Relations and their Consequences," in Jehne and Pfeilschifter (2006), 139–53.

Patterson, J. R. (2006b). "Colonization and Historiography: The Roman Republic," in Wilson and Bradley (2006), 189–218.

Patterson, J. R. (2012), "Contact, Co-operation, and Conflict in Pre-Social War Italy," in Roselaar (2012), 215–26. Leiden.

Pelgrom, J. (2014), "Roman Colonization and the City-State Model," in Stek and Pelgrom (2014), 73–86.

Pfeilschifter, R. (2005), *Titus Quinctius Flamininus: Untersuchungen zur römischen Griechenlandpolitik*. Göttingen.

Pfeilschifter, R. (2007), "The Allies in the Republican Army and the Romanization of Italy," in R. Roth and J. Keller (eds.), *Roman by Integration: Dimensions of Group Identity in Material Culture and Text*. 27–42. Portsmouth.

Phang, S. E. (2001), *The Marriage of Roman Soldiers (13 BC–AD 235): Law and Family in the Imperial Army*. Leiden.

Phang, S. E. (2008), *Roman Military Service: Ideologies of Discipline in the Late Republic and Early Principate*. Cambridge.

Philipp, H. and W. Koenigs (1979), "Zu den basen des L. Mummius in Olympia," *MDAI(A)* 94, 93–116.

Pierson, D. S. (1999), "Natural Killers-Turning the Tide of Battle," *Military Review* 79(3), 60–65.

Pina Polo, F. (1995), "Procedures and Functions of Civil and Military *contiones* in Rome," *Klio* 77, 203–16.

Pina Polo, F. (2003), "¿Por qué fue reclutada la turma Salluitana en Salduje?" *Gerión* 21, 197–204.

Pina Polo, F. (2008), "Hispania of Caesar and Pompey: A Conflict of *clientelae*?" in M. Paz García-Bellido, A. Mostalac, and A Jiménez (eds.), *Del imperium de Pompeyo a la auctoritas de Augusto: Homenaje a Michael Grant*. 41–48. Madrid.

Pinault, G.-J. (1987), "*Bellum*: la guerre et la beauté," in G. Freyburger (ed.), *De Virgile à Jacob Balde: hommage à Mme Andrée Thill*. 151–56. Mulhouse.

Pinsent, J. (1954), "The Original Meaning of Municeps," *CQ* 4, 158–64.

Pinsent, J. (1957), "Municeps, II," *CQ* 7, 89–97.

Pinsent, J. (1975), *Military Tribunes and Plebeian Consuls: The Fasti from 444 V to 342 V.* Wiesbaden.

Piper, D. J. (1988), "*The ius adispiscendae ciuitas Romanae per magistratum* and its Effects on Roman-Italian Relations," *Latomus* 47, 59–68.

Pitassi, M. (2011), *Roman Warships*. Rochester.

Pittenger, M. R. P. (2008), *Contested Triumphs: Politics, Pageantry, and Performance in Livy's Republican Rome*. Berkeley, CA.

Platner-Ashby, T. (1927), *The Roman Campagna in Classical Times*. London.

Poccetti, P. (1982), "Sulle dediche tuscolane del tribuno militare M. Furio," *MÉFRA* 94, 657–74.

Pohl, R., and M. Roock (2011), "Sozialpsychologie des Krieges: Der Krieg als Massenpsychose und die Rolle der militärisch-männlichen Kampfbereitschaft," in T. Jäger and R. Beckmann (eds.), *Handbuch Kriegstheorien*. 45–53. Wiesbaden.

Popkin, M. L. (2016), *The Architecture of the Roman Triumph: Monuments, Memory, and Identity*. Cambridge.

Powell, J. G. F. (1990), "The Tribune Sulpicius," *Historia* 39, 446–60.

Prag, J. R. W. (2007), "*Auxilia* and *Gymnasia*: A Sicilian Model of Roman Imperialism," *JRS* 97, 68–100.

Prag, J. R. W. (2011a), "Provincial Governors and Auxiliary Soldiers," in N. Barrandon, and F. Kirbihler (eds.), *Les gouverneurs et les provinciaux sous la République romaine*. 15–28. Rennes.

Prag, J. R. W. (2011b), "Troops and Commanders: *Auxilia Externa* under the Roman Republic," in D. Bonanno, R. Marino, and D. Motta (eds.), *Truppe e Comandanti nel mondo antico (Hormos: Ricerche di storia antica)*. 101–13. Palermo.

Prag, J. (2013), "Sicily and Sardinia-Corsica: The First Provinces," in Hoyos (2013), 53–65.

Prag, J. R. W. (2015), "*Auxilia* and *Clientelae*: Military Service and Foreign *Clientelae* Reconsidered," in Jehne and Pina Polo (2015), 281–94.

Premerstein, A. von (1937), *Vom Werden und Wesen des Prinzipats*. Munich.

Preston, S. H., A. McDaniel, and C. Grushka (1972), *Causes of Death: Life Tables for National Populations*. New York.

Pryke, L. M. (2017), *Ishtar*. London.

Quesada Sanz, F. (1997), "*Gladius Hispaniensis*: An Archaeological View from Iberia," *JRMES* 8, 251–70.

Quesada Sanz, F. (2006), "Not So Different: Individual Fighting Techniques and Small Unit Tactics of Roman and Iberian Armies Within the Framework of Warfare in the Hellenistic Age," *Pallas* 70, 245–63.

Raaflaub, K. A. (1996), "Born to be Wolves? Origins of Roman Imperialism," in Wallace and Harris (1996), 273–314.

Raaflaub, K. A. (2005a), "The Conflict of the Orders in Archaic Rome: A Comprehensive and Comparative Approach," in Raaflaub (2005b), 1–46.

Raaflaub, K. A. (ed.) (2005b), *Social Struggles in Archaic Rome: New Perspectives on the Conflict of the Orders*, Expanded and Updated Edition. Oxford.

Raaflaub, K. A. (2010), "Between Myth and History: Rome's Rise from Village to Empire (the Eighth Century to 264)," in Rosenstein and Morstein-Marx (2006), 125–46.

Raaflaub, K. A., and N. S. Rosenstein (1999a), "Introduction," in Raaflaub and Rosenstein (1999b), 1–6.

Raaflaub, K. A., and N. S. Rosenstein (eds.) (1999b), *War and Society in the Ancient and Medieval Worlds: Asia, the Mediterranean, Europe, and Mesoamerica*. Cambridge, MA.

Rankov, B. (2007), "Military Forces," in P. Sabin, H. van Wees, and M. Whitby (eds.), *The Cambridge History of Greek and Roman Warfare, Vol. 2: Rome from the Late Republic to the Late Empire*. 30–75. Cambridge.

Rathbone, D. W. (1993), "The Census Qualifications of the *Assidui* and the *Prima Classis*," in H. Sancisi-Weerdenburg (ed.), *De Agricultura: In Memoriam Pieter Willem de Neeve (1945–1990)*. 121–52. Amsterdam.

Rathbone, D. W. (2008), "Poor Peasants and Silent Sherds," in de Ligt and Northwood (2008), 305–32.

Rawlings, L. (1999), "Condottieri and Clansmen: Early Italian Raiding, Warfare and the State," in K. Hopwood (ed.), *Organised Crime in Antiquity*. 97–127. London.

Rawlings, L. (2007), "Army and Battle During the Conquest of Italy (350–264 BC)," in Erdkamp (2007c), 45–62.

Rawlings, L. (2016), "The Significance of Insignificant Engagements: Irregular Warfare during the Punic Wars," in Armstrong (2016d), 204–36.

Rawson, B. (2006), "Finding Roman Women," in Rosenstein and Morstein-Marx (2006), 324–41.

Rawson, E. (1971a), "The Literary Sources for the Pre-Marian Army," *PBSR* 39, 13–31.

Rawson, E. (1971b), "Prodigy Lists and the Use of the *Annales Maximi*," *CQ* 21, 158–69.

Rawson, E. (1979), "L. Cornelius Sisenna and the Early First Century B.C.," *CQ* 29, 327–46.

Rawson, E. (1990), "The Antiquarian Tradition: Spoils and Representations of Foreign Armour," in Eder (1990), 158–173.

Reese, P. (1992), *Homecoming Heroes: An Account of the Reassimilation of British Military Personnel into Civilian Life*. London.

Rich, J. W. (1976), *Declaring War in the Roman Republic in the Period of Transmarine Expansion*. Brussels.

Rich, J. W. (1983), "The Supposed Roman Manpower Shortage of the Later Second Century B.C.," *Historia* 32, 287–331.

Rich, J. W. (1993), "Fear, Greed, and Glory: The Causes of Roman War-Making in the Middle Republic," in J. Rich and G. Shipley (eds.), *War and Society in the Roman World*. 38–68. London.

Rich, J. W. (2005), "Valerius Antias and the construction of the Roman past," *BICS* 48, 137–61.

Rich, J. W. (2007a), "Warfare and the Army in Early Rome," in Erdkamp (2007c), 7–23.

Rich, J. W. (2007b), "Tiberius Gracchus, Land and Manpower," in O. Hekster, G. de Kleijn and D. Slootjes (eds.), *Crisis and the Roman Empire*. 155–66. Leiden.

Rich, J. W. (2008), "Treaties, Allies and the Roman Conquest of Italy," in P. de Souza and J. France (eds.), *War and Peace in Ancient and Medieval History*. 51–75. Cambridge.

Rich, J. W. (2011), "The Fetiales and Roman International Relations," in Richardson and Santangelo (1014b), 187–242.

Rich, J. W. (2012), "Roman Attitudes to Defeat in Battle Under the Republic," in Marco Simón, Pina Polo, and Remesal Rodriguez (2012), 83–112.

Rich, J. W. (2013), "Roman Rituals of War," in Campbell and Tritle (2013), 542–68.

Rich, J. W. (2014), "The Triumph in the Roman Republic: Frequency, Fluctuation and Policy," in Lange and Vervaet (2014), 197–258.

Rich, J. W., and G. Shipley (eds.) (1993), *War and Society in the Roman World*. London.

Richard, J. C. (1987), "Recherches sur le interprétation populaire de la figure du roi Servius Tullius," *RPh* 61, 205–225.

Richard, J.-C. (1990), "Historiographie et histoire: L'expédition des Fabii à la Crémère," in Eder (1990), 174–99.

Richardson, J. H. (2007), "On the Location of the *Urbs* and *Tribus Scaptia*," *Hermes* 135, 166–73.

Richardson, J. H., and F. Santangelo (2014a), "Introduction," in Richardson and Santangelo (2014b), 1–15.

Richardson, J. H., and F. Santangelo (eds.) (2014b), *The Roman Historical Tradition: Regal and Republican Rome*. Oxford.

Richardson, J. S. (1975), "The Triumph, the Praetors and the Senate in the Early Second Century B.C.," *JRS* 65, 50–63.

Richardson, L. (1992), *A New Topographical Dictionary of Ancient Rome*. Baltimore, MD.

Richlin, A. (1992), *The Garden of Priapus: Sexuality and Aggression in Roman Humor*. Revised edn. Oxford.

Richlin, A. (2017), *Slave Theater in the Roman Republic: Plautus and Popular Comedy*. Cambridge.

Rickman, G. (1980), *The Corn Supply of Ancient Rome*. Oxford.

Ridley, R. T. (1979), "The Origin of the Roman Dictatorship: An Overlooked Opinion," *RhM* 122, 303–9.

Ridley, R. T. (1986), "The 'Consular Tribunate': The Testimony of Livy," *Klio* 68, 444–65.

Ridley, R. (2014) "The Enigma of Servius Tullius," in Richardson and Santangelo (2014b), 83–128. Originally published in *Klio* 57 (1975), 147–77.

Robinson, H. R. (1975), *The Armour of Imperial Rome*. New York.

Rocco, R. (2017), "The Origin and Evolution of the Roman Manipular Legion," Dissertation, University of Auckland, Auckland.

Roller, M. B. (2004), "Exemplarity in Roman Culture: The Cases of Horatius Cocles and Cloelia," *CPhil.* 99, 1–56.

Roller, M. B. (2009), "The Exemplary Past in Roman Historiography and Culture," in Feldherr (2009b), 214–30.

Roselaar, S. T. (2009), "*Assidui* or *Proletarii*? Property in Roman Citizen Colonies and the *Vacatio Militiae*," *Mnemosyne* 62, 609–23.

Roselaar, S. T. (2010), *Public Land in the Roman Republic: A Social and Economic History of Ager Publicus in Italy, 396–89 BC*. Oxford.

Roselaar, S. T. (ed.) (2012), *Processes of Integration and Identity Formation in the Roman Republic*. Leiden.

Roselaar, S. T. (2013a), "The Concept of *Commercium* in the Roman Republic," *Phoenix* 66, 381–413.

Roselaar, S. T. (2013b), "The Concept of *Conubium* in the Roman Republic," in P. Du Plessis (ed.), *New Frontiers: Law and Society in the Roman World*. 102–22. Edinburgh.

Roselaar, S. T. (2019), *Italy's Economic Revolution: Integration and Economy in Republican Italy*. Oxford.

Roselaar, S. T., and M. Helm (eds.) (forthcoming), *Spoils in the Roman Republic – a Re-evaluation*. Berlin.

Rosenstein, N. S. (1990), *Imperatores Victi: Military Defeat and Aristocratic Competition in the Middle and Late Republic*. Berkeley, CA.

Rosenstein, N. S. (1993), "Competition and Crisis in Mid-Republican Rome," *Phoenix* 47, 313–38.

Rosenstein, N. S. (1999), "Republican Rome," in Raaflaub and Rosenstein (1999b), 193–216.

Rosenstein, N. S. (2002), "Marriage and Manpower in the Hannibalic War: *Assidui*, *Proletarii* and Livy 24.18.7–8," *Historia* 51, 163–91.

Rosenstein, N. S. (2004), *Rome at War: Farms, Families, and Death in the Middle Republic*. Chapel Hill, NC.

Rosenstein, N. S. (2006), "Aristocratic Values," in Rosenstein and Morstein-Marx (2006), 365–82.

Rosenstein, N. S. (2007), "Military Command, Political Power, and the Republican Elite," in Erdkamp (2007c), 132–47.

Rosenstein, N. S. (2010), "Phalanges in Rome?" in G. Fagan and M. Trundle (eds.), *New Perspectives on Ancient Warfare.* 289–303. Leiden.

Rosenstein, N. (2011), "War, Wealth and Consuls," in Beck, Duplá, Jehne, and Pina Polo (2011), 133–59.

Rosenstein, N. S. (2012a), "Integration and Armies in the Middle Republic," in Roselaar (2012), 85–104.

Rosenstein, N. S. (2012b), *Rome and the Mediterranean 290 to 146 BC: The Imperial Republic.* Edinburgh.

Rosenstein, N. S. (2016a), *"Tributum* in the Middle Republic," in Armstrong (2016d), 80–97.

Rosenstein, N. S. (2016b), *"Bellum se ipsum alet?* Financing Mid-Republican Imperialism," in Beck, Jehne, and Serrati (2016), 114–44.

Rosenstein, N. S., and R. Morstein-Marx (eds.) (2006), *A Companion to the Roman Republic.* Malden, MA.

Roth, J. P. (1999), *The Logistics of the Roman Army at War (264 B.C. – A.D. 235).* Leiden.

Roth, J. P. (2009), *Roman Warfare.* Cambridge.

Rouland, N. (1977), *Les esclaves romains en temps de guerre.* Brussels.

Rowe, G. (2002), *Princes and Political Cultures: The New Tiberian Senatorial Decrees.* Ann Arbor, MI.

Roy, A. M. (2017), "Engineering Power: The Roman Triumph as Material Expression of Conquest, 211–55 BCE," PhD dissertation, University of Washington, Seattle, WA.

Rubio-Campillo, X, P. Valdés Matías, and E. Ble (2015), "Centurions in the Roman Legion: Computer Simulation and Complex Systems," *Journal of Interdisciplinary History* 46, 245–63.

Rudán, B., and U. Brandl (2008), "… *intrare castra feminis non licet* – Tatsache oder literarische Fiktion? Ein kritischer Literaturüberblick," in U. Brandl (ed.), *Frauen und Römisches Militär: Beiträge eines Runden Tisches in Xanten vom 7. bis 9. Juli 2005.* 1–19. Oxford.

Rüpke, J. (1990), *Domi militiae: die religiöse Konstruktion des Krieges in Rom.* Stuttgart.

Rüpke, J. (1995), "Wege zum Töten, Wege zum Ruhm: Krieg in der römischen Republik," in Stietencron and Rüpke (1995), 213–40.

Rüpke, J. (2006), "Triumphator and Ancestor Rituals between Symbolic Anthropology and Magic," *Numen* 53, 251–89.

Rüpke, J. (2008), "Neue Perspektiven auf alte Statuenrituale: Überlegungen zu Res Gestae Divi Augusti 4," in H. Krasser, D. Pausch, and I. Petrovic (eds.), *Triplici invectus triumpho: Der römische Triumph in augusteischer Zeit.* Stuttgart, 11–26.

Rüpke, J. (2016), *On Roman Religion: Lived Religion and the Individual in Ancient Rome.* Ithaca, NY.

Rzepka, J. (2008), "The Units of Alexander's Army and the District Divisions of Late Argead Macedonia," *GRBS* 48, 39–56.

Sabin, P. (1996), "The Mechanics of Battle in the Second Punic War," in T. Cornell, B. Rankov, and P. Sabin (eds.), *The Second Punic War: A Reappraissal. BICS* Supp. 67. 59–79. London.

Sabin, P. (2000), "The Face of Roman Battle," *JRS* 90, 1–17.

Saddington, D. B. (1982), *The Development of the Roman Auxiliary Forces from Caesar to Vespasian (49 B.C. – A.D. 79).* Harare.

Sage, M. M. (2008), *The Republican Roman Army: A Sourcebook*. London.

Sage, M. M. (2013), "The Rise of Rome," in Campbell and Tritle (2013), 216–35.

Sage, M. M. (2018), *The Army of the Roman Republic: From the Regal Period to the Army of Julius Caesar*. Barnsley.

Sallares, R. (2002), *Malaria and Rome. A History of Malaria in Ancient Italy*. Oxford.

Saller, R. (1994), *Patriarchy, Property and Death in the Roman Family*. Cambridge.

Salmon, E. T. (1958), "Notes of the Social War," *TAPA* 89, 159–84.

Salmon, E. T. (1963), "The *Coloniae Maritimae*," *Athenaeum* 41, 3–33.

Salmon, E. T. (1967), *Samnium and the Samnites*. Cambridge.

Salmon, E. T. (1982), *The Making of Roman Italy*. Ithaca, NY.

Sandberg, K. (2001), *Magistrates and Assemblies: A Study of Legislative Practice in Republican Rome*. Rome.

Sandberg, K., and C. Smith (eds.) (2017), *Omnium Annalium Monumenta: Historical Writing and Historical Evidence in Republican Rome*. Leiden.

Santangelo, F. (2008), "The Fetials and Their *Ius*," *BICS* 51, 63–93.

Scanlon, T. (2015), *Greek Historiography*. Malden, MA.

Shean, J. F. (1996), "Hannibal's Mules: The Logistical Limitations of Hannibal's Army and the Battle of Cannae, 216 B.C.," *Historia* 45, 159–87.

Scheid, J. (1983), "G. Dumézil et la méthode experimentale," *Opus* 2, 343–54.

Scheid, J. (1985), *Religion et piété à Rome*. Paris.

Scheid, J. (1998), *La Religion des Romains*. Paris.

Scheid, J. (2011), *Pouvoir et religion à Rome*. With the collaboration of Jean-Maurice de Montrémy. Paris.

Scheid, J. (2016), *The Gods, the State, and the Individual: Reflections on Civic Religion in Rome*. Translated by Clifford Ando. Philadelphia, PA.

Scheidel, W. (1994), "Seasonal Mortality and Endemic Disease in Rome," *AncSoc* 25, 151–75.

Scheidel, W. (1996a), "Finances, Figures and fiction," *CQ* 46, 222–38.

Scheidel, W. (1996b), "Seasonal mortality in the Roman Empire," in W. Scheidel (ed.), *Measuring Sex, Age, and Death in the Roman Empire: Explorations in Ancient Demography*, *JRA Suppl.* 21. 139–63. Ann Arbor, MI.

Scheidel, W. (1999a), "The Slave Population of Roman Italy. Speculation and Constraints," *Topoi* 9, 129–44.

Scheidel, W. (1999b), "Emperors, Aristocrats, and the Grim Reaper: Towards a Demographic Profile of the Roman Elite," *CQ* 49, 254–81.

Scheidel, W. (2001a), "Roman Age Structure: Evidence and Models," *JRS* 91, 1–26.

Scheidel, W. (2001b), "Progress and Problems in Roman Demography," in W. Scheidel (ed.), *Debating Roman Demography*. 1–82. Leiden.

Scheidel, W. (2003), "Germs for Rome," in C. Edwards and G. Woolf (eds.), *Rome the Cosmopolis*. 158–76. Cambridge.

Scheidel, W. (2005), "Human Mobility in Roman Italy, II: The Slave Population," *JRS* 95, 64–79.

Scheidel, W. (2008), "Roman Population Size: The Logic of the Debate," in de Ligt and Northwood (2008), 17–70.

Schneider, H.-C. (1977), *Das Problem der Veteranenversorgung in der späteren römischen Republik*. Bonn.

Schweighäuser, J. (ed.) (1795), *Lexicon Polybianum*. Leipzig.

Scopacasa, R. (2015), *Ancient Samnium: Settlement, Culture, and Identity Between History and Archaeology*. Oxford.

Scopacasa, R. (2016), "Rome's Encroachment on Italy," in A. E. Cooley (ed.), *A Companion to Roman Italy*. 33–56. Malden, MA.

Seager, R. (2002), *Pompey: A Political Biography*[2]. Oxford.

Sealey, R. (1959), "Consular Tribunes Once More," *Latomus* 18, 521–30.

Segal, E. (1975), "Perché *Amphitruo*," *Dioniso* 46, 247–67.

Segalen, M., and F. Zonabend (eds.) (1996), *Geschichte der Familie. Altertum. Vol. 1.* Frankfurt.

Sen, A. (1993), "The Causation and Prevention of Famines: A Reply," *Journal of Peasant Studies* 21, 29–40.

Serrati, J. (2011), "The Rise of Rome to 264 BC," in Hoyos (2011), 9–27.

Serrati, J. (2013), "Imperialism and the Fall of the Republic: *Post hoc ergo propter hoc?*" in Hoyos (2013), 155–68.

Serrati, J. (2016), "The Financing of Conquest: Roman Interaction with Hellenistic Tax Laws," in Beck, Jehne, and Serrati (2016), 97–113.

Sewell, J. (2014), "Gellius, Philip II and a Proposed End to the 'Model-Replica' Debate," in Stek and Pelgrom (2014), 125–39.

Shanin, T. (1972), *The Awkward Class: Political Sociology of Peasantry in a Developing Society: Russia 1910–1925*. Oxford.

Shay, J. (2002), *Odysseus in America: Combat Trauma and the Trials of Homecoming*. New York.

Shatzman, I. (1972), "The Roman General's Authority over Booty," *Historia* 21, 177–205.

Shatzman, I. (1975), *Senatorial Wealth and Roman Politics*. Brussels.

Sherwin-White, A. N. (1973), *The Roman Citizenship*[2]. Oxford.

Sherwin-White, A. N. (1980), "Rome the Agressor?" *JRS* 70, 177–81.

Shochat, Y. (1980), *Recruitment and the Programme of Tiberius Gracchus*. Brussels.

Slavik, J. F. (2018), "*Pilum* and *Telum*: The Roman Infantryman's Style of Combat in the Middle Republic," *CJ* 113, 151–71.

Small, A. (2000), "The Use of Javelins in Central and South Italy in the 4th Century BC," in D. Ridgway, F. R. Serra Ridgway, M. Pearce, E. Herring, R. D. Whitehouse, and J. B. Wilkins (eds.), *Ancient Italy in its Mediterranean Setting: Studies in Honour of Ellen MacNamara*. 221–34. London.

Smith, C. J. (1996), *Early Rome and Latium: Economy and Society c. 1000 to 500 BC*. Oxford.

Smith, C. J. (2006a), *The Roman Clan: The Gens from Ancient Ideology to Modern Anthropology*. Cambridge.

Smith, C. J. (2006b), "*Adfectatio Regni* in the Roman Republic," in Lewis (2006), 49–64.

Smith, C. J. (2011), "The Magistrates of the Early Roman Republic," in Beck, Duplá, Jehne, and Pina Polo (2011), 19–40.

Smith, C. J., and L. M. Yarrow (2012), "Introduction," in C. J. Smith and L. M. Yarrow (eds), *Imperialism, Cultural Politics, and Polybius*. 1–14. Oxford.

Smith. R. E. (1944), "The Sources of Plutarch's Life of Titus Flamininus," *CQ* 38, 89–95.

Sordi, M. (1972), "L'arruolamento dei capite censi nel pensiero e nell'azione politica di Mario," *Athenaeum* 50, 379–85.

Speidel, M. A. (2016), "Actium, Allies, and the Augustan Auxilia: Reconsidering the Transformation of Military Structures and Foreign Relations in the Reign of Augustus," in Wolff and Faure (2016), 79–95.

Speidel, M. P. (1982), "Legionary Cohorts in Mauretania: The Role of Legionary Cohorts in the Structure of Expeditionary Armies," *ANRW* 2, 850–60.

Speidel, M. P. (1990), "The Names of Legionary Centuriae," *Arctos* 24, 135–37.

Spurr, M. S. (1986), *Arable Cultivation in Roman Italy: c.200 B.C.–c.A.D. 100*. London.

Staveley, E. S. (1953), "The Significance of the Consular Tribunate," *JRS* 43, 30–36.

Staveley, E. S. (1972), *Greek and Roman Voting and Elections*. London.

Staveley, E. S. (1989), "Rome and Italy in the Early Third Century," *CAH²* 7.2, 420–55.

Stein, C. (2007), "Qui sont les aristocrates romains a 'la fin de la re' publique?" in H.-L. Fernoux and C. Stein (eds.), *Aristocratie antique: mod'eles et exemplarit'e sociale*. 127–59. Dijon.

Stek, T. D. and J. Pelgrom (eds.) (2014), *Roman Republican Colonization: New Perspectives from Archaeology and Ancient History*. Rome.

Stemmler, M. (1997), *Eques Romanus – Reiter und Ritter: Begriffsgeschichtliche Untersuchungen zu den Entstehungsbedingungen einer römischen Adelskategorie im Heer und in den comitia centuriata*. Frankfurt.

Stemmler, M. (2000), "Die römische Manipularordnung und der Funktionswandel der Centurien," *Klio* 82, 107–25.

Stewart, O. (2017), "Citizenship as a Reward or Punishment? Factoring Language into the Latin Settlement," *Antichthon* 51, 186–201.

Stewart, R. (1998), *Public Office in Early Rome: Ritual Procedure and Political Practice*. Ann Arbor, MI.

Stibbe, C. M., G. Colonna, C. de Simone, and H. S. Versnel (1980), *Lapis Satricanus: Archaeological, Epigraphical, Linguistic and Historical Aspects of the New Inscription from Satricum*. The Hague.

Stietencron, H. von (1995), "Töten im Krieg. Grundlagen und Entwicklungen," in Stietencron and Rüpke (1995), 17–56.

Stietencron, H. von, and J. Rüpke (eds.) (1995), *Töten im Krieg*. Freiburg, Munich.

Stockton, D. (1979). *The Gracchi*. Oxford.

Stone, A. M. (2005), "*Optimates*: An Archaeology," in K. Welch, and T. W. Hillard (eds.), *Roman Crossings: Theory and Practice in the Roman Republic*. 59–94. Swansea.

Stroup, S. C. (2007), "Making Memory: Ritual, Rhetoric, and Violence in the Roman Triumph," in J. K. Wellman (ed.), *Belief and Bloodshed: Religion and Violence across Time and Tradition*. 29–46. Lanham, MD.

Sumi, G. S. (2005), *Ceremony and Power: Performing Politics in Rome between Republic and Empire*. Ann Arbor, MI.

Sumner, G. V. (1970), "The Legion and the Centuriate Organization," *JRS* 60, 67–78.

Suolahti, J. (1955), *The Junior Officers of the Roman Army in the Republican Period: A Study on Social Structure*. Helsinki.

Swank, R. L. and W. E. Marchand (1946), "Combat Neuroses: Development of Combat Exhaustion," *Archives of Neurology & Psychiatry* 55, 236–47.

Takács, S. A. (2008), *Vestal Virgins, Sibyls, and Matrons: Women in Roman Religion*. Austin.

Talbert, R. (1987), *The Senate of Imperial Rome*. Princeton, NJ.

Tan, J. (2017), *Power and Punlic Finance at Rome, 264–49 BCE*. New York.

Tatum, W. J. (1999), *The Patrician Tribune: Publius Clodius Pulcher*. Chapel Hill, NC.

Taylor, D. (2017), *Roman Republic at War: A Compendium of Battles from 498 to 31 BC*. Barnsley.

Taylor, L. R. (2013), *The Voting Districts of the Roman Republic: The Thirty-Five Urban and Rural Tribes.* With Updated Material by Jerzy Linderski. Ann Arbor, MI.

Taylor, M. J. (2014), "Roman Infantry Tactics in the Mid-Republic: A Reassessment," *Historia* 63, 301–22.

Taylor, M. J. (2015), "Finance, manpower, and the Rise of Rome," PhD dissertation, University of California, Berkeley, CA.

Taylor, M. J. (2017a), "State Finances in the Middle Roman Republic: A reevaluation," *AJP* 138, 143–80.

Taylor, M. J. (2017b), "Etruscan Identity and Service in the Roman Army: 300-100 BCE," *AJA* 1212, 275–92.

Tellegen-Couperus, O. E. (ed.) (2011), *Law and Religion in the Roman Republic.* Leiden.

Terrenato, N. (2001), "The Auditorium Site in Rome and the Origins of the Villa," *JRA* 14, 5–32.

Terrenato, N. (2011), "The Versatile Clans: Archaic Rome, and the Nature of the Early City-States in Central Italy," in Terrenato and Haggis (2011), 231–44.

Terrenato, N. (2019), *The Early Roman Expansion into Italy: Elite Negotiation and Family Agendas.* Cambridge.

Terrenato, N., and D. C. Haggis (eds.) (2011), *State Formation in Italy and Greece: Questioning the Neoevolutionist Paradigm.* Oxford.

Thein, A. (2016), "Booty in the Sullan Civil War of 83–82 B.C.," *Historia* 65, 450–72.

Thomas, Y. (1996), "Rom: Väter als Bürger in einer Stadt der Väter," in A. Burguière, C. Klapisch-Zuber, M. Segalen, and F. Zonabend (eds.), *Geschichte der Familie, Bd. 1: Altertum.* 277–326. Frankfurt.

Thommen, L. (2012), *An Environmental History of Ancient Greece and Rome.* Translated by Philip Hill. Cambridge.

Thomsen, R. (1980), *King Servius Tullius: A Historical Synthesis.* Copenhagen.

Tibiletti, G. (1953), "La politica delle colonie e città Latine nella Guerra Sociale," *RIL* 86 45–63.

Timpe, D. (1972), "Fabius Pictor und die Anfänge der römischen Historiographie," *ANRW* 1, 928–69.

Torelli, M. (1968), "Il Donario di M. Fulvio Flacco nell'Area Sacra di S. Omobono," *Quaderni dell'Istituto di Topografia antica della Università de Roma* 5, 71–6.

Torelli, M. (2012), "The Early Villa: Roman Contributions to the Development of a Greek Prototype," in J. A. Becker and N. Terrenato (eds.), *Roman Republican Villas: Architecture, Context, and Ideology.* 8–31. Ann Arbor, MI.

Toynbee, A. J. (1965), *Hannibal's Legacy: The Hannibalic War's Effects on Roman Life.* Oxford.

Tränkle, H. (1977), *Livius und Polybios.* Basel.

Treggiari, S. (1969), *Roman Freedmen during the Late Republic.* Oxford.

Trundle, M. (2010), "Light Troops in Classical Athens," in D. M. Pritchard (ed.), *War, Democracy and Culture in Classical Athens.* 139–60. Cambridge.

Tucholsky, K. (1932), *Lerne Lachen ohne zu Weinen.* Berlin.

Turney-High, H. H. (1949), *Primitive War: Its Practices and Concepts.* Columbia, SC.

Ungern-Sternberg, J. von (1990), "Die Wahrnehmung des Ständekampfes in der römischen Geschichtsschreibung," in Eder (1990), 92–102.

Ungern-Sternberg, J. von (2000), "Eine Katastrophe wird verarbeitet: Die Gallier in Rom," in Bruun (2000), 207–22.

Urso, G. (2011), "The Origin of the Consulship in Cassius Dio's *Roman History*," in Beck, Duplá, Jehne, and Pina Polo (2011), 41–60.

Vaan, M. de (2008), *Etymological Dictionary of Latin and Other Italic Languages*. Leiden.

Valgaeren, J. H. (2012), *The Jurisdiction of the Pontiff in the Roman Republic: A Third Dimension*. Nijmegen.

VanDerPuy, P. J. (2017), "'Uis Ingens Aeris Alieni': Agriculture and Debt in the Early Roman Republic, c. 450-287 BC," PhD dissertation, The Ohio State University, Columbus, OH.

van der Vliet, E. C. L. (2011), The Early Greek Polis: Regime Building, and the Emergence of the State," in Terrenato and Haggis (2011), 119–34.

van Sickle, J. (1987), "The Elogia of the Cornelii Scipiones and the Origin of Epigram at Rome," *AJPhil*. 108, 41–55.

van Wees, H. (2004), *Greek Warfare: Myths and Realities*. London.

Vasaly, A. (2015), *Livy's Political Philosophy: Power and Personality in Early Rome*. Cambridge.

Vernole, V. E. (2002), *Servius Tullius*. Rome.

Versnel, H. S. (1970), *Triumphus: An Inquiry into the Origin, Development and Meaning of the Roman Triumph*. Leiden.

Versnel, H. S. (1976), "Two Types of Roman *Devotio*," *Mnemosyne* 29, 365–410.

Versnel, H. S. (1980), "Historical Implications," in C. M. Stibbe, G. Colonna, C. de Simone and H. S. Versnel (eds.), *Lapis Satricanus: Archaeological, Epigraphical, Linguistic and Historical Aspects of the New Inscription from Satricum*. 97–150. The Hague.

Vervaet, F. J. (2014), *The High Command in the Roman Republic: The Principle of the summum imperium auspiciumque from 509 to 19 BCE*. Stuttgart.

von Arnim, H. (1892), "Ineditum Vaticanum," *Hermes* 27, 118–130.

von Fritz, K. (1950), "The Reorganisation of the Roman Government in 366 BC and the So-Called Licinio-Sextian Laws," *Historia* 1, 3–44.

Wachter, R. (1987), *Altlateinische Inschriften: Sprachliche und epigraphische Untersuchungen zu den Dokumenten bis etwa 150 v. Chr.* Bern.

Walbank, F. W. (1957), *A Historical Commentary on Polybius, Vol. I: Commentary on Books I–VI*. Oxford.

Walbank, F. W. (1967), *A Historical Commentary on Polybius, Vol. II: Commentary on Books VII–XVIII*. Oxford.

Walbank, F. W. (2002), *Polybius, Rome, and the Hellenistic World: Essays and Reflections*. Cambridge.

Wallace, R. W., and E. M. Harris (eds.) (1996), *Transitions to Empire: Essays in Greco-Roman History, 360–146 BC, in Honor of E. Badian*. Norman, OK.

Wallace-Hadrill, A. (1989a), "Patronage in Roman Society: From Republic to Empire," in Wallace-Hadrill (1989b), 63–87.

Wallace-Hadrill, A. (ed.) (1989b), *Patronage in Ancient Society*. London.

Wallace-Hadrill, A. (2008), *Rome's Cultural Revolution*. Cambridge.

Waller, M. (2011), "Victory, Defeat and Electoral Success at Rome, 343–91 B.C.," *Latomus* 70, 18–38.

Walter, U. (2004), *Memoria und res publica: Zur Geschichtskultur im republikanischen Rom*. Frankfurt.

Walters, C. F., and R. S. Conway (1918), "Restorations and Emendations in Livy VI–X," *CQ* 12, 1, 1–14.

Ward, G. (2016), "The Roman Battlefield: Individual Exploits in Warfare of the Roman Republic," in W. Riess and G. Fagan (eds.), *The Topography of Violence in the Greco-Roman World*. 299–324. Ann Arbor, MI.

Warren, L. B. (1970), "Roman Triumphs and Etruscan Kings: The Changing Face of the Triumph," *JRS* 60, 49–66.

Warrior, V. M. (2006), *Roman Religion*. Cambridge.

Watson, G. R. (1969), *The Roman Soldier*. Ithaca, NY.

Webster, G. (1969), *Roman Imperial Army*. London.

Weinstock, S. (1971). *Divus Julius*. Oxford.

Welch, K. E. (2006), "*Domi Militiaeque*: Roman Domestic Aesthetics and War Booty in the Republic," in Dillon and Welch (2006), 91–161.

Welch, K. E. (ed.) (2015), *Appian's Roman History: Empire and Civil War*. Swansea.

Wellesley, K. (2000), *The Year of the Four Emperors*[3]. London.

Westall, R. (2015), "The Sources for the *Civil Wars* of Appian of Alexandria," in Welch (2015), 125–67.

Wheeler, E. L. (1996), "The Laxity of Syrian Legions," in D. L. Kennedy (ed.), *The Roman Army in the East*. 229–76. Ann Arbor, MI.

Wiedemann, T. (1986), "The *Fetiales*: A Reconsideration," *CQ* 36, 478–90.

Willems, P. (1878), *Le Sénat de la République Romaine: Sa Composition et Ses Attributions*. Louvain.

Wilson, J.-P., and G. Bradley (eds.) (2006), *Greek and Roman Colonization: Origins, Ideologies and Interactions*. Swansea.

Wintjes, J. (2012), "'Keep the Women Out of the Camp!': Women and Military Institutions in the Classical World," in B. C. Hacker, and M. Vining (eds.), *A Companion to Women's Military History*. 17–59. Leiden.

Wiseman, T. P. (1964), "Some Republican Senators and the Their Tribes," *CQ* 14, 122–33.

Wiseman, T. P. (1970), "The Definition of 'Eques Romanus' in the Late Republic and Early Empire," *Historia* 19, 67–83.

Wiseman, T. P. (1971), *New Men in the Roman Senate: 139 B.C. – A.D. 14*. Oxford.

Wiseman, T. P. (1995), *Remus: A Roman Myth*. Cambridge.

Wiseman, T. P. (2004), *The Myths of Rome*. Exeter.

Wolff, C. (2007), "Le refus du service militaire à Rome à l'époque républicaine," *RÉMA* 4, 17–58.

Wolff, C. (2009), *Déserteurs et transfuges dans l'armée romaine à l'époque républicaine*. Naples.

Wolff, C., and P. Faure (eds.) (2016), *Les auxiliaires de l'armée romaine: Des alliés aux fédérés*. Lyon.

Woods, R. (2007), "Ancient and Early Modern Mortality: Experience and Understanding," *The Economic History Review* 60(2), 373–99.

Wrangham, R. W., and L. Glowacki (2012), "Intergroup Aggression in Chimpanzees and War in Nomadic Hunter-Gatherers: Evaluating the Chimpanzee Model," *Human Nature* 23, 5–29.

Yarrow, L. (2006), "Lucius Mummius and the Spoils of Corinth," *SCI* 25, 57–70.

Yoshimura, T. (1961), "Die Auxiliartruppen und die Provinzialklientel in der römischen Republik," *Historia* 10, 473–95.

Zack, A. (2001), *Studien zum "Römischen Völkerrecht": Kriegserklärung, Kriegsbeschluss, Beeidung und Ratifikation zwischenstaatlicher Verträge, internationale*

Freundschaft und Feindschaft während der römischen Republik bis zum Beginn des Prinzipats. Göttingen.

Zevi, F. (1968–9), "Considerazioni sull'elogio di Scipione Barbato," *Studi miscellanei* 15, 65–73.

Zhmodikov, A. (2000), "Roman Republican Heavy Infantrymen in Battle (IV – II Centuries B.C.)," *Historia* 49, 67–78.

Ziółkowski, A. (1986), "The Plundering of Epirus in 167 B.C.: Economic Considerations," *PBSR* 54, 69–80.

Ziółkowski, A. (1992), *The Temples of Mid-Republican Rome and Their Historical and Topographical Context*. Rome.

Ziółkowski, A. (2011), "The Capitol and the 'Auspices of Departure,'" in S. Ruciński, C. Balbuza, and C. Królczyk (eds.), *Studia Lesco Mrozewicz ab amicis et discipulis dedicata*. 465–71. Poznań.

Zollschan, L. (2011), "The Longevity of the Fetial College," in Tellegen-Couperus (2011), 119–44.

Index

Note: Page numbers followed by "n" denote endnotes.

Taylor & Francis Group
an **informa** business

Taylor & Francis eBooks

www.taylorfrancis.com

A single destination for eBooks from Taylor & Francis
with increased functionality and an improved user
experience to meet the needs of our customers.

90,000+ eBooks of award-winning academic content in
Humanities, Social Science, Science, Technology, Engineering,
and Medical written by a global network of editors and authors.

TAYLOR & FRANCIS EBOOKS OFFERS:

A streamlined
experience for
our library
customers

A single point
of discovery
for all of our
eBook content

Improved
search and
discovery of
content at both
book and
chapter level

REQUEST A FREE TRIAL
support@taylorfrancis.com

 Routledge
Taylor & Francis Group

 CRC Press
Taylor & Francis Group